American Institute of Chemical Engineers

# $11^{th}$ Conference on Refinery Processing

Topical Conference at the
2008 AIChE Spring National Meeting

April 6-10, 2008
New Orleans, Louisiana, USA

**Printed from e-media with permission by:**

Curran Associates, Inc.
57 Morehouse Lane
Red Hook, NY 12571
www.proceedings.com

**ISBN: 978-1-60560-213-4**

**Some format issues inherent in the e-media version may also appear in this print version.**

ISBN: 978-1-60560-213-4



American Institute of Chemical Engineers
Proceedings
Three Park Avenue
New York, NY 10016-5991
Phone: 212-591-8100

www.aiche.org

American Institute of Chemical Engineers

11th Conference on Refinery Processing
2008

# TABLE OF CONTENTS

# Mapping Hydroprocessing Reactor Operating Regimes

*Jinwen Chen and Neil Wang*
*National Centre for Upgrading Technology (NCUT)*
*One Oil Patch Drove, Devon, AB T9G 1A8, Canada*

*Fabian Mederos and Jorge Ancheyta*
*Instituto Mexicano del Petroleo (IMP), Eje Central Lazaro Cardenas Norte 152*
*Col. San Bartolo Atepehuacan, Mexico D.F. 07730*

Prepared for Presentation at the 2008 AIChE National Spring Meeting, Advances in Hydroprocessing, April 6-11, 2008, New Orleans, LA, USA



February 2008

## Abstract

In pilot plant hydrotreating studies, it is important to operate the hydrotreater under a desired operating regime – such as plug flow, full catalyst wetting, operation that is free of reactor wall effects, etc. – to generate reliable and reproducible data for commercial scale up. In this paper, the mapping of operating conditions and parameters was conduced to investigate the conditions that such desired operating regimes could be met for two different middle distillates feeds (straight run LGO and FCC LCO) with similar boiling ranges. It was found that the LCO feed required either higher/lower pressure, lower gas/oil ratio, longer catalyst bed, or smaller diluent particles than the LGO feed to maintain plug flow of the liquid in the reactor. However, the two feeds required similar operating parameters to have full catalyst wetting.

## 1. Introduction

During hydroprocessing (hydrotreating and hydrocracking), petroleum distillates and hydrogen pass through trickle-bed catalytic reactors at high temperatures and pressures with relatively large hydrogen flow. These conditions result in partial vaporization of oil and partial dissolving of hydrogen in liquid to form a vapor-liquid equilibrium (VLE) system with both vapor and liquid phases containing oil and hydrogen. Our previous related studies (Chen et al. 2005, 2006) with middle distillates have shown that as the temperature and hydrogen flow rate increase and pressure decreases, the VLE results in a significant decrease in liquid phase flow rates as well as significant changes in liquid phase properties. This could significantly affect the flow dynamics in the reactors and, therefore, the catalyst performance. For pilot plant studies, this becomes especially important since any discrepancies from the desired ideal operating regime (plug flow, full catalyst wetting, operation that is free of reactor wall effect, etc) could result in unreliable and unrepresentative data for commercial scale-up. Therefore, it is necessary to correctly predict the VLE in hydroprocessing reactors and, furthermore, predict the operating regimes under which reliable and representative pilot plant results can be generated.

There have been only a few studies published considering VLE in three phase reactors (Akgerman et al., 1985; Smith and Satterfield, 1986; LaVopa and Satterfield, 1988; Jong, 1994; Tesser et al., 2003). There have been even fewer studies of the VLE effect on hydroprocessing (Kocis and Ho, 1986; Bellos and Papayannakos, 2003). Experimental data on VLE of hydrogen-petroleum fraction systems obtained under commercial hydroprocessing conditions are rarely published in the open literature because of the poor definition of the problem, stemming from the complex composition of those systems. Instead, systems consisting of hydrogen and a single hydrocarbon or a simulated hydrocarbon mixture are often studied (Simnick et al., 1978; Laugier et al., 1983; Ramanujam et al., 1985; Connolly and Kandalic, 1986; Wiegand et al., 1989). These studies have been briefly reviewed in our previous paper (Chen et al., 2005).

In this study, VLEs in middle distillates pilot plant hydrotreaters are predicted using an in-house calibrated flash calculation program under various operating conditions and other operating parameters. The calculated VLE results are further used for mapping reactor operating regimes (plug flow, full catalyst wetting, operation that is free of reactor wall effects, etc.). Detailed mapping of these hydrotreater operating regimes are discussed based on two typical but different middle distillates feeds and under various operating conditions and parameters.

## 2. Simulation Basis

Our discussion mainly focuses on pilot plant hydrotreaters which are frequently used in hydrotreating catalyst development, evaluation, and activity stability studies to generate reliable data for commercial scale-up. The pilot plant specifications used in this study are:

Reactor ID: 2.53 cm
Catalyst volume: 100 to 200 ml
Catalyst particle size (average): 1.3X5 mm
Diluenting particle size: 0.20 to 0.4 mm
Catalyst-to-diluents volumetric ratio: 1:1

The operating conditions used were typical for commercial middle distillates hydrotreating:

Temperature: 300, 320, 340, 360 and 380°C
Pressure: 40, 50, 60, 70, and 80 bars
Liquid Hourly Space Velocity (LHSV): 1.5/h
Gas/oil ratio: 500, 800, and 1000 NL/h

Two typical middle distillates feeds were used, one was straight run light gas oil (LGO) and the other was light cycle oil (LCO) from a local refinery. The properties of the two feeds are shown in Table 1:

Table 1. Physical properties and chemical compositions of the feeds

| | LGO | LCO |
|---|---|---|
| Density (15.6°C), g/mL (ASTM D4052) | 0.8392 | 0.9360 |
| Saturates (wt%) | 73.2 | 15.7 |
| Aromatics (wt%) | 26.8 | 83.8 |
| SimDis, °C (ASTM D2887) | | |
| IBP | 126.5 | 128.4 |
| 10wt% | 214.5 | 225.0 |
| 30wt% | 262.5 | 255.3 |
| 50wt% | 295.0 | 288.5 |
| 70wt% | 326.5 | 326.4 |
| 90wt% | 370.5 | 373.9 |
| FBP | 435.0 | 430.8 |

In Table 1, the two feeds are quite close to each other in terms of boiling range. However there are significant difference in density and chemical composition. The LGO feed has lower density and low aromatics content while the LCO feed has higher density and aromatics content.

The vapor-liquid phase equilibrium calculations were performed using a modified flash calculation program that has been described in detail in our previous studies (Chen et al., 2005, 2006).

## 3. Theoretical Background

In a trickle-bed pilot plant hydrotreating study, in order to generate reliable and consistent experimental data, it is critical to maintain the reactor operation in the desired regime, which includes plug flow of the liquid phase, complete catalyst wetting, and operation that is free of reactor wall effects, etc. There have been many published papers that discuss these important issues. We will review these studies in a separate paper.

### 3.1 Plug Flow Criteria

In a pilot plant trickle-bed reactor, attention should always be paid to the axial dispersion, or back mixing in the liquid phase, especially in a short bed with a high conversion. Otherwise, the catalyst evaluation results, based on assumed plug flow, could cause deviation in reactor scale-up. An empirical criterion frequently used to verify the plug flow has the simplest form (Carberry, 1963):

$$\frac{L_B}{d_{pe}} > 100 \qquad (1)$$

Mears (1971) derived a criterion from the perturbation solution of an axial dispersion model with N-series mixers

$$\frac{L_B}{d_{pe}} > \frac{20n}{Pe_L} ln \frac{1}{(1-x)} \qquad (2)$$

which was based on the assumption that the minimum bed length for freedom of axial diffusion should have less than 5% deviation from the bed required by plug flow. In this equation, an implicit assumption that conversion is less than 90% was applied. Gierman (1988) considered that Mears' criterion was too conservative and, therefore, derived a relaxed criterion

$$\frac{L_B}{d_{pe}} > \frac{8n}{Pe_L} \ln \frac{1}{(1-x)} \qquad (3)$$

Equation 3 was based on the 10% deviation from plug flow. In our ultra low sulphur diesel (ULSD) studies, it was found that meeting this criterion might still be too strict due to the very high conversions (99.5% or even higher). Accordingly, an even more relaxed criterion with 15% deviation from plug flow was used in our computation:

$$\frac{L_B}{d_{pe}} > \frac{\sqrt{20}n}{Pe_L} \ln \frac{1}{(1-x)} \qquad (4)$$

It should be pointed out that the above equations contain the Peclet number $P_{eL}$ based on catalyst particle diameter. In this work, we used the Hochman and Effron (1969) correlation

$$Pe_L = 0.034 \, \mathrm{Re}_L^{0.53} \qquad (5)$$

in which the particle Reynolds Number

$$\mathrm{Re}_L = \frac{d_{pe} u_L \rho_L}{\mu_L} \qquad (6)$$

Gierman (1988) reported that, in their radioactive tracer experiments, the particle Peclet number tended to be constant at 0.04 when the Reynolds number was less than 1.0 without a sound reason. In our opinion, there is no reason to assume a constant Peclet number even when the Reynolds number is less than 1.0. Therefore we used Hochman and Effron's correlation (Equation (5)) throughout for calculation of Peclet number.

3.2 Full catalyst wetting criteria

In a trickle-bed reactor, it is desirable to have all the catalyst particles in the reactor completely covered by the flowing liquid, which gives maximum catalyst utilization. Any partial catalyst wetting would cause by-passing of the feed and, therefore, reduce catalyst performance. If the liquid superficial velocities are very low, which is common in a pilot plant reactor due to relatively small volume of catalyst loading, partial catalyst wetting may occur. In this work, criterion for wetting efficiency proposed by Gierman (1988) was used:

$$\frac{\mu_L u_L}{\rho_L d_{pe}^2 g} > 5.0 \times 10^{-6} \qquad (7)$$

3.3 Free of Reactor Wall Effects Criteria

Normally reactor wall effect on reactor performance is not a concern for a commercial hydrotreating reactor, since the reactor diameter is much larger than the catalyst particle size. However in a pilot plant reactor, a concern arises when a small diameter reactor and relatively large catalyst particles are used. To keep the wall effect at a minimum, a trickle bed reactor should meet the following criterion (Chu and Ng, 1989)

$$\frac{D_R}{d_{pe}} > 25 \qquad (8)$$

In our study, the calculated $\frac{D_R}{d_{pe}} = 58.2 > 25$. Therefore, the reactor wall effect was negligible and no further discussion is given in this paper.

## 4. Computation Results and Discussion

4.1 Plug Flow Mapping

The calculated operating regimes that meet the plug flow criterion are shown in Figures 1(a) and 1(b) for the LGO feed and LCO feed, respectively. Each curve in the figures represents, under certain gas/oil ratios, the boundary of pressure and temperature at which plug flow is met. The zone above the curve represents the operating temperatures and pressures that meet the plug flow criterion and the zone below the curve represents the operating temperatures and pressures that cannot meet the plug flow criterion. Normally, under fixed gas/oil ratio, increased reactor temperature requires increased pressure in order to ensure plug flow of the liquid phase. Under the same temperature, increased gas/oil ratio also required increased pressure to maintain liquid plug flow.

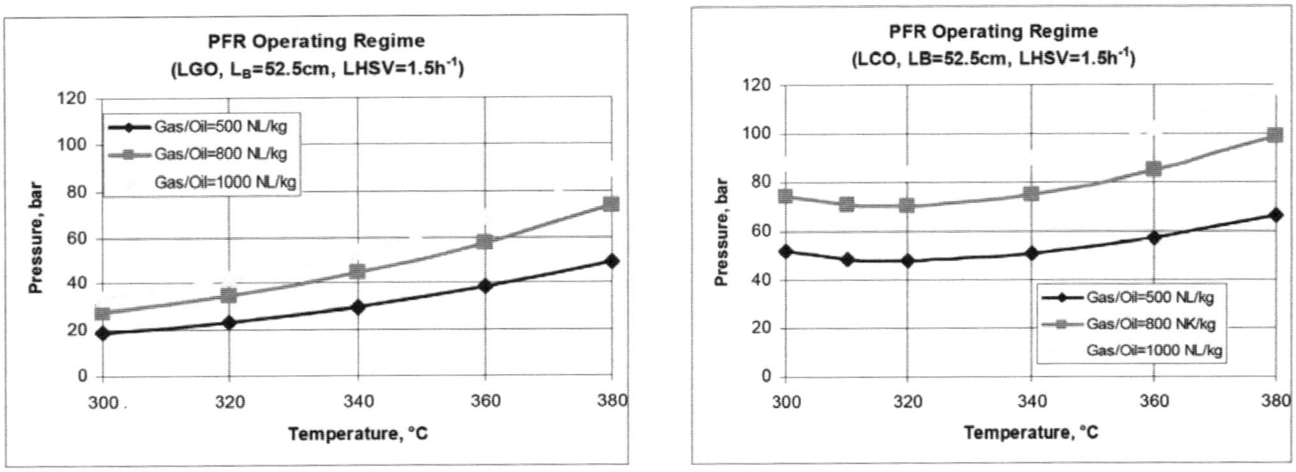


(a) LGO feed (2) LCO feed

Figure 1. Plug flow mapping, temperature-pressure-gas/oil ratio

Some differences were observed between the LGO feed (Figure 1(a)) and LCO feed (Figure 1(b)) as seen in the two figures. Generally, under the similar reactor temperature and gas/oil ratio, the LCO feed required higher pressure than the LGO feed in order to maintain liquid phase plug flow. This was even more pronounced at a lower temperature range (below 340°C). As seen in Figure 1(b), in the low temperature range, there was even an inverse trend that reduced the temperature required to increase pressure to maintain plug flow. This was due to the significant increase in viscosity for the LCO feed with the decrease in temperature. To maintain the Reynolds number and Peclet number, higher liquid velocity/flowrate was needed, which required higher pressure to increase the amount of oil in the liquid phase. This important finding indicates that even for similar middle distillates, such as LCO and LGO with similar boiling ranges, different liquid densities and chemical compositions could result in different flow dynamics and, therefore, reactor performance.

During a pilot plant study, sometimes the operating conditions, catalyst particle size and diluenting particle size are fixed according to process requirements. Therefore, to maintain plug flow of the liquid inside the reactor, catalyst bed length has to meet a certain value under a particular set of operating conditions. Figures 2(a) and 2(b) show the requirement for catalyst bed length at different reactor temperatures and pressures with fixed LHSV of 1.5/h and gas/oil ratio of 800 NL/kg for the LGO and LCO feed, respectively. Each curve in the figures represents, under certain pressure, the boundary of temperature and catalyst bed length at which plug flow is met. Again, for each curve in the figures, the zone above the curve represents the operating temperatures and catalyst bed lengths that meet the plug flow criterion and the zone below the curve represents the operating temperatures and catalyst bed lengths that cannot meet the plug flow criterion. As seen in these two figures, the increased temperature required an increased catalyst bed length in order to meet plug flow of the liquid. At higher temperature and lower pressure, the required catalyst bed length increased sharply. Under similar operating conditions, the LCO feed (Figure 2(b)) required longer catalyst bed length than the LGO feed (figure 2(a)) to maintain plug flow of the liquid in the reactor.

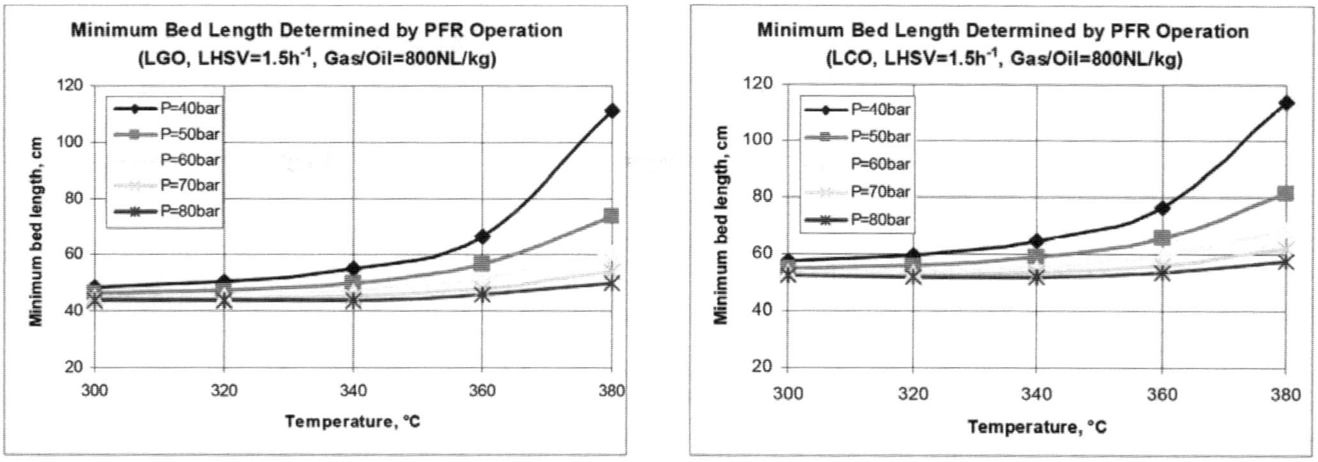


(a) LGO feed (2) LCO feed

Figure 2. Plug flow mapping, temperature-pressure-catalyst bed length

To achieve better liquid and gas distribution in the catalyst bed and to have better heat transfer, it is a common practice to use smaller diluent particles in pilot plant reactors. Using smaller diluent particles also improves the flow dynamics and helps in meeting the plug flow criterion. Therefore if operating conditions, catalyst size and volume are fixed, changing diluent particle size (with fixed catalyst-to-diluent ratio of 1:1) can also achieve plug flow of the liquid phase. Figures 3(a) and 3(b) show the requirement for diluent particle size at different reactor temperatures and pressures with fixed LHSV of 1.5/h and gas/oil ratio of 800 NL/kg for the LGO and LCO feed, respectively. Each curve in the figures represents, under certain pressure the boundary of temperature and diluent particle size at which plug flow is met or. Again, for each curve in the figures, the zone above the curve represents the operating temperatures and diluent particle size that meet the plug flow criterion and the zone below the curve represents the operating temperatures and diluent particle size that cannot meet the plug flow criterion. In these figures, the increased in temperature required reduced diluent particle size. Under similar operating conditions, the LCO feed (Figure 3(b)) required smaller diluent particles than the LGO feed to maintain plug flow of the liquid in the reactor.

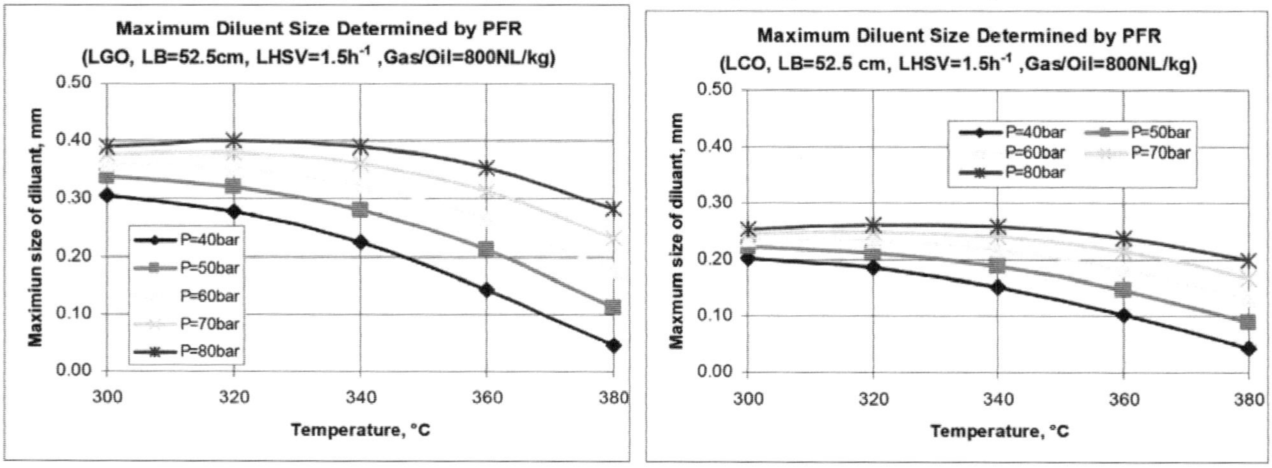


(a) LGO feed (2) LCO feed

Figure 3. Plug flow mapping, temperature-diluent particle size-pressure

4.2 Full catalyst wetting mapping

As discussed earlier, the VLE in the pilot plant hydrotreater could significantly change the liquid phase flow rate due to partial vaporization of the oil into vapor phase. Such a change in liquid flow rate could result in incomplete catalyst wetting, leading to by-passing of the feed and, therefore, reduced catalyst performance. The extent of the partial catalyst wetting is normally random and differs from one condition to another. Therefore, it is important to maintain full/complete catalyst wetting in a pilot plant hydrotreater in order to generate reliable and reproducible data. Unlike the plug flow criterion, the full catalyst wetting criterion is not related to the catalyst bed length. It is only related to the operating conditions, and catalyst and diluent particle sizes. Figures 4(a) and 4(b) show the mapping of full catalyst wetting for the LGO and LCO feeds respectively. Each curve in the figures represents, under certain gas/oil ratios, the boundary of pressure and temperature at which full catalyst wetting is met. The zone above the curve represents the operating temperatures and pressures that meet the full catalyst wetting criterion and the zone below the curve represents the operating temperatures and pressures that cannot meet full catalyst wetting. Normally, under fixed gas/oil ratios, increased reactor temperature requires increased pressure in order to ensure full catalyst wetting. Under the same temperature, increased gas/oil ratio also requires increased pressure to maintain full catalyst wetting.

It is interesting to see in Figures 4(a) and 4(b) that under the same temperatures and gas/oil ratios the LGO feed and the LCO feed required similar pressures to have complete catalyst wetting, which is different from the plug flow mapping discussed earlier where the LCO feed required higher pressure than the LGO feed to maintain plug flow in the liquid under the same temperatures and gas/oil ratios (see Figures 1(a) and 1(b)).

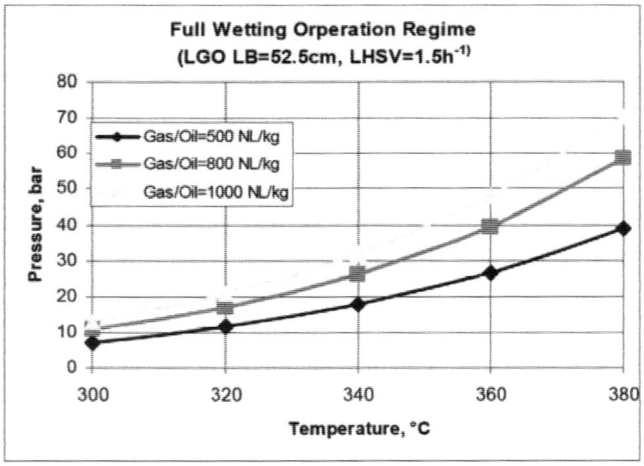


(a) LGO feed

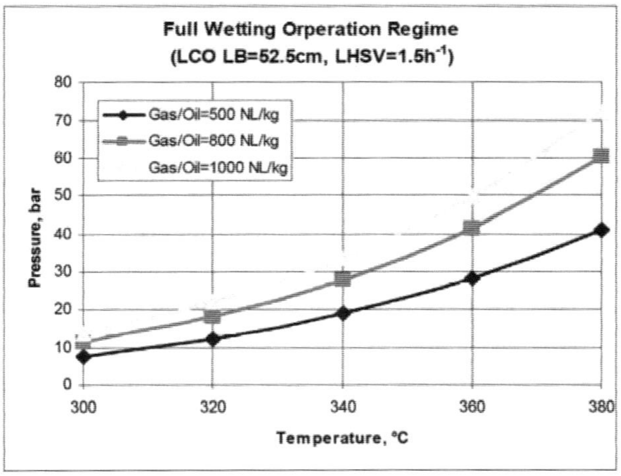


(2) LCO feed

Figure 4. Full catalyst wetting mapping, temperature-pressure-gas/oil ratio

## 5. Concluding Remarks

In this paper, VLEs in middle distillate pilot plant hydrotreaters were predicted using an in-house calibrated flash calculation program under various operating conditions. Two typical middle distillates with similar boiling ranges were used – a straight run LGO and a FCC LCO. The calculated VLE results were further used to predict the flow dynamics in the hydrotreaters to provide a mapping of operating parameters under which the desired operating regime (plug flow, full catalyst wetting, operation that is free of reactor wall effect, etc.) could be maintained. It was found that the LCO feed required higher or lower pressure, , or lower gas/oil ratio, or longer catalyst bed, or smaller diluent particles than the LGO feed to maintain plug flow of the liquid in the reactor. However, the two feeds required similar operating parameters to have full catalyst wetting.

## 6. Acknowledgements

Partial funding for NCUT has been provided by the Canadian Program for Energy Research and Development (PERD), the Alberta Research Council (ARC) and the Alberta Energy Research Institute (AERI). Partial funding for this study was also provided by Instituto Mexicano del Petroleo (IMP). The authors are grateful to Norm Sacuta for editing this paper.

## 7. References

1. Akgerman, A., Collins, G. M., Hook, B. D., "Effect of Feed Volatility on Conversion in Trickle-bed reactors", Ind. Engng. Chem. Fundam., 24, 398-401, 1985.
2. Bellos, G. D., and Papayannakos, N. G., "The Use of a Three Phase Microreactor to Investigate HDS Kinetics", Catal. Today, 79&80, 349-355, 2003.
3. Carberry, J. J., and M. M. Wendel, "A Computer of the Fixed Bed Catalytic Reactor: The Adiabatic and Quasi-adiabatic Cases," AIChE J, 9, 129 (1963).
4. Chen, J., Yang H., and Ring, Z., "Experimental Study on Vapour-Liquid Phase Equilibrium of LCO Hydrotreating Systems at High Temperatures and Pressures", Proceedings of 2005 AIChE National Spring Meeting, April 10-14, 2005, Atlanta, GA, USA
5. Chen, J., Jiang, W., Yang H., Hawkins, R., and Ring Z., "Comparative Study of Vapor-Liquid Equilibrium during Hydroprocessing of Different Petroleum Feedstocks", Proceedings of 2006 AIChE National Spring Meeting, April 23-27, 2006, Orlando, FL, USA
6. Chu, C. F. and K. M. Ng, "Flow in Packed Tubes with a Small Tube to Particle Diameter Ratio", AIChE J., 35, 148, 1989.
7. Connolly, J. and Kandalic, G., "Gas Solubilities, Vapour-Liquid Equilibria, and Partial Molar Volumes in Some Hydrogen-Hydrocarbon Systems", J. Chem. Eng. Data, 31, 396-406, 1986.
8. Gierman, H., "Design of Laboratory Hydrotreating Reactors Scaling Down of Trickle-flow Reactors", Appl. Catal., 43, 277, 1988.
9. Hochman J. and Effron E., Two-phase Cocurrent Downflow in Packed Beds", Ind. Engng. Chem. Funds., 8, 63, 1969.
10. de Jong, K. P., "Effect of Vapour-Liquid Equilibrium on Coke Deposition in Trickle-Bed Reactors during Heavy Oil Processing. 2. Modeling", Ind. Engng. Chem. Res. 33, 3141-3145, 1994.
11. Kocis, G. R. and Ho, T. C., "Effects of Liquid Evaporation on the Performance of Trickle-Bed Reactors", Chem. Engng. Res. Des., 64, 288-291, 1986.
12. Laugier, S., Richon, D., and Renon, H., "Vapour-Liquid Equilibria of Hydrogen-Hydrocarbon Mixtures Up to 40 MPa, 400°C: Measurements and Representation", Chemical Engineering Thermodynamics, paper presented at the Thermodynamics

Sessions of the 2nd World Congress on Chemical Engineering, 1983, Montreal, Que., Canada.
13. LaVopa, V., and Satterfield, C. N., "Some Effects of Vapour-Liquid Equilibria on Performance of a Trickle-Bed Reactor", Chem. Engng. Sci., 43, 2175-2180, 1988.
14. Mears, D. E., "The Role of Axial Dispersion in Trickle-Flow Laboratory Reactors", Chem. Engng. Sci., 26, 1361,1971.
15. Ramanujam, S., Leipziger, S., and Well, S., "Vapour-Liquid Equilibrium for a Hydrogen/ Simulated Coal-Derived Liquid System", Ind. Engng. Chem. Proc. Des. Dev., 24, 364-368, 1985.
16. Simnick, J., Sebastian, H., Li, H-M, and Chao, K-C, "Solubility of Hydrogen in Toluene at Elevated Temperatures and Pressures", J. Chem. Eng. Data, 23, 339-340, 1978.
17. Smith, C. M. and Satterfield, C. N., "Some Effects of Vapour-Liquid Flow Ratio on Performance of a Trickle-bed reactor", Chem. Engng. Sci. 41, 839-843 1986.
18. Tesser, R., Di Serio, M., and Santacesaria, E., "Influence of the Vapour-Liquid Equilibrium (VLE) on the Kinetics in Gas-Liquid and Gas-Liquid-Solid Systems", Catal. Today, 79&80, 323-331, 2003.
19. Wiegand, K., Strobel, B. and Hofmann, H., "Vapour-Liquid Equilibrium of Hydrogen/Coal Liquid and Methane/Coal Liquid Systems at Elevated Temperatures and Pressures", Chem. Engng. Technol., 12, 280-288, 1989.

# HYDROCONVERSION OF FISCHER-TROPSCH WAXES OVER PLATINUM LOADED CATALYST: ASSESSMENT OF PRODUCT DISTRIBUTION

*Chiara Gambaro, Gianni Pederzani, Vincenzo Calemma*
*Eni R&M, San Donato Milanese, Italy*

## Introduction

Since the 1990s the European Union has established emission standards to set limits for vehicles exhaust gasses, in terms of NOx, CO, HC and particulate emissions. The current fuel regulations require a minimum diesel cetane number of 51 and a maximum sulfur content for gasoline and diesel of 50 ppm (2005), while "sulfur free" fuels (≤ 10 ppm) must be available since 2009. Natural gas conversion to liquid fuels (GTL) via Fischer-Tropsch technology provides middle distillates characterized by very low concentrations of sulfur and nitrogen and, due to their paraffinic nature, high cetane numbers (>75) but poor cold flow properties. The current cobalt catalyzed low temperature slurry FT technologies are aimed to the synthesis of heavy products (waxes), which are subsequently hydrocracked to produce high quality isomerized middle distillates [1]. Isoparaffins have a deep influence on both cold flow properties and cetane number [2, 3].

## Results and discussion

Hydroconversion tests were carried out in a continuous bench scale fixed bed reactor operated in down flow mode, on a Pt/amorphous $SiO_2$-$Al_2O_3$ catalyst. The operating conditions were changed according to a Central Composite Design in the following range: T=319-351°C; P=35-60 bar; $H_2$/wax=0.06-0.15 g/g; WHSV=1-3 $h^{-1}$. The feed used throughout the experiments was a light paraffinic FT wax, characterized by a C22+ weight fraction of 36%.
The distribution of the feed and the products at different conversion degrees, shown in figure 1, points out that during hydroconversion there is a progressive decrease of the heavier components with a contemporary increase of the lighter compounds mainly in the $C_{10}$-$C_{20}$ range. The selectivity for naphtha (i.e. $C_5$-$C_9$) formation is significantly lower and the production of $C_1$-$C_3$ compounds is negligible till $C_{22+}$ fraction conversion value up to 85-90%.
The low production of $C_1$-$C_3$ compounds and the constancy of the $C_{10}$-$C_{14}$/$C_{15}$-$C_{22}$ ratio are consistent with a situation where the consecutive reaction of the first formed products is absent or scarce and hydrocracking of each normal paraffin occurs according to "ideal hydrocraking", where with the exception of the three terminal C-C bonds, the probability of scission of the C-C bonds is similar. [4].

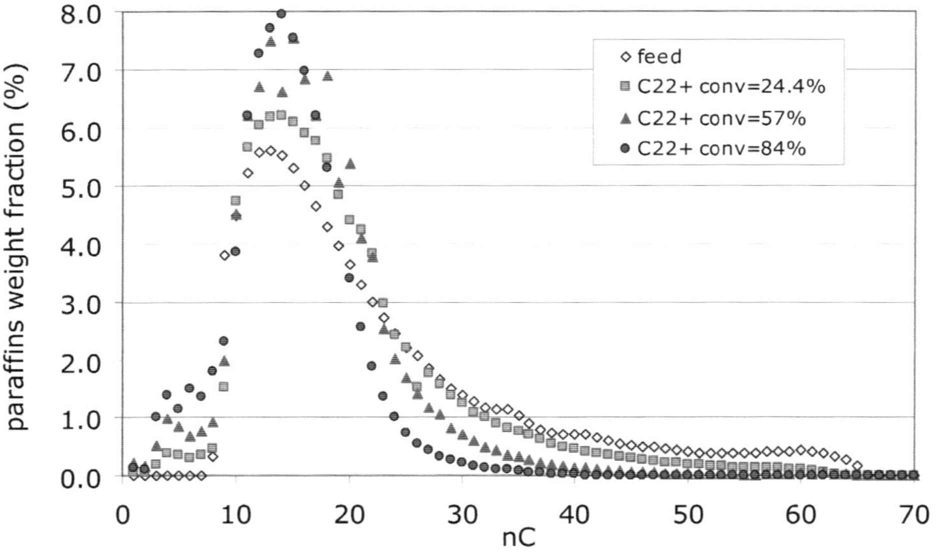


**Figure 1.** Products yields distribution at different conversions.

Plots reported in figure 2 show that normal paraffins present a distribution as a function of chain length shifted towards lower molecular weights in comparison with the distribution displayed by the iso-paraffins. Particularly, the peaks of normal and iso-paraffins are centered at $C_{12}$ and $C_{20}$ respectively. This shift of the distribution maximum for the branched products can be explained in the light of the hydroconversion reaction pathway and reactivity of n-alkanes as a function of chain length. In this regard the higher isomer content of heavier aliphatic chains would merely reflect the increasing conversion degree of the single n-paraffin due to its higher reactivity in comparison with the lighter ones [5]. This can be accounted for by the fact that the reactivity is expected to be proportional to the number of secondary carbon atoms per molecule or, as proposed by Sie, to $C_{n-6}$ and $C_{n-4}$ for hydrocracking and hydroisomerization respectively [6]. However, as demonstrated by several works [7, 8], the higher reactivity of heavier n-paraffins can be also ascribed to their stronger physisorption, which in turn is the consequence of the direct relationship between heat of adsorption and chain length. However, the overall reactivity of normal paraffins could be also significantly affected by the vapor liquid equilibrium (VLE) inside the reaction environment. In this regard, it as been recently shown that in hydroconversion of alkanes the physisorption behavior in liquid phase is very different from that observed in vapor phase [9, 10]: more specifically, both in liquid and vapor phase the heaviest alkanes are more reactive, but the difference in reactivity is much more pronounced in vapor phase than in liquid phase. Those changes in operating conditions that lead to an increase in conversion, also favor the vaporization of the hydrocarbon mixture and thus the liquid phase is enriched in heavier paraffins: the heavier and more reactive n-paraffins, in the liquid phase, are almost completely converted into the corresponding iso-paraffins, which are in turn cracked to lighter products. This can explain the difference in the two distributions shown in figure 2.

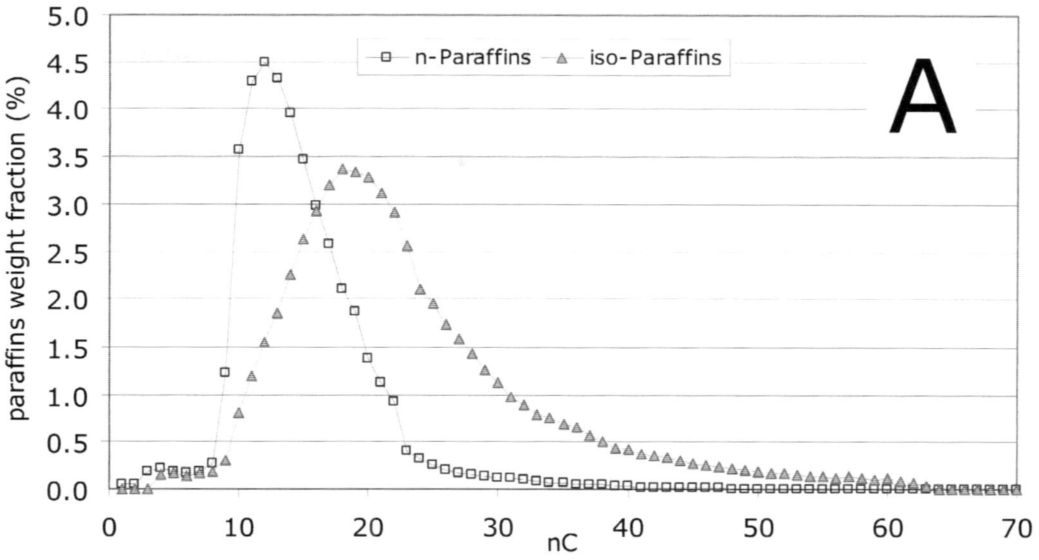


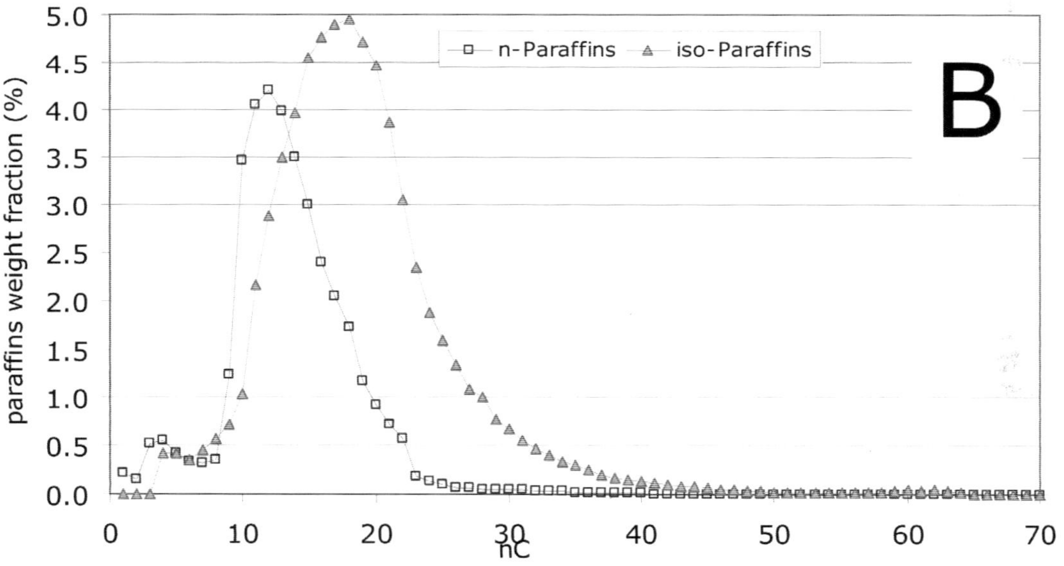


**Figure 2.** Normal and isoparaffins yields distribution at different $C_{22+}$ conversions (A: conversion=24.3%; B: conversion=57%), for the tests presented in figure 1

Kerosene and gasoil specifications require a freezing point of -47°C and a cold filter plug point of -5÷ -20 °C respectively. In this regard, one of the major drawbacks of middle distillates derived from FT synthesis is strictly connected with their paraffinic nature which confers very poor cold flow properties. In this circumstances the isomerization reaction holds a fundamental

role in FT wax upgrading because it allows to lower the melting point of paraffinic compounds with a consequent improvement of the cold flow properties of middle distillates.
The data reported in figure 3 clearly show for both kerosene and gasoil fractions a speedy increase of the isomer content at the lower bound of conversion followed by a slower increase reaching values of ~0.85 and ~0.6 for $C_{15}$-$C_{22}$ and $C_{10}$-$C_{14}$ respectively.

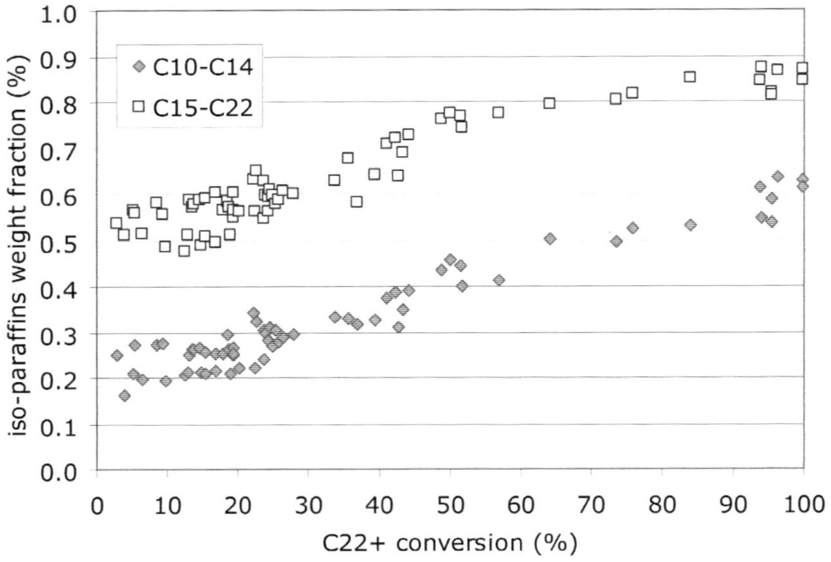


**Figure 3.** Isoparaffins yields for kerosene (C10-C14) and diesel (C15-C22) cuts vs. C22+ conversion.

Although a direct relationship between isomer content and melting point is from a qualitative stand point, is expected the extent of the decrease significantly depends on the degree, the position and the length of the branching. Branching positions in the middle of the chain have a higher effect than those near the end of the chain, while longer side chains lead to a higher decrease of melting point. It is therefore important to get a detailed characterization of the isomer fraction in order to achieve a reliable esteem the cold flow properties of middle distillates.
The results reported in figure 4, show that both mono and multibranched isomers concentrations in the $C_{22-}$ fraction increase with conversion, the concentration of mono-branched is always higher than multibranched but the latter are progressively more important with the increase of conversion degree till to became the major component at high conversion levels. The observed trend is in agreement with the reaction pathway of n-alkane occurring through a series of consecutive reactions, where the n-alkane is first isomerized into monobranched isomers which undergo subsequent isomerization steps and cracking reaction [11].
A more detailed analysis reveals that monobranched isomers lump is a mixture of methyl-paraffins, with lower amounts of ethyl-paraffins and traces of propyl-paraffins. Results reported in figure 5 display that monomethyl paraffins are by far the most abundant isomers and the

ratio mono-methyl/mono-ethyl/mono-propyl up to 50% conversion of C22+ fraction is approximately 100/10/1. Furthermore, the concentration of mono-methyl isomers increases with conversion up to about 60% and remains constant thereof, while ethyl isomers weight fraction increases in the whole range.

Formation of methyl branching can be readily explained according to protonated cyclo propane mechanism (PCP) [12], while, by analogy, the formation of ethyl and propyl isomers may result by the formation of protonated cyclobutanes and cyclopentanes as intermediate stage respectively. With increasing the number of carbon atoms of the protonated rings their formation probability rapidly decreases and consequently the formation of ramifications longer than methyl [13]. Alternatively, as proposed by Weitkamp [14], ethyl and propyl branching may be the result of consecutive isomerization reaction of methyl isomers via classical isomerization of type A. A clear distinction between the two routes is not possible presently. However we notice that the increase of ethyl branching concentration at the higher bond of conversion supports the view that at least part of ethyl branching may be formed through consecutive reaction of mono methyl isomers.

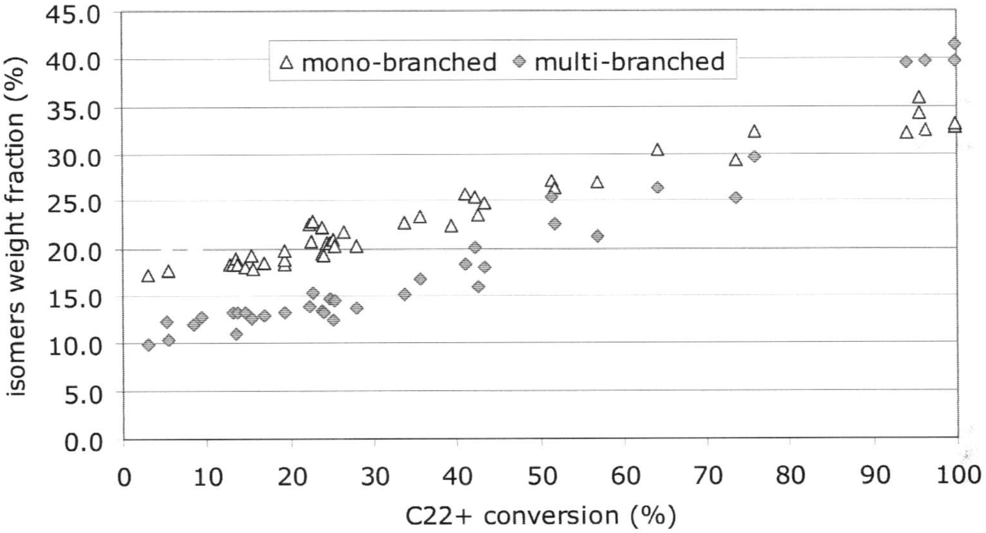


**Figure 4.** Mono and multi-branched isomers yields in the C22- fraction vs. C22+ conversion.

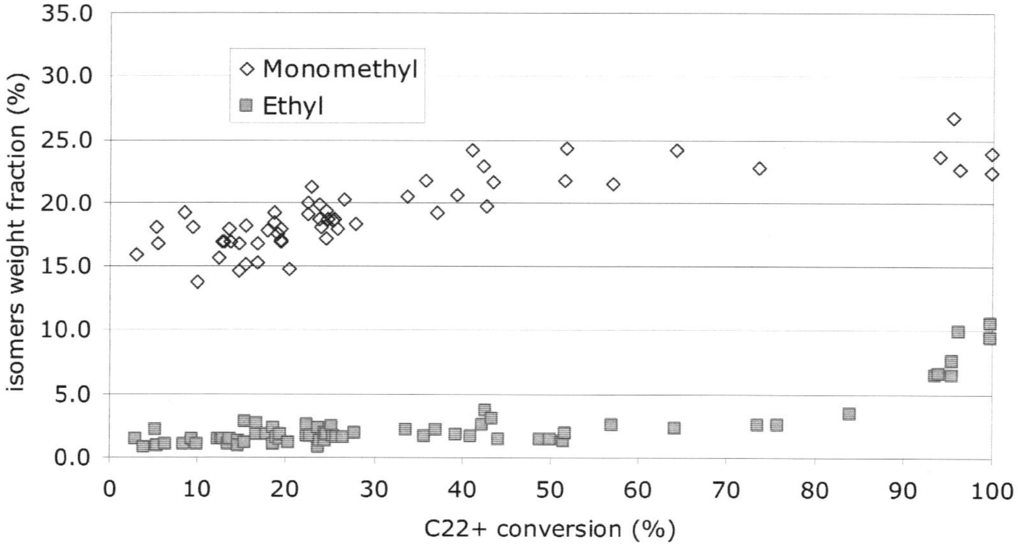


**Figure 5.** Monomethyl and ethyl isomers yields in the C22- fraction vs. C22+ conversion.

## Conclusion

The hydrocracking behavior of a $C_{10}$-$C_{70}$ Fischer-Tropsch derived wax was investigated: experimental results show that during hydrocracking longer aliphatic chains are hydrocracked preferentially leading to an increase of middle distillate cut. Furthermore, the conversion was also characterized by a low naphtha selectivity and scarce formation of $C_1$-$C_4$ fraction. The latter data together with the evolution of Kerosene/Gasoil ratio indicate an "ideal hydrocraking" mechanism, where consecutive reactions of the primary products are negligible. The isomer content of reaction products shows a strong increase with conversion reaching values of 80% and 60% for the $C_{15}$-$C_{22}$ and $C_{10}$-$C_{14}$ fractions respectively. A detailed analysis of the $C_{22-}$ fraction in the product mixture evidenced that isomers are mainly made up of monobranched paraffins, while multibranched are more abundant at the highest conversion; this behavior is consistent with the accepted hydroconversion pathway of n-parafins.

### *References*

1. Dry, M.E. *Catal. Today,* **2002**, *71, 227.*
2. Murphy, M.J., Taylor, J.D., Mc Cormic, R.I. Compendium of Experimental Cetane Number Data, NREL/SR-540-36805 (2004).
3. Handbook of Chemistry and Physics, D.R. Lide (Ed.), 85$^{th}$ Edition 2004-2005, CRC Press
4. Weitkamp, J., *Erdöl, Kohle, Erdgass, Petrochem*, **1978**, *31, 13.*
5. Weitkamp, J. "ACS Symposium Series"; J.W. Ward, S.A. Quader, Eds.; American Chemical Society Washinton DC, 1975, vol. 20, p. 1.

6. Sie, S. T. *Ind Eng. Chem. Res.*, **1993**, *32, 403.*
7. Denayer, J.F.; Baron, G.V.; Jacobs, P.A.; Martens, J.A. *Phys. Chem. Chem. Phys.*, **2000**, *2, 1007.*
8. Denayer, J.F.; Baron, G.V.; Martens, J.A.; Jacobs, P.A. *J Phys. Chem. B*, **1998**, *102, 3077.*
9. Denayer, J.F.M.; De Jonckheere, B.; Hioch, M.; Marin, G.B.; Vanbustele, G.; Martens, J.A.; Baron, G.V. *J. Catal,.* **2002**, *210, 445.*
10. Denayer, J.F.M.; Ocakoglu, R.A.; Huybrechts, W.; De Jonckheere, B.; Jacobs, P.; Calero, S.; Krishna, R.; Smit, B.; Baron, G.V.; Martens, J.A. *J. Catal.*, **2003**, *220, 66.*
11. Steijns, M.; Froment, G.; Jacobs, P.; Uytterhoeven, J.; Weitkamp, J. *Ind. Eng. Chem. Prod. Res. Dev.,* **1981**, *20, 654.*
12. Sie, S. T. *Ind Eng. Chem. Res.*, **1992**, *31, 1881*
13. Marcilly, C. "Catalyse acido-basique. Application au raffinage et à la pétrochimie", Editions Technip 2003, vol.1, chapter 3, p. 168.
14. Weitkamp, J. *Ind. Eng Chem Prod. Res. Dev.*,**1982**, *21, 550.*

**ExxonMobil MIDW™ Process – Technology and Commercial Experiences for Production of Low-Cloud Point ULSD**

Stuart S. Shih and Anna B. Gorshteyn
ExxonMobil Research and Engineering Company, Fairfax, Virginia, USA

## INTRODUCTION

The worldwide demand on middle distillates including diesel, kerosene, and jet fuels are increasing. Meanwhile, specifications of the distillate fuels also are rapidly changing and become more stringent in response to the environmental concerns and increased fuel efficiency. The production of ultra-low sulfur diesel (ULSD) is the current focus. Other fuel qualities, such as cetane,aromatics, density and cold flow property are parts of the matrix for production of premium diesel fuels. The cold flow property is generally expressed as cloud point, pour point, or cold filter plugging point (CFPP). Most specifications require the maximum diesel cloud point be equal or lower than the lowest expected ambient temperature. Consequently, the clod flow property specifications can be country-specific, regional, and seasonal.

In 1996, ExxonMobil announced a new fuels dewaxing process, Mobil Isomerization Dewaxing or more simply MIDW™ [1,2] to lower the pour point of waxy gas oil and distillates by hydroisomerization of linear paraffins to branched paraffins. High yields of the premium, low pour point distillates are obtained by retaining the isomerized paraffins in the distillate boiling range. The process is very flexible. It can be either a dedicated process or combined with hydrotreating and hydrocracking to produce high yields of premium low-cloud point ULSD from a wide range of feed stocks. The paper will focus on commercial MIDW experiences of different process configurations to illustrate the technology flexibility.

## PROCESS CHEMISTRY

The MIDW™ Process chemistry is quite simple: linear paraffins are isomerized to branched paraffins for improving the cold flow property while hydrogenation reaction is optimized to improve the fuel quality but side reaction, such as hydrocracking, is minimized to preserve a high distillate yield.

It is known that melting point of paraffins depends on molecular weight (carbon number) and degree of branching (isomerization). Table 1 lists the melting points of n-$C_{14}$, n-$C_{18}$, n-$C_{19}$, and selected isomers which are in the diesel boiling point range to illustrate the importance of isomerization for the cold flow property improvement.

Table 1: Effects of Branching (Isomerization) on Melting Point

| Hydrocarbon | Melting Point, °C |
| --- | --- |
| n-Tetradecane | 5.0 |
| 7-Methyltridecane | -37.2 |
| | |
| n-Octadecane | 28.2 |
| 2-Methylheptadecane | 5.7 |
| | |
| n-Nonadecane | 32.1 |
| 7-Hexyltridecane | -28.3 |

## PROCESS CONFIGURATIONS

The MIDW™ Process is a flexible process. Different process configurations have been developed and commercialized to meet various objectives from various feed qualities. The flow diagram of the MIDW™ Process is very similar to the conventional desulfurization or hydrotreating processes. The MIDW catalyst, the backbone of the process, is highly tolerant to ammonia and hydrogen sulfide. Consequently, it can operate in either "sweet" or "sour" conditions depending on the presence of sulfur and nitrogen in the feed ammonia and hydrogen sulfide in the treat gas. Depending on feed quality (sulfur, nitrogen, and boiling point range), at least four process configurations have been developed and commercialized:

**Configuration I**
MIDW alone: Clean feeds which are very low in sulfur and nitrogen.

**Configuration II**
Cascaded HDT/MIDW: Feeds with moderate to high S and N

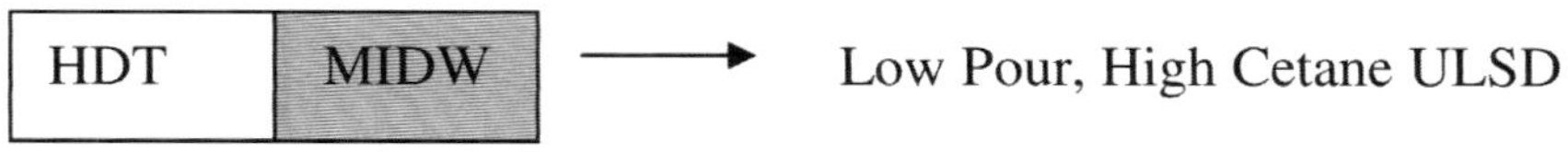


**Configuration III**
Two-stage HDT/MIDW: Feeds with high S and N

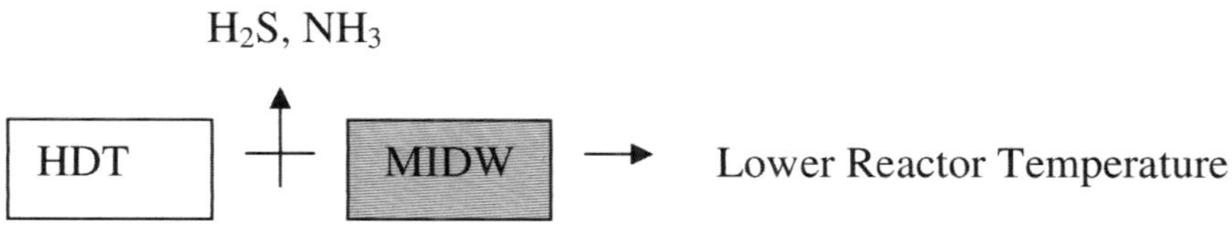

**Configuration IV**
Combined with VGO MPHC to increase premium ULDSD yields

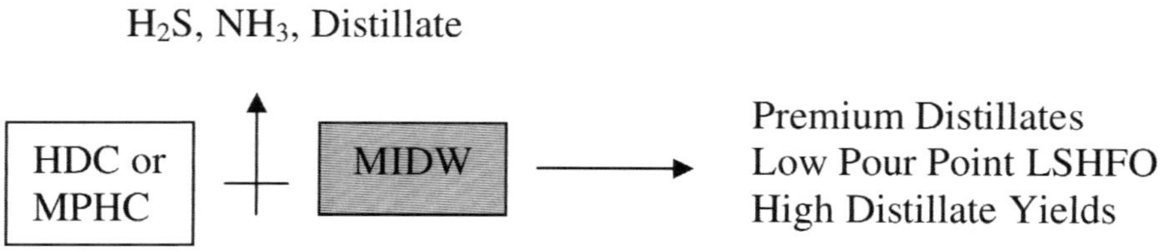


Depending on treat gas composition, the MIDW™ process can be operated under "sour" conditions, such as the Configuration II, and "sweet" conditions, such as the Configurations I, II, and IV. Feed to the MIDW process is not limited to the middle distillate boiling range material; it can be VGO, coker gas oil, light cycle oil (LCO) derived from FCC, and feed components separated from synthetic crude oils. The difference between the Configurations III and IV is the feed boiling range only (VGO vs distillate). Feed to the Configuration IV is heavier than that for the Configuration II and III. Except for the Configuration I, commercial experiences for the Configurations II, III, and IV will be discussed in more details.

## COMMERCIAL EXPERIENCES

### Configuration II: Cascaded HDT/MIDW

Refiners can take advantages of the MIDW™ process by combining it with hydrotreating technology offered by ExxonMobil to produce low-cloud ULSD in one unit. Currently, there are three MIDW units in operational using the Configuration II approach. For simplicity, only one commercial unit will be discussed.

Refinery A

This MIDW unit consists of one reactor filled with desulfurization and MIDW catalyst. Feed to the MIDW process is primarily straight-run diesel boiling range material. The sulfur and nitrogen contents are 0.2-0.3 wt% S and 90-130 ppmw N, respectively. The unit does not have an amine scrubber. Consequently, hydrogen sulfide in the recycle gas was quite high (>0.5 vol%). The feed with 4-8°C cloud points was dewaxed to a delta cloud point (feed CP minus product CP) of 10-20°C depending on the seasonal cloud point requirements. The unit produced <10 ppmw S ULSD in the entire cycle.

As shown in Fig. 1, the MIDW catalyst performed well under these conditions. The MIDW catalyst lost only 30°C in terms of operating window. However, most of the operating temperature increase occurred during the lining out period in the first 150 days on stream.

Fig. 1: MIDW Performance: Refinery A

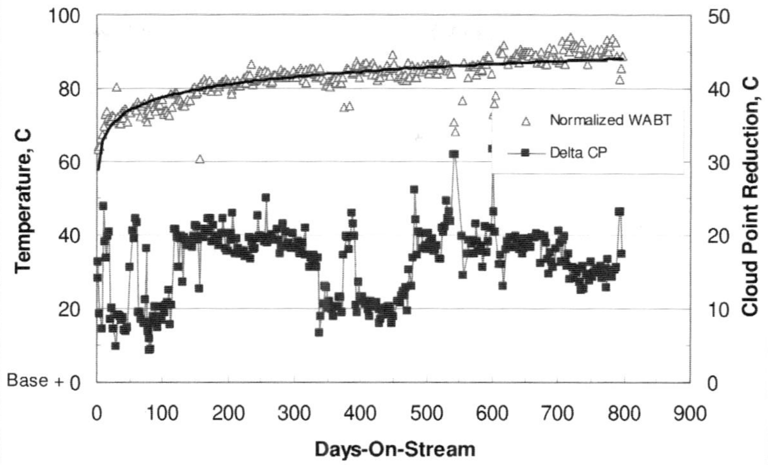


## Configuration III (and Configuration I) Commercial Experience

The Configuration III performance operating under the "sweet" conditions is expected to be better than that demonstrated for the Configuration II under the "sour" conditions. At least, the cycle length of the MIDW catalyst would be much longer. The Configuration III (and IV) has potential to improve diesel quality by the enhanced hydrogenation activity of the MIDW catalyst.

Refinery B

It is a revamped unit for the Refinery B. An upstream process provides high-cloud but ultra-low sulfur (<10 ppmw S) heavy (nominal $275^{o}C^{+}$) diesel as the feed to the MIDW unit. Consequently, the MIDW unit is operating under the "sweet" conditions.

The average cloud point of the feed is $-3.5^{o}C$. To maximize the blending flexibility, the refinery further separates the low-cloud products into light diesel (nominal $320^{o}C$ cut point) and heavy diesel (nominal $320^{o}C^{+}$ bottoms). The operating conditions are originally targeted at producing $-40^{o}C$ cloud point for the dewaxed light and heavy diesel fractions (Fig.2). The actual cloud points for the first winter cycle was much lower (less than $-70^{o}C$ cloud point) for both the light and heavy diesel fractions. Note that the cloud point measurement limit is $-70^{o}C$.

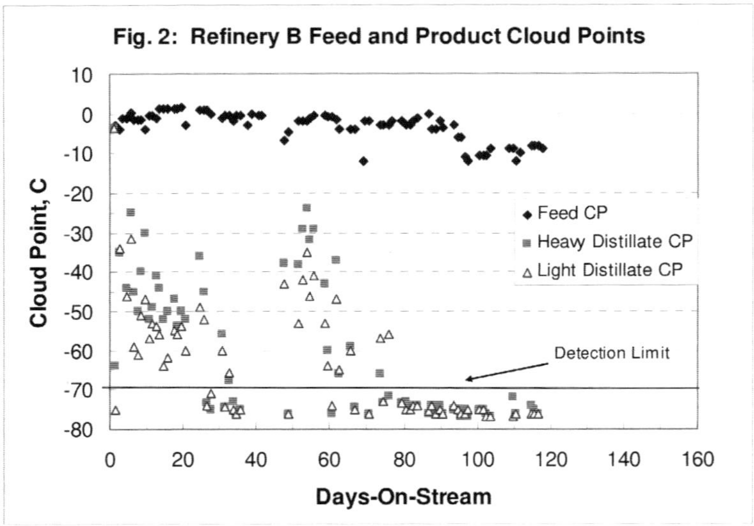


As expected, the catalyst was stable for the first winter season operation. The ultra-low cloud point makes the reactor normalization data scattered somewhat. The raw reactor temperature (or WABT, weighted average bed temperature) appeared to be lining out at 70°C above the temperature base after 40 days on stream before a temporary shutdown. After the re-commission, the catalyst was unexpectedly lost about 10°C in its activity. The cause of this activity loss was not clear. The catalyst appeared to line out at 83°C above the base and stable while the cloud point of the heavy diesel fraction reached less than -70°C.

The diesel selectivity was very high. The refinery reported that the total diesel yields (light plus heavy diesels) were greater than 97 vol% at -40°C cloud point.

The capability to produce very low cloud point (<-70°C) by the MIDW™ process allows the refinery to further increase the low-cloud diesel throughput by blending some high-cloud point components into the diesel pool. Alternatively, if the kerosene (or jet fuels) price is favorable, the MIDW™ process also allows the refinery to market more kerosene instead of blending the kerosene into the low-cloud diesel pool.

The unit is designed for the production of artic diesels and will be operating during the winter season only. Due to the excellent stability under the "sweet" conditions, the cycle length is predicted to be greater than three years on the actual operation days or greater than 6 years on the calendar days.

## Configuration IV Commercial Experiences

The feed to the Configuration IV is also low in sulfur and nitrogen but its end point is much higher. The objective of this process configuration is to enhance premium kerosene and diesel production from heavy vacuum gas oil (VGO) by combination with a moderate pressure hydrocracker (MPHC) or high pressure hydrocracker (HDC).

Refinery C

The Refinery C implemented MPHC and MIDW to convert more than 65% of a 337-510°C VGO into kerosene and diesel. For simplicity, only the MIDW™ process will be discussed. Note that ExxonMobil also designed the MPHC Process for this refinery.

The Refinery C took the capability of the MIDW process to produce premium kerosene (jet fuel) and diesel by operating the MIDW conditions at slightly more severe than the diesel deaxing applications which have been discussed earlier. Feed to the MIDW reactor is a 350-510°C MPHC bottoms available in the refinery. The MIDW feed contains about 200 ppmw sulfur and its pour point is greater than 38°C. Table 2 summarizes process yields and product quality.

Table 2: MIDW Process Yields and Product Quality

| | Feedstock | Naphtha | Kerosene | Diesel | LSHFO |
|---|---|---|---|---|---|
| Boiling Range, °C | 337-510 | $C_5$ - 150 | 150-260 | 260-387 | $387^+$ |
| Yields, wt% | 100.0 | 9.7 | 13.6 | 24.7 | 50.7 |
| Gravity, °API | 32.0 | 73.0 | 49.5 | 34.7 | 29.5 |
| Sulfur, ppmw | 200 | <1 | <2 | <15 | 40 |
| Smoke Point, mm | | | 32 | | |
| Freeze Point, °C | | | <-54 | | |
| Pour Point, °C | >38 | | | -43 | -7 |
| Cetane Index | | | 52 | 56 | |
| P/N/A, wt% | 44/39/17 | | -/-/10 | 45/31/24 | 36/42/22 |

As shown in Table 2, the MIDW™ process produces high yields of premium kerosene and diesel. The kerosene product has a high smoke point (32 mm) and is suitable for jet fuel blending. The low pour diesel has an excellent cold flow property and high cetane index. The 387°C$^+$ bottom is used as a heavy fuel oil blending component for this refinery but it is also an excellent FCC feed.

As shown in Fig. 3, the MIDW catalyst was very stable. It had accumulated more than 4 years on-stream. The MIDW catalyst took about one year to line out for this unit. After the lining-out, the catalyst aging rate was minimal. The volumetric distillate (kerosene plus diesel) yields were constant, varying between 35-45 vol% during the whole 4-year cycle

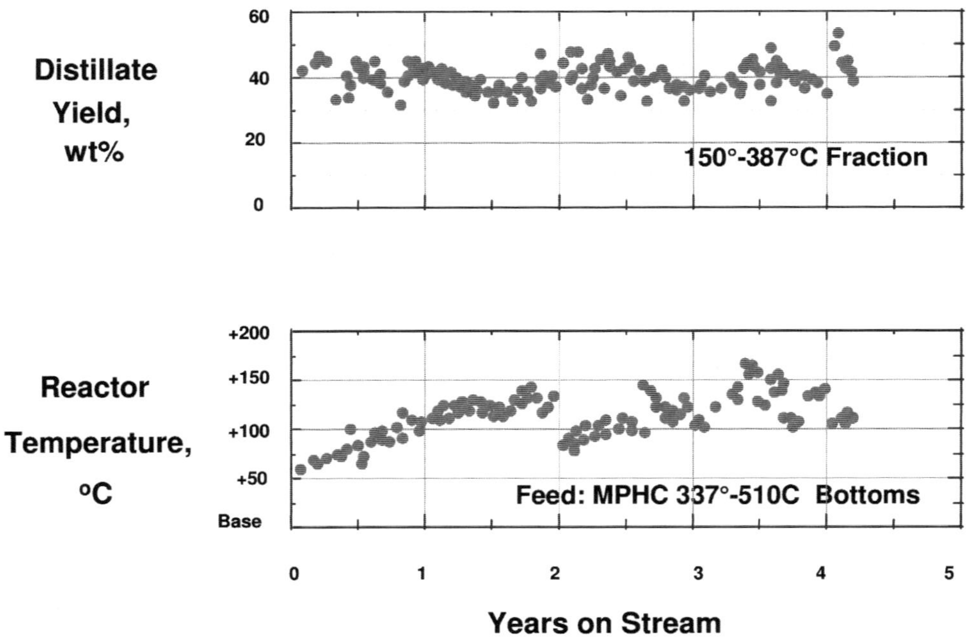


Fig. 3: Distillate Yield and MIDW Catalyst Stability for the Refinery E.

## SUMMARY

The MIDW™ Process is a commercially proven technology. It is a flexible and attractive process for the production of low-cloud diesels. The process can operate under either "sweet" or "sour" conditions. By combining with ExxonMobil hydrotrating techlogies[3,4], it can produce low-cloud, <10 ppmw S ULSD at a high yield. It also can be combined with MPHC or hydrocracking process to produce high yields of premium distillates from VGO boiling range materials.

Depending on feed quality and refinery needs, there are four process configurations have been developed and commercialized. The commercial experiences demonstrate that the MIDW catalyst is stable and produces high diesel yields are consistent with the original predictions.

## REFERENCES

1. P. J. Angevine, F. M. Buyan, D. A. Pappal, and R. D. partridge, "MIDW – Mobil's New Flexible Dewaxing Process", The European Refining Technology Conference, London, England, October 28-30, 1996.
2. W. J. Tracy, G. K. Chitnis, W. J. Novak, and T. E. Helton, "ExxonMobil MIDW Process: Using Selective Dewaxing and Advanced Hydrotreating Catalysts", The 3rd European Catalyst Technology Conference, Amsterdam, The Netherlands, February 26-27, 2002.

## 20d Ecofining Process Overview

Peter Kokayeff[1], **T.L. Marker**[1], Chris Gosling[1], Giovanni Faraci[2], Franco Baldiraghi[2], Silvia Guanziroli[2], and Luigina Sabatino[2]. (1) Refining Process Development, UOP LLC, P.O. Box 163, Riverside, IL 60546, (2) Refining, Eni, Milan, Italy

The EcofiningTM process, jointly developed by UOP and ENI, converts vegetable oil, grease, tallow and algae oil to paraffins at mild conditions yielding an excellent diesel blending stock. This new process, utilizes decarboxylation, hydrodeoxygenation and hydroisomerization reactions to produce a high quality diesel fuel from bio-oils plus hydrogen at mild conditions. The EcofiningTM process is currently offered for licensing and two EcofiningTM process units are in construction with expected start-up in 2009.

The talk will include data on the following:

1 Yields and effect of alternative feeds

2 Green diesel product properties and blending synergisms

3 Life cycle analysis

4 Economics and capital costs

## 20e Feed-Flexible Processing of Oil-Rich Crops to Jet Fuel

**Michael J. McCall**[1], Joseph Kocal[2], Alak Bhattacharyya[1], and Amar Anumakonda[3]. (1) R&D, UOP LLC, 25 E. Algonquin Rd., Des Plaines, IL 60017, (2) UOP, 50 East Algonquin Road, Des Plaines, IL 60017, (3) Renewable Energy, UOP LLC, 25 E. Algonquin Rd., Des Plaines, IL 60017

UOP is leading a project to demonstrate the efficient conversion of vegetable and other natural oils to jet fuel. The goal of the project, which is funded by the Defense Advanced Research Projects Agency (DARPA), is to develop and commercialize a process to produce Jet Propellant 8 (JP-8), which is used by U.S. and NATO militaries.

Approximately 4.5 billion gallons of JP-8 fuel are used by the U.S. Air Force, U.S. Army and NATO annually. The kerosene-based, high-performance fuel is less flammable and less hazardous than other fuel options, allowing for better safety and combat survivability. In addition to jets, JP-8 is also used to fuel heaters, stoves, tanks, and other vehicles in military service. Commercial airliners use Jet A and Jet A-1, which is also kerosene-based.

UOP is working with Honeywell Aerospace, Cargill Inc., Arizona State University, Sandia National Laboratories and Southwest Research Institute on the 18-month project which will be completed by the end of 2008. Fuel produced by the new process will have to meet stringent military specifications and is expected to achieve high energy efficiency for maximum conversion of feed to fuel, reduced waste and reduced production costs. The technology will be viable for future use in the production of jet fuel for commercial jets.

The jet fuel technology builds upon the UOP/Eni EcofiningTM process, which was developed by UOP and European energy company Eni to convert vegetable oils and waste into a high-cetane green diesel fuel with low emissions and high efficiency. The new jet fuel process produces a paraffinic fuel that meets JP-8 specifications, and therefore aligns with today's standard refinery practices, a key focus of UOP's Renewable Energy and Chemicals business unit. The presentation will discuss the progress towards the project goals.

## 21a Novel Reverse-Selective Membrane for Refinery Applications

**Alvin Ng**, Douglas E. Gottschlich, and David Johnstone. Membrane Technology & Research, Inc, 1360 Willow Road, Suite 103, Menlo Park, CA 94025

Over the last decade MTR has successfully developed and commercialized an innovative "reverse-selective" membrane. This membrane preferentially permeates the larger, more condensible hydrocarbon components over the smaller, lighter compounds. To date the largest use of these membranes has been for petrochemical applications, most notably in olefin and nitrogen recovery from polyolefin vent streams. However, this innovative membrane is also an attractive solution for treating various refinery streams. Refineries produce a variety of purge streams from hydrotreating and other cracking/conversion processes. These streams, while containing valuable compounds are too dilute for direct reuse, and therefore are typically used as fuel or flared. Traditionally a size-selective membrane has been used to treat these purge streams, primarily to recover hydrogen. However, in some situations the conventional hydrogen membrane is not suitable, particularly when retrofitting an existing hydroprocsssing unit with limited compression capacity. In addition, the size-sieving hydrogen membrane does not perform well when treating low-pressure, low concentration hydrogen streams, such as fuel gas streams. In contrast, MTR's membrane offers an innovative and economic solution capable of providing hydrogen enrichment as well as the potential to recover C3+ hydrocarbons valuable LPG.

## Benzene Reduction from Gasoline by Pervaporation Using Cation-Exchanged Membrane Containing Metal Ions

Maoqi Feng, Division of Chemistry and Chemical Engineering, Southwest Research Institute, San Antonio, TX 78238, email: mfeng@swri.org.

**Abstract**: The EPA's new regulations under Mobile Sources Air Toxics Phase 2 (MSAT2) is forcing refiners to reduce the annual average benzene content in gasoline to 0.62 vol% effective Jan. 11, 2011. After July 1, 2012, a maximum benzene content in gasoline will be 1.3 vol%. The regulations allow refineries to trade credits to meet the 0.62 vol% specification. Typical sources of benzene contributions to gasoline include reformate, FCC naphtha, light straight run naphtha, and VGO hydrocracked naphtha, etc. Differences on sorption, diffusion, and pervaporation properties between benzene and substituted benzene on cation-exchanged membranes are the driving force for benzene separation. Reduction of benzene by membrane pervaporation is more attractive than by hydrogenation. The paper describes different types of polymer membranes and composite membranes based on polyelectrolytes that could be synthesized by interfacial reaction with oppositely charged ionic surfactants, or by cross-linking with metal ions. A series of cations, Cu(II), Ag(I), and Cu(I) were discussed.

Keywords: benzene, membrane, pervaporation, separation

### Introduction

The current level of benzene in gasoline is 0.97% - 1.3% by volume. As we know, benzene is carcinogenic, and the toxic benzene emission renders a big health problem in urban area. Since EPA announced a new rule on Feb. 9, 2007, which limits the benzene content of gasoline to an annual refinery average of 0.62% volume by 2011, there is an increasing demand for benzene reduction in petroleum products, and especially in gasoline, diesel and kerosene. The current benzene elimination route is by hydrogenation on transition metal supported catalysts under high pressure and temperature. Unfortunately, those catalysts are very susceptible to poisoning through the small amounts of sulfur present in the feed stream.

The typical benzene-reduction process, hydrotreating, which requires high temperature and pressure, is energy-intensive, and big investment. According to EPA, the new rule will cost an average of 0.27 cents per gal of gasoline for benzene removal by hydrotreating, reflecting the average $14 million in capital investment needed at each refinery to reduce benzene levels in gasoline.

Membrane pervaporation provides an alternative method for benzene reduction. Compared with the classical separation methods, e.g., distillation, rectification, extraction and crystallization, membrane processes are faster, more economical, energy saving and ecologically friendly. Membrane pervaporation can also be used in combination with distillation (membrane-distillation column hybrid processes) in order to lower operating costs and produce higher throughput. In addition, membranes can separate azeotropic and close-boiling mixtures, which would be difficult or impossible to separate by distillation alone. However, the separation of aromatics/aliphatics by pervaporation is not yet accomplished on industrial scale.

## Membranes Selection

Polymers are the most used membrane materials. The driving forces for mass transport of aromatics/aliphatics through polymer membranes are temperature, pressure, chemical potential gradients, external force field, and the nature of polymer material and penetrating component, etc [1]. The mechanism for membrane pervaporation is complicated and still unclear.

For separation of aromatics/aliphatics, cyclohexane is commonly used as an aliphatic model compound. Benzene and cyclohexane have very similar molecular size and mass. For separation of benzene-cyclohexane binary mixture by pervaporation, the diffusion selectivity of benzene for 7 nonporous polymer membranes is listed in Table 1 [2].

In pervaporation, mass transfer through a nonporous membrane occurs by a solution-diffusion mechanism. The mass transport process of gases or vapors through nonporous polymer membranes can be described as a sequence of consecutive processes of sorption of the components on the one side, activated diffusion through membrane and desorption of the components on the other side of the membrane. Selection of a suitable membrane partly depends on the relative solubility of the membrane polymer in the organic components to be separated.

Table 1. Comparison of the flux and selectivity of different membranes for pervaporative separation of a 1:1 benzene-cyclohexane mixture [2]

| Membrane | Diffusion selectivity for benzene | Flux, $kg/m^2$ h | Temperature, °C |
|---|---|---|---|
| Polyethylene | 1.63 | 1.25 | 25 |
| Treated Poly(vinylidene fluoride) | 5.45 | 0.148 | 56 |
| Polyphosphonate | 13.3 | 0.08 | 30 |
| Styrene-acrylic block copolymer | 9.57 | 4.836 | 20 |
| Poly(methyl methacrylate) grafted on HDPE | 7.5 | 2.55 | 50 |
| Polyphenylquinoxalane | 8.5 | N/A | N/A |
| Poly(acrylonitrile-*co*-styrene) | 22.5 | 0.263 | 30 |

Preferential sorption of benzene compared to cyclohexane is observed particularly in the polyurethane (PU)-membrane and less pronounced in the polyether-polyamide block-copolymer (PEBA)-membrane [3], which is not shown in Table 1.

It's well known that swelling of the membrane can be avoided by crosslinking as it improves the chemical stability and resistance. Therefore, crosslinking of the above polymers is preferred.

Polyimides have been found to be very attractive as membrane materials for benzene/aliphatics separation, as they have better separation characteristics compared to other glassy polymers, e.g. polycarbonates, and they also offer good thermal and chemical resistance and are easy to process. In order to improve the separation characteristics and to assure applications over 100° C in aromatic/aliphatic systems, cross-linking of the polymer is necessary [4]. Cross-linked copolyimides are very stable, they can be used for aromatic/aliphatic mixtures up to 150°C with an aromatic feed concentration of around 30%.

Different polymer membrane may be used for separation of benzene from other aromatics, such as toluene and xylene. Nitrile rubber membrane (Aparene-N 553 NS) with a bound acrylonitrile content of 34% and blended with 5% ZnO was reported to separate benzene from toluene and xylene [5].

During the pervaporation, the sorption coefficient that is related to the equilibrium sorption of the penetrant is calculated using the equation S =Ms/Mp, where Ms is the mass of the solvent at equilibrium swelling and Mp is the mass of the polymer sample. These values are shown in Table 2.

The sorption coefficient value decreases with increasing nitrile rubber membrane content in the blend composition because the solvent resistance increases with increasing nitrile rubber membrane content. The values of permeability coefficient P, which implies the net effect of sorption and diffusion, are given in Table 3.

Table 2. Values of sorption coefficient (S) (g/g) [5]

| Membrane | Benzene | Toluene | p-Xylene |
|---|---|---|---|
| Nitrile rubber | 3.48 | 3.45 | 3.39 |

Table 3. Values of permeability coefficient ($P \times 10^5$) ($cm^2 s^{-1}$) [5]

| Membrane | Benzene | Toluene | p-Xylene |
|---|---|---|---|
| Nitrile rubber | 14.70 | 11.90 | 11.25 |

For separation of benzene from di- and trinuclear aromatics, polyimide membrane was reported to permeate di- and trinuclear aromatics with almost no loss of selectivity, if a certain temperature is exceeded [6]. The permeability can also be influenced by other parameters i.e. the chain length of the cross-linker, the permeate pressure and pre-treatment of the membranes.

## Composite Membranes

To improve the separation performance of the polymer membranes, hybrid membrane can be used. A hybrid membrane was prepared by filling carbon graphite (CG) into poly (vinyl alcohol) (PVA) and chitosan (CS) blending mixture [7]. Homogenous distribution of graphite particles is important, because it will alter the hydrogen bonding interaction, decrease the crystallinity degree, dramatically enhance mechanical properties, and significantly increase free volume in CG-PVA/CS hybrid membranes.

The pervaporation properties of the hybrid membranes for separation of benzene/cyclohexane mixtures were investigated through varying graphite content and mass ratio of PVA to chitosan in hybrid membranes. Sorption studies indicated that the degree of swelling of CG-PVA/CS hybrid membranes increased with graphite content increasing and the separation factor of benzene/cyclohexane was dominated by solubility selectivity rather than diffusivity selectivity. The concentration-averaged diffusion coefficients of benzene were larger than those of cyclohexane in all CG-PVA/CS hybrid membranes with different graphite loading.

Compared to PVA and PVA/CS membranes, higher permeation flux and separation factor of CG-PVA/CS hybrid membranes were obtained. The highest separation factor achieved was 59.8 with a permeation flux of 124.2 g/($m^2$ h) at 50 °C, 1 KPa. The diffusion coefficient of benzene decreased as graphite content increased from 0 to 10 wt.%.

The effect of graphite filling on diffusivity of permeant may be explained as follows: the incorporation of graphite resulted in lower crystallinity and higher swelling ratio of membranes, which enhanced the diffusion of benzene and cyclohexane; the benzene adsorbed on to graphite particles could not diffuse freely due to the interaction between benzene and graphite, which decreased the diffusion rate of benzene. In addition, the non-porous graphite particle served as the barrier for the diffusion of permeant. The diffusion of cyclohexane decreased as graphite content ranged from 6 to 10 wt.% because of the barrier effect.

Carbon nanotubes (CNTs) can also be used as filling material to make hybrid membranes. PVA/CNTs hybrid membranes were prepared in [8] and β-cyclodextrin (β-CD) was used for dispersion of CNTs. Compared with pure PVA and β-CD/PVA membranes, there was significant improvement in Young's modulus and thermal stability. In pervaporation separation of benzene/cyclohexane mixtures, the PVA/CNTs membranes showed good pervaporation properties. The permeation flux of benzene reached 61.0 g/($m^2$ h) and separation factor was 41.2.

## Ion-Exchanged Membranes

In pervaporation, penetrant molecules are first adsorbed by the membrane polymer, then diffusion through the polymer. The diffusivity value decreases with increase in molecular size of the penetrant. As the size of the permeant molecule increases, the solvent uptake decreases, i.e., the low molecular weight solvent benzene shows the highest uptake while xylene, the high molecular weight solvent shows lowest uptake.

However, the permeance induced by molecular size difference is not enough for efficient separation. Some metal cations which will interact with the permeant are added into the membrane. This is called solid carrier membranes. There are two types of solid carrier membranes: one is ion-exchange membrane impregnated carrier [9], and the other is carrier salt and polymer blend type membrane [10]. Silver ion or salt can reversibly react with olefin and make complexes by electrostatic interactions, and the ionic carrier or complex can diffuse through the membrane, thus the carrier membrane can show high selectivity for olefin due to its facilitation effect. Solution of stannic iodide was reported to form complex formation with aromatic hydrocarbons as evidenced by spectrophotometric measurements [11].

PVA membrane containing $AgNO_3$ was studied [12]. Here $Ag^+$ ion is used as the carrier. This type of membrane can selectively transport benzene from a benzene/cyclohexane mixture. Experiments were performed using benzene/cyclohexane feed mixtures with volume ratios of 8/2 and 2/8 (v/v). The permeation rates for benzene was between 0.5 and 5.5 kg $\mu m\ m^{-2}\ h^{-1}$ and the benzene/cyclohexane selectivity ranged from 51 to 84.

$Ag^+$-Nafion 117 membrane and $Na^+$-Nafion 117 membrane were used for pervaporation of benzene and cyclohexane [13]. Nafion-117 membranes were treated with NaOH, ion-exchanged with $AgNO_3$, and swollen in glycerol yielded higher benzene fluxes and better separations. Increasing operating temperatures enhanced the benzene and cyclohexane fluxes, and the activation energy was 50.1 kJ/mol for benzene and 55.1 kJ/mol for cyclohexane. Pervaporation selectivity was induced by the sorption and/or the diffusion phenomena depending diffusivity on the membrane-permeant characteristics. The pure benzene uptake by the $Ag^+$ membranes was only 1.5 times that of the pure cyclohexane uptake, however, the sorption selectivity in the binary solutions was increased to 15-19. The extent of solvent sorption is greatly influenced by competition between benzene and cyclohexane.

Benzene uptake in the $Ag^+$-membrane was much higher than in the $Na^+$-membrane. This demonstrates that in addition to the hydrophilic interaction between benzene and ionic groups in the membrane described above, specific interaction between benzene and $Ag^+$ ions must play a crucial role in benzene sorption. Silver ions can form coordination complexes with olefin double bounds [14]. Benzene molecules might form complexes with silver ions similar to alkene molecules.

The $Ag^+$-Nafion membranes needs some medium, such as water, to allow the transport of the silver salt-benzene complexes. To use silver form ion-exchange membranes, humidification must be supplied to the membrane during operation to maintain its permeability and selectivity. $AgBF_4$/Nafion blend membrane showed benzene selectivity over cyclohexane at low humidity. The humidity changes the membrane conductance, thus changes the benzene flux [14]. Another reason is, the aqueous solubility of benzene is significantly greater than cyclohexane as shown in Table 4, due to the $\pi$ electrons.

Table 4. Mole fraction solubility of benzene and cyclohexane in water

| Temperature | Benzene | Cyclohexane |
|---|---|---|
| 70 °C | $5.75 \times 10^{-4}$ | $19.7 \times 10^{-6}$ |
| 100 °C | $9.22 \times 10^{-4}$ | $35.0 \times 10^{-6}$ |
| 130 °C | $16.20 \times 10^{-4}$ | $61 \times 10^{-6}$ |
| 150 °C | $23.2 \times 10^{-4}$ | $131 \times 10^{-6}$ |

Table 5 shows that the aqueous solubility of benzene is greater than substituted benzenes. With substitution by alkyl groups, the increasing length of the non-polar alkyl substituent decreases the solubility. This facilitates the separation of benzene from other aromatics.

Table 5. Solubility of a series of aromatics in water

| Compound | Benzene | Toluene | o-Xylene | m-Xylene | p-Xylene | Ethylbenzene |
|---|---|---|---|---|---|---|
| Solubility, mol/L | 0.0228 | 0.0055 | 0.0017 | 0.0014 | 0.0017 | 0.0015 |

As we can see, silver ions ($Ag^+$) are effective in benzene–$Ag^+$ coordinate complex formation. The resulting $Ag^+$–Nafion membrane exhibits a preferential sorption behavior toward benzene rather than cyclohexane. However, it was observed that the $Ag^+$-membrane darkened upon exposure to light and its pervaporation performance decayed. This was probably due to silver ion reduction under daylight. Therefore, this membrane containing $Ag^+$ must be shielded from light during storage and preparation.

Copper ions are more stable and less expensive than silver ions. Copper ions can also be used for facilitating transport of benzene. Pervaporation of benzene/cyclohexane mixtures on cation-exchange membranes containing copper ions ($Cu^{2+}$) was investigated [15]. The Neosepta-CMX cation exchange membrane contains 45–65% sulfonated styrene/divinylbenzene random copolymer and 45–55% polyvinylchloride. For the Cu(II)-membrane, pure solvent sorption ratio of benzene/cyclohexane was 2.59, which is higher than Na(I)-membrane of 1.53. $Cu^{2+}$ membrane had better separation efficiency and higher benzene flux throughout the entire benzene concentration range. The facilitating capability for benzene transport for $Cu^{2+}$ membrane is better than $Na^+$ membrane due to the preferential sorption property of $Cu^{2+}$ toward benzene.

## Zeolite Membranes

Most polymer membranes can only be operated under 100 °C. Other problems, e.g., swelling, deformation and limited lifetime, are also disadvantages for polymer membranes. Zeolite membrane can overcome these problems.

For zeolite membrane, the permeant mainly diffuses through the zeolitic pores by a sorption–diffusion mechanism. Table 6 shows that FAU-type zeolite membrane selectively adsorbs benzene rather than cyclohexane. The permeance of benzene and a series of hydrocarbons are listed in Table 7. n-Butane, n-pentane, n-hexane, and n-heptane are all more permeable than benzene. The selective adsorption of benzene is in part responsible for the low permeance of benzene.

Table 6. Calculated number of hydrocarbon molecules adsorbed by FAU-type zeolites [16]

| Feed | Saturated adsorption ($mol\ kg^{-1}$) | Molecules per unit cell | Molecules per unit supercage structure |
|---|---|---|---|
| Benzene | 3.14 | 35.89 | 4.49 |
| Cyclohexane | 2.32 | 26.51 | 3.31 |

Table 7. Permeances through the FAU-type zeolite membrane at 100 °C [16]

| Feed | Molecular width, nm | Molecular length, nm | Permeance ($mol\ m^{-2}\ s^{-1}\ Pa^{-1}$) |
|---|---|---|---|
| Benzene | 0.585 | 0.75 | $9.5 \times 10^{-8}$ |
| Cyclohexane | 0.60 | 0.76 | $2.0 \times 10^{-7}$ |
| *n*-Butane | 0.43 | 0.78 | $5.3 \times 10^{-7}$ |
| *n*-Pentane | 0.43 | 0.90 | $3.8 \times 10^{-7}$ |
| *n*-Hexane | 0.43 | 1.02 | $2.3 \times 10^{-7}$ |
| *n*-Heptane | 0.43 | 1.14 | $1.0 \times 10^{-7}$ |

Note: The total pressure of hydrocarbon(s) on the feed side = 7.5–9.5 kPa.

Zeolites Cu(I)-Y and Ag(I)-Y had been used for desulfurization of liquid fuels by adsorption via $\pi$-complexation [17]. Cu(I)-Y and Ag(I)-Y adsorbent were prepared by ion exchange of zeolites using known ion-exchange procedures [18]. Zeolite Y is highly hydrophilic, and it uptakes > 20 wt% water at ambient conditions. Both Ag(I)-Y and Cu(I)-Y turned black after exposure to pure benzene, which indicates that there is interaction and complex formation between the aromatics and the solids. This was not observed for Na-Y.

The calculated and measured adsorption bond energies were listed in Table 8. The experimental and theoretical values are in good agreement. These results indicate that $Cu^+$ is stronger than $Ag^+$ in bonding ($\pi$-complexation) with benzene [19]. The natural bond orbital analysis showed that the bonding followed the classical picture of $\pi$-complexation: there is some donation of electron charges from the $\pi$ orbital of benzene to the vacant *s* orbital of metals (known as $\sigma$ donation) and, simultaneously, back-donation of electron charges from the *d* orbitals of metals to the $\pi^*$ orbital (i.e., the antibonding $\pi$

orbital) of benzene (known as π back-donation). The results also showed that the π back donation is more important for benzene. $Cu^+$ is more favorable for π-complexation than $Cu^{2+}$. Hopefully, membranes of Cu(I)-Y and Ag(I)-Y are effective for benzene removal from liquid fuel.

Table 8. Adsorption bond energies (kcal/mol) for benzene on zeolites from ab initio theory MO theory and from the experiment [19]

| Adsorbent | From MO theory | From experiment |
|---|---|---|
| Cu(I)-Y | 20.5 | 20.5 |
| Ag(I)-Y | 19.1 | 19.5 |

**The Current Status of Benzene Reduction by Membrane Pervaporation**

Currently there is no commercial plant for benzene separation by membrane pervaporation in the world. However, since 2004, PolyAn, a German company is running a pilot plant using functionalized, aromatics-selective, polyimide membranes to produce nearly benzene-free fuel [20]. An economical feasibility study of zeolite membranes for industrial scale separation of aromatics from naphtha cracker feed can be found in [21]. As EPA's benzene limit phase-in date is drawing near, more and more cost-effective membrane pervaporation will be emerged in the future. It will be interesting to see what the pilot plant work will reveal.

**Summary**

As the benzene limitation in gasoline is getting tighter, membrane pervaporation provides an alternative way for benzene reduction. Four kinds of membranes are reviewed:

1) Polymer membranes are typically used for benzene pervaporation. Cross-linking polymers with a combination of two or three polymers are desirable. Polyimide membranes are very attractive.

2) Composite membranes with carbon graphite or carbon nanotubes filling have better separation factor, thermal stability as well as mechanical properties.

3) The metal ions $Ag^+$ and $Cu^{2+}$ in ion-exchanged membranes facilitate the transport of benzene, and thus improve the separation capability.

4) Zeolite membranes with $Ag^+$ and $Cu^+$ ions can also be used for benzene separation by π-complexation with benzene.

Among the above membraneconcepts, composite membranes are very promising for future development

**References**

[1] M. Mulder, 1998, Basic Principles of Membrane Technology, Kluwer Academic Publishers, Dordrecht.

[2] Samit K. Ray, Sudhir B. Sawant, Jyestharaj B. Joshi, and Vishwas G. Pangarkar, 1997, "Development of New Synthetic Membranes for Separation of Benzene-Cyclohexane Mixtures by Pervaporation: A Solubility Parameter Approach", *Ind. Eng. Chem. Res.*, Vol. 36, p. 5265-5276.

[3] L. Enneking, A. Heintz, R.N. Lichtenthaler, 1996, "Sorption equilibria of the ternary mixture benzene/cyclohexene/cyclohexane in polyurethane- and PEBA-membrane polymers", *Journal of Membrane Science*, Vol. 115, p. 161 – 170.

[4] D. Katarzynski, C. Staudt, 2006, "Permeation properties of different aromatic substances in multicomponent aromatic/aliphatic pervaporation experiments", *Desalination*, Vol. 200, p. 23–25.

[5] Asha Elizabeth Mathai, R.P. Singh, Sabu Thomas, 2002, "Transport of substituted benzenes through nitrile rubber/natural rubber blend membranes", *Journal of Membrane Science*, Vol. 202, p. 35–54.

[6] D. Katarzynski, C. Staudt-Bickel, 2006, "Separation of multi component aromatic/aliphatic mixtures by pervaporation with copolyimide membranes", *Desalination*, Vol. 189, p. 81–86.

[7] Lianyu Lu, Honglei Sun, Fubing Peng, Zhongyi Jiang, 2006, "Novel graphite-filled PVA/CS hybrid membrane for pervaporation of benzene/cyclohexane mixtures", *Journal of Membrane Science*, Vol. 281, p. 245–252.

[8] Peng, F., Hu, C., Zhongyi Jiang, 2007, "Novel ploy(vinyl alcohol)/carbon nanotube hybrid membranes for pervaporation separation of benzene/cyclohexane mixtures", *Journal of Membrane Science*, Vol. 297, p. 236–242.

[9] LeBlanc, O. H.; Ward, W. J.; Matson, S. L.; Kimura, S. G., 1980, *Journal of Membrane Science*, Vol. 6, p. 339.

[10] Bryant, D. L.; Noble, R. D.; Koval, C. A., 1997, *Journal of Membrane Sci*ence, Vol.127, p.161.

[11] J. F. Murphy and D. E. Baker, 1965, "Complexes of stannic iodide with aromatic hydrocarbons", *Canadian Journal of Chemistry*, Vol. 43, No. 5, p.1272–1278.

[12] Debra L. Bryant, Richard D. Noble, Carl A. Koval, 1997, "Facilitated transport separation of benzene and cyclohexane with poly(vinyl alcohol)-$AgNO_3$ membranes", *Journal of Membrane Science*, Vol. 127, p. 161-170.

[13] Sumin Tanya Kao, Fengjan Joshua Wang, Shingjiang Jessie Lue, 2002, "Sorption, diffusion, and pervaporation of benzenekyclohexane mixtures on silver-Nafion membranes", *Desalination*, Vol.149, p. 35-40.

[14] Takeo Yamaguchi, Hayato Kurita, and Shin-ichi Nakao, 1999, "Transport Mechanism of Aromatic Vapor through Silver Salt Carrier/Polymer Blend Membrane and Its Humidity Effect", *J. Phys. Chem. B,* Vol.103, p.1831-1835.

[15] S. Jessie Lue, F. Joshua Wang, Shian-Yiaw Hsiaw, 2004, "Pervaporation of benzene/cyclohexane mixtures using ion-exchange membrane containing copper ions", *Journal of Membrane Science*, Vol. 240, p.149–158.

[16] Byeong-Heon Jeong, Yasuhisa Hasegawa, Ken-Ichiro Sotowa, Katsuki Kusakabe, Shigeharu Morooka, 2003, "Permeation of binary mixtures of benzene and saturated C4–C7 hydrocarbons through an FAU-type zeolite membrane", *Journal of Membrane Science*, Vol. 213, p.115–124.

[17] Arturo J. Hernandez-Maldonado and Ralph T. Yang, 2003, "Desulfurization of Liquid Fuels by Adsorption via $\pi$-Complexation with Cu(I)-Y and Ag-Y Zeolites", *Ind. Eng. Chem. Res.*, Vol. 42, p.123-129.

[18] A. J. Hernandez-Maldonado, R. T. Yang, 2003, *Ind. Eng. Chem. Res.*, Vol. 42, p.123.

[19] Ralph T. Yang, Arturo J. Hernandez-Maldonado, Frances H. Yang, 2003, "Desulfurization of Transportation Fuels with Zeolites Under Ambient Conditions", *Science*, Vol. 301, No. 4, p.79-81.

[20] The website of PolyAn company: http://www.poly-an.de/.

[21] Meindersma, G.W., de Haan, A.B., 2002, "Economical feasibility of zeolite membranes for industrial scale separation of aromatic hydrocarbons", *Desalination*, Vol. 149, p.29-34.

## 21c The Synthesis of Separation Networks with Complex Column

**Libin Zhang**, University of Illinois at Chicago, 851 S. Morgan St., SEO 218, Chicago, IL 60607, Jeonghwa Moon, Chemical Engineering, University of Illinois at Chicago, 851 S. Morgan St., Room 218, Chicago, IL 60607, and Andreas A. Linninger, Bioengineering, University of Illinois at Chicago, 851 S. Morgan St., Room 218, Chicago, IL 60607.

**Summary:**

Separation processes make up 40%-70% of capital and operating costs of chemical manufacturing. Distillation accounts for more than 60% of the total process energy for the manufacture of commodity chemicals. Complex column configurations are estimated to harness energy savings up to 70%. Therefore, distillation synthesis with complex is a meaningful target for energy improvements on an industry-wide scale. However, design of complex separation system requires (i) structural decisions such as sequencing of distillation columns for solvent recovery, and (ii) determination of associated operating conditions (e.g. distillate/bottoms compositions, reflux ratio etc). For computer tools this process synthesis constitutes a very formidable challenge. Technically MINLP incorporating binary as well as continuous variables can address synthesis problem. However, infeasible operation following from ineffective structural decision may jeopardize the robustness and convergence of MINLP. In this presentation, we demonstrate a hybrid algorithm combining stochastic search techniques with novel formulation of complex distillation configuration. An evolutionary algorithm will construct automatically structurally different complex separation network with the complex basic distillation configuration. Feasibility of the design will be delegated to an advanced feasibility tests based on temperature collocation of finite elements. Our search algorithm is capable of solving complex distillative synthesis problem robustly and reliably without user interactions. Application will demonstrate the algorithms performance in complex column sequencing problems for solvent recovery.

**Scope:**

We propose a novel bi-level hybrid algorithm for the synthesis of complex column sequences. The master problem will make structural decisions using evolutionary algorithms, while the subproblem will rigorously assess the feasibility of each of these structures using our novel MInimum Bubble Point DIstance (MIDI) Algorithm. The minimum distance algorithm computes the section temperature profiles by orthogonal finite-element collocation technique. This minimum bubble point distance is obtained by solving a gradient based NLP optimization problem. A minimum distance of zero ensures feasibility, whereas positive values indicate infeasible specifications. Penalizing the genetic algorithm's objective for infeasible specifications ensures that the chromosomes are feasible after a few generations.

One of challenge for the determination of pinch points will also be addressed in the synthesis and design of complex separation systems. We develop the new format homotopy continuous methods which can locate all the fixed points starting from pure components. We also present interval methods, global terrain method, niche genetic algorithm for validating the results. However, continuous method has to carefully construct the proper homotopy function. The classification and order of the fixed points also are given by calculating the eigenvalue and the entropy.

***Significance:***

The major innovation of the proposed approach is an evolutionary program combined with a novel temperature collocation algorithm to systematically build and optimize complex column configurations based on column sections. Massive problem size reductions due to temperature collocation ensure the realistic composition profiles of each column in the network without sacrificing the computational and thermodynamic rigor. The evolutionary algorithm constructs automatically different separation sequences; feasibility of the synthesized sequences is ascertained by means of the MIDI algorithm. In contract to existing approaches, this method synthesizes several clusters of designs solutions, each one corresponding to regions of local optimality.

# DIESEL DESULFURIZATION TO MAKE ULSD – OVERCOMING NITROGEN INHIBITION

*Manuela Serban, Joseph Kocal, Peter Kokayeff, and Chris Gosling*
**UOP LLC, a Honeywell Company**


## Abstract

It has been extensively reported that the nitrogen compounds present in typical diesel hydrotreater feedstocks are the strongest inhibitors in hydrodesulfurization. A reduction of the nitrogen content in diesel feedstocks results in reducing the cost associated with producing ultra-low sulfur fuels. We have developed a method of reversibly extracting the organic nitrogen (N) compounds using ionic liquids (ILs) prior to the catalytic hydrodesulfurization step. Pilot plant runs aimed at determining the benefit of denitrogenation of the feed prior to desulfurization, have shown that it is possible to use 50% less catalyst for a treated feed vs. non-treated feed for the same desulfurization level. This translates into significantly improved economic performance for sulfur removal through a combination of lower catalyst requirement, longer cycle length, and reduced utility consumption.

## Introduction

The importance of improved desulfurization technologies to the refining industry for the production of clean fuels has been clearly established. As the specification of the final fuel products approaches "zero sulfur", traditional hydrodesulfurization economics become increasingly less attractive. The achievement of very low sulfur levels is dependent on the ability to remove the most refractory sulfur species at minimal cost and degradation of the hydrocarbon fuel components. The optimal solution to achieve these specifications is refinery specific and depends on the plant configuration, crude quality, market requirements, etc. Options available to refiners include improvements of the existing hydrodesulfurization catalysts and processes, novel process concepts such as oxidative desulfurization, and pretreatment of the feed to make hydrodesulfurization more effective.

The catalytic hydrodesulfurization (HDS) of the most refractive sulfur containing molecules in diesel i.e., dibenzothiophene (DBT) and especially 4,6-dimethyl-dibenzothiophene (DMDBT) is inhibited at different degrees by the components in the reaction mixture such as organic heterocompounds and polyaromatic hydrocarbons. It has been extensively reported that the nitrogen compounds present in the diesel feedstock are the strongest inhibitors in catalytic HDS [1-7]. In general, the following order of inhibition has been noticed: saturated and mono-aromatic hydrocarbons < condensed aromatics ~ oxygen compounds ~ $H_2S$ < organic sulfur compounds < nitrogen compounds [1, 2]. It has been shown [3] that the hydrogenation (HYD) of the aromatic ring followed by the C-S bond hydrogenolysis route, which is the predominant reaction pathway for the hydrodesulfurization of 4 and 6 alkyl-substituted dibenzothiophenes, is more strongly suppressed by the nitrogen compounds than the direct hydrogenolysis (DDS) route, the predominant desulfurization route for DBT. The DBT conversion remained nearly unaffected by the presence of 50 ppm quinoline and carbazole in the feed, while the conversion of 4,6-DMDBT was significantly retarded even by the presence of such small quantities of nitrogen compounds [2]. The DBT conversion decreased when larger amounts of nitrogen compounds, i.e., 500-650 ppm, were present in the reaction mixture.

Kasztelan et al. [8] proposed, in the case of unpromoted $MoS_2$, that corner sites with higher unsaturation are active for HYD while the edge sites with relatively low unsaturation are active for DDS. This is also in accordance with the observation that basic nitrogen compounds exhibit a stronger poisoning effect compared to non-basic compounds. The basic nitrogen compounds, e.g., quinoline, would be expected to preferentially donate electrons to sites of high unsaturation, thus poisoning the corner sites active for the HYD route. However, even carbazole species, despite their low basicity are poisons for the HDS of refractory sulfur compounds over commercial catalysts. Ho et al. [6,7] who studied the poisoning effect of 3-ethylcarbazole (3-ECBZ) on the HDS of 4,6-diethyldibenzothiophene (4,6-DEDBT) showed that even a trace amount of 3-ECBZ could have a huge impact on the HDS of 4,6-DEDBT since the alkylcarbazoles are highly adsorptive and refractive and could eventually block off all the HYD sites.

We have developed a method of reversibly extracting the organic N-compounds using ionic liquids (ILs) prior to the catalytic hydrodesulfurization step. Pilot plant runs aimed at determining the benefit of denitrogenation of the feed prior to desulfurization, have shown that it is possible to use 50% less catalyst for a treated feed vs. non-treated feed for the same desulfurization level. This translates into significantly improved economic performance for sulfur removal through a combination of lower catalyst requirement, longer cycle length, and reduced utility consumption. An extractive process that is able to selectively remove the organic nitrogen compounds from diesel feedstocks is desirable since it results in reducing the cost associated with producing ultra-low sulfur clean fuels.

## Experimental

All extraction experiments were performed on two high nitrogen containing diesel blends consisting of straight run diesel (SR), light cycle oil (LCO) and light coker gas oil (LCGO) in a 4:3:3 and 1:1:1 weight ratios simulating the components in a typical refinery diesel fuel pool. The liquid-liquid extraction experiments were performed at room temperature and atmospheric pressure.

The total nitrogen content was determined with a chemiluminescence analysis using ASTM D4629 method, the basic nitrogen was determined as total base number (mg KOH/g sample) using ASTM D2896 and total sulfur was determined by energy-dispersive X-ray fluorescence spectroscopy (ASTM 4294). The nitrogen species concentration and distribution was determined using a GCxGC N-comprehensive method. The high boiling point nitrogen species were identified with positive electrospray mass spectrometry. The concentration of ILs in the diesel blend was determined by liquid anion chromatography, while C-13 and H-1 liquid NMR analyses were used to measure the amount of hydrocarbons in the 'spent' ILs.

The hydrodesulfurization pilot plant runs were performed with a commercial high activity Ni/Mo catalyst.

## Results and Discussion

Ideally any extracting agent should have the following properties: the partition coefficient for the N-compounds should be high, the extracting agent should be insoluble in the feed, the N-free hydrocarbons should not be soluble in the extracting agent, and the regeneration of the agent should be relatively easy. We found that specific ionic liquids were able to fulfill all these process requirements.

Ionic liquids, the so-called "green-solvents", are nonaqueous, aprotic solvents, with low melting points, no vapor pressure and good chemical and thermal stability. Since the melting points are low, ILs act as solvents in which reactions/extractions can be performed, and because the liquids are made of ions rather than molecules, such reactions/extractions often give distinct reactivities/selectivities when compared to conventional organic solvents. The complete absence of vapor pressure is another major advantage over organic solvents.

A high nitrogen containing diesel blend consisting of straight run diesel (SR), light cycle oil (LCO) and light coker gas oil (LCGO) in a 4:3:3 weight ratio was denitrogenated by liquid-liquid extraction using an ionic liquid (IL). The diesel blend contained 650 ppm total nitrogen, 220 ppm basic nitrogen and 1.35 wt% S. The extraction experiments were done at low temperature and pressure for a diesel feed to IL weight ratio of 1:0.5. Equilibrium was reached in less than 5 minutes and the two distinct liquid layers formed after mixing were separated by decantation. As shown in Table 1, IL2 was able to extract 90% of both total nitrogen and basic nitrogen from the diesel blend. Due to availability reasons, all experiments were however performed with IL1.

**Table 1. Percent Total Nitrogen and Basic Nitrogen Removed from a Diesel Blend**

| *Ionic Liquid* | *Total N Removal (%)* | *Basic N Removal (%)* |
|---|---|---|
| IL1 | 70 | > 90 |
| IL2 | 90 | > 90 |

The GCxGC N-comprehensive chromatographic analysis of the diesel blend (Figure 1) indicates that the distribution of the nitrogen species in the feed ranges from $C_0$-$C_5$+ carbazoles, $C_1$-$C_6$+ indoles; high boiling point N-compounds and traces of aniline and quinoline. Positive electrospray mass spectrometry (+ESI-MS) identified the high boiling point nitrogen compounds as a mixture of 17 classes of basic N compounds and 23 classes of basic di-N compounds. The GCxGC N-comprehensive chromatographic analysis of the diesel blend after extraction with IL1 (Figure 2) shows that indeed the majority of the basic N-compounds (quinoline and high boiling point basic N-compounds) were extracted. A direct comparison of the level of denitrogenation of the IL1 treated feed vs. the non-treated feed is shown in Figure 3.

**Figure 1. GCxGC-N Comprehensive Chromatogram of the Diesel Blend.**

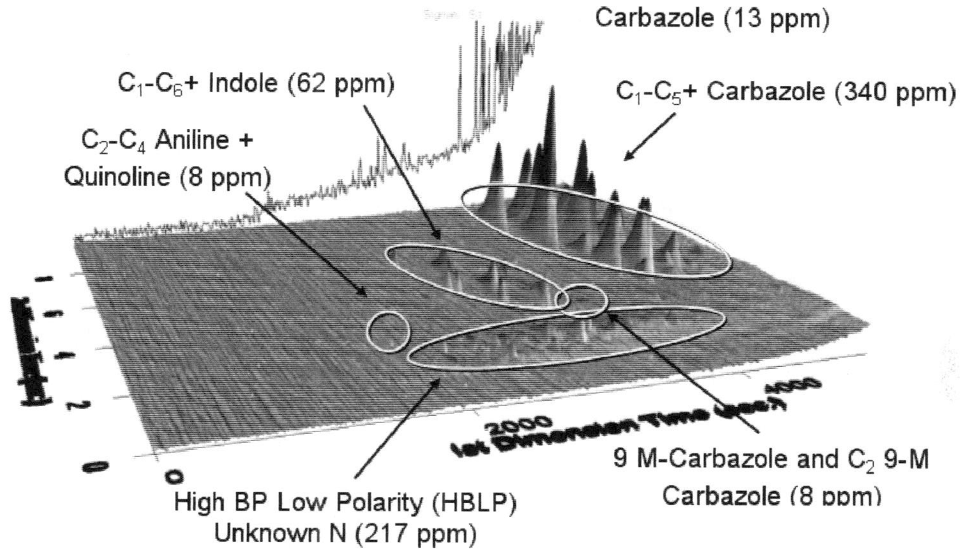


**Figure 2. GCxGC-N Comprehensive Chromatogram of the Diesel Blend after Nitrogen Extraction with IL1.**

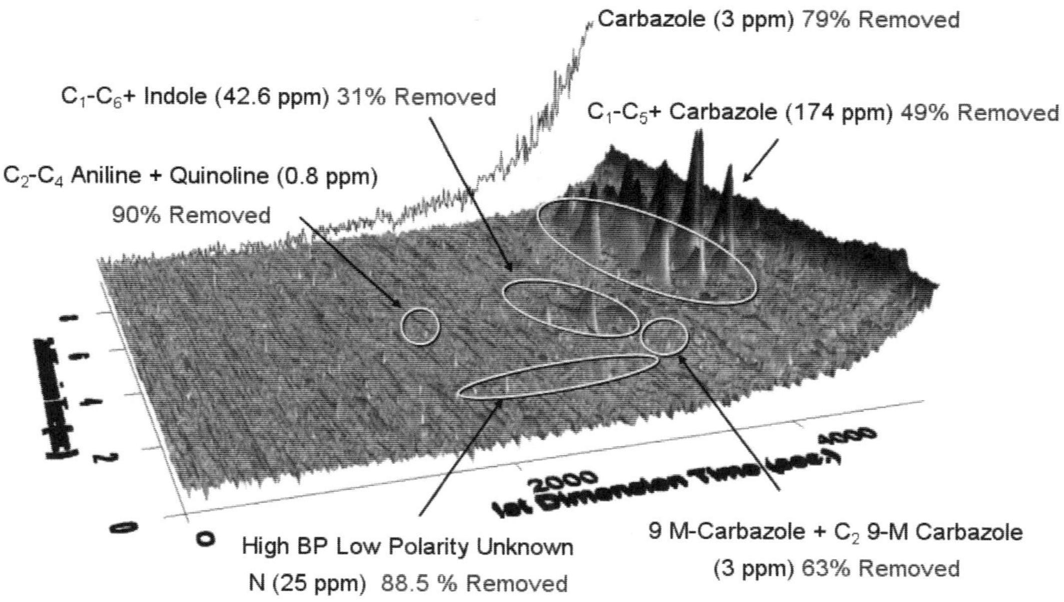

**Figure 3. Comparison of the N-Compounds in Untreated vs. IL1 Treated Diesel Blend.**

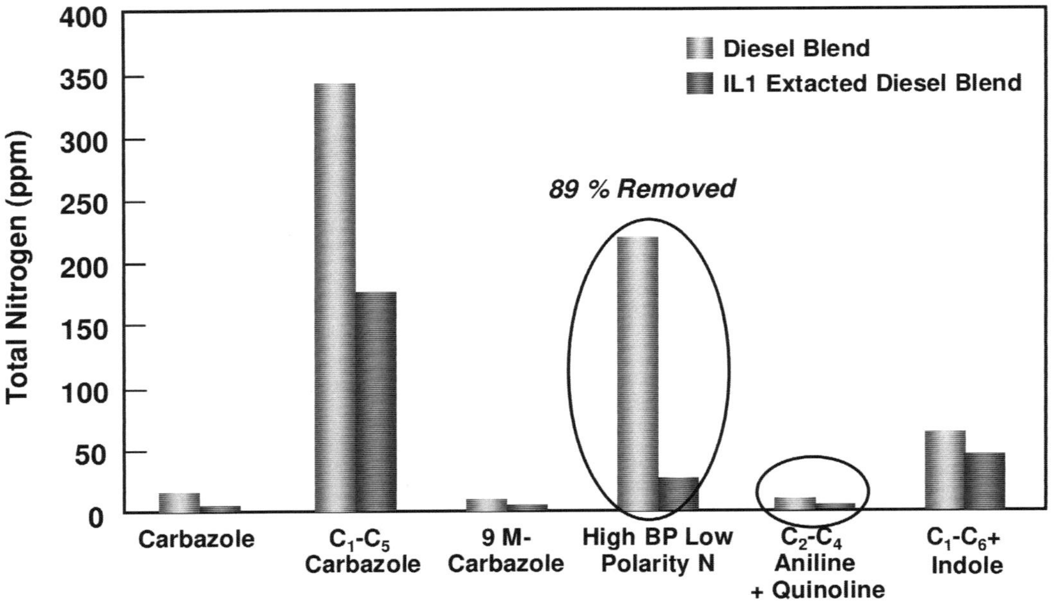


The desulfurization level achieved by the ionic liquid extraction process was insignificant. The simultaneous removal of the S-containing species would be desirable, but with these particular ILs, this proved to be unsuccessful. Most importantly, the C-13 and H-1 NMR analyses of the fresh and 'spent' IL1, the extraction step did not affect the aromatic hydrocarbon content suggesting that the hydrocarbons are not soluble in the extracting agent. Also, the low temperature of the IL pretreatment process suppresses any dissociation, disproportionation and degradation reactions, such that the fuel components remain structurally unmodified. The liquid chromatographic anion analysis detected <0.0088% IL1 in the diesel blend after the nitrogen extraction, showing that IL1 is highly selective toward the nitrogen compounds and it does not cross-contaminate the diesel blend.

Single stage extractions with increasingly higher IL1:diesel blend weight ratios (Figure 4) indicated that even with a 0.25:1 IL1:diesel blend ratio, the level of denitrogenation is above 65%, while at higher weight ratios, i.e., 1.5-2.3, the denitrogenation level reaches a plateau. The nitrogen species that are not soluble in IL1 and remain un-extracted are mainly the highly substituted carbazole species (Figure 3). Several experiments aimed at determining the effect of staged extractions vs. a single step extraction showed that there is no significant benefit in performing staged extractions for the same total cumulative IL1:diesel blend weight ratios as for a single stage extraction, e.g., ~83% total nitrogen removed in a Single Stage Extraction for IL:Blend=1.5:1 vs. Step 3 (total cumulative IL:Blend = 1.5). This data is shown in Figure 4. However, the benefit of a staged extraction process vs. a single stage extraction step would be the lower IL inventory required to achieve a desired denitrogenation level.

**Figure 4. Single Stage vs. Staged Extractions with IL1 at Different Weight Ratios.**

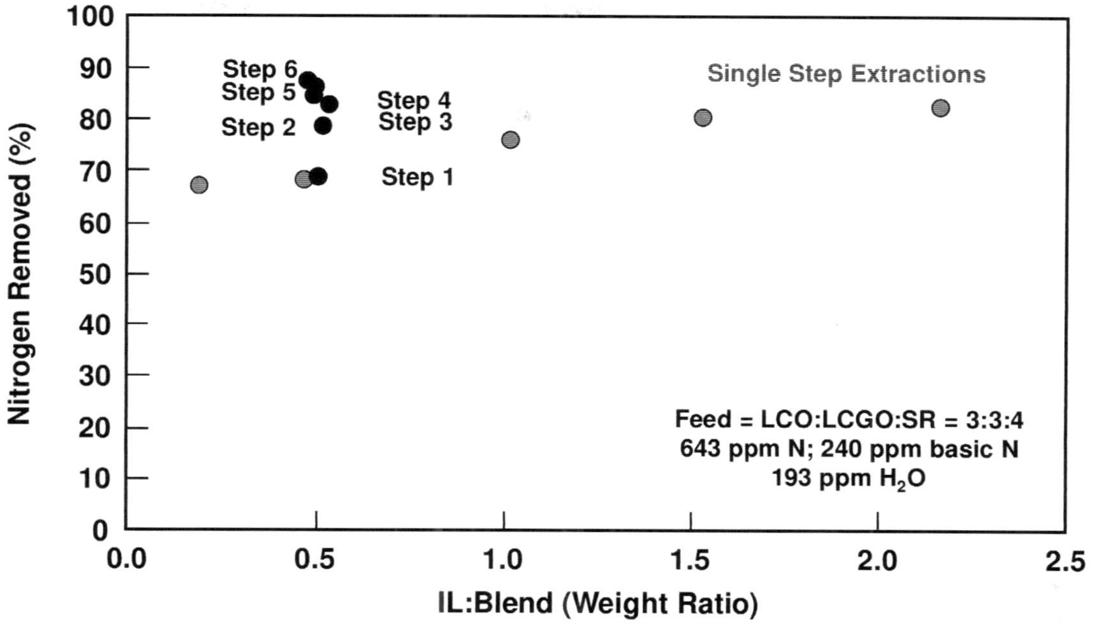


It is theorized that the nitrogen species are extracted through a π-π interaction between the ionic liquid aromatic ring and the aromatic rings of the basic and non-basic nitrogen-containing species. Even further, the acidic ionic liquids target specifically the basic N-compounds (Table 2). These types of interactions are reversible, such that the spent ionic liquid solvent can be regenerated and recovered without any contamination and with minimal losses. This is imperative for making the selective extractive denitrogenation process an economical one.

**Table 2. Total Nitrogen Removal Capacity of Acid vs. Basic ILs.**

| *Ionic Liquid* | *#C in the Aromatic Ring Side Chain* | *% Total N Removed from Diesel Feed* |
|---|---|---|
| Acidic | 0 | 50 |
| Acidic | 2 | 50 |
| Acidic | 4 | 70 |
| Basic | 4 | 10 |
| Neutral | N/A | Cross-contaminates the feed |

Recovery by distillation is not suitable for low volatility and thermally unstable products, but is applicable for products that have a high vapor pressure below the decomposition temperature of the ionic liquid. Extraction with water, works for hydrophilic products that are easily separated from hydrophobic ionic liquids, although the cross-contamination between the streams must be taken into account. The extracted nitrogen-containing species (carbazoles, indoles and quinolines) have a very low water solubility and vapor pressure and the IL used here is fully miscible with water, such that a water extraction method would not be appropriate for the quantitative recovery of the IL.

Steam-stripping the spent IL is a viable recovery method. Critical to the success of this regeneration method is the intimate contact between the steam flow and the IL, as well as the efficient displacement of the nitrogen species from the spent IL. The water contamination is minimized as long as the water is in the vapor phase during contact with the IL. Laboratory scale experiments have proved the feasibility of this concept for a total of four consecutive extraction/regeneration cycles. Ionic Liquid 1 used to denitrogenate the diesel blend was regenerated by steam stripping it at 150°C (302°F). After the first regeneration, the ionic liquid lost only 2.5 and 4.5% of its extraction capacity for total nitrogen and basic nitrogen, respectively compared to the 1$^{st}$ cycle. The performance in the 2$^{nd}$, 3$^{rd}$ and 4$^{th}$ cycles is identical (Figure 5) and the results were reproducible.

**Figure 5. Regeneration of IL1 in Four Consecutive Extraction-Regeneration Cycles.**

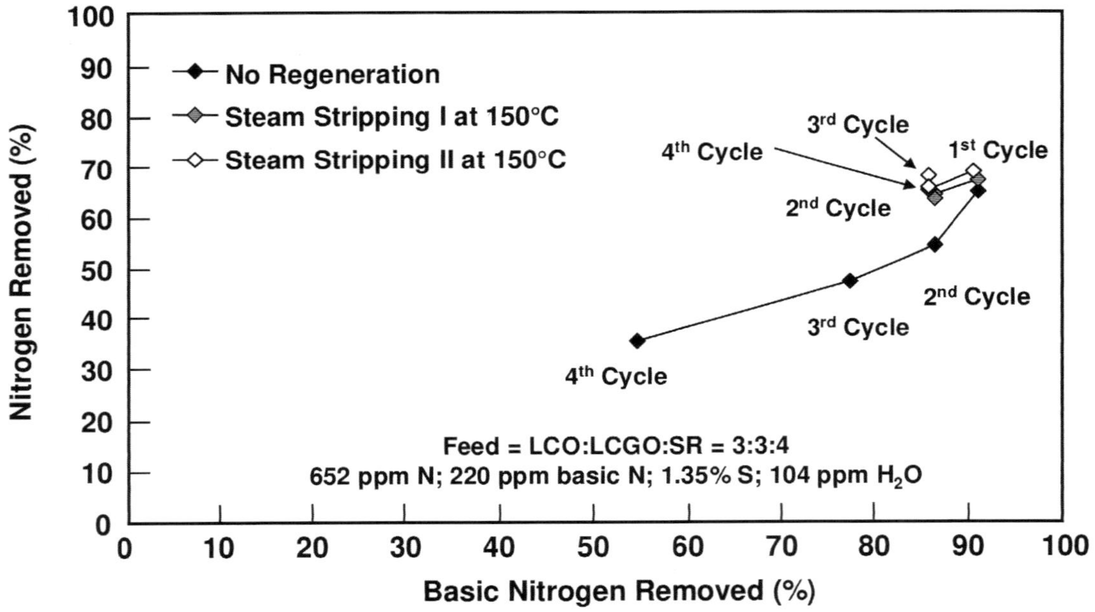


In order to determine and quantify the benefit of denitrogenating the feed prior to desulfurization, a large amount of diesel blend consisting of straight run diesel, light cycle oil and light coker gas oil in a 1:1:1 weight ratio was extracted with IL1 at low temperature and pressure. The pilot plant runs were done at 800 psig and at a LHSV of 1 $hr^{-1}$. The IL1 treated feed was run at 655°F while the untreated feed was run at both 655 and 680°F. The data was taken at varying contact times with the catalyst, allowing the possibility of tracking the conversion as a function of catalyst bed depth (a bed depth of 100% corresponding to a LHSV=1 $h^{-1}$).

The pilot plant data showed that it is possible to use 50% less catalyst for an IL1 treated feed vs. non-treated feed for the same desulfurization level (Figure 6). The S conversion for the treated feed at 655°F is similar to that for the untreated feed at 680°F. Also, the impact of N removal is observed even at very short contact times, at 15% of the catalyst bed depth. This implies that even the non-sterically hindered S species are impacted by N poisoning, although the biggest impact of N removal is observed for the sterically hindered S species (higher contact times corresponding to >40% of the catalyst bed depth), which is in accordance with literature data [2-5, 9].

**Figure 6. Product Sulfur vs. Catalyst Bed Depth.**

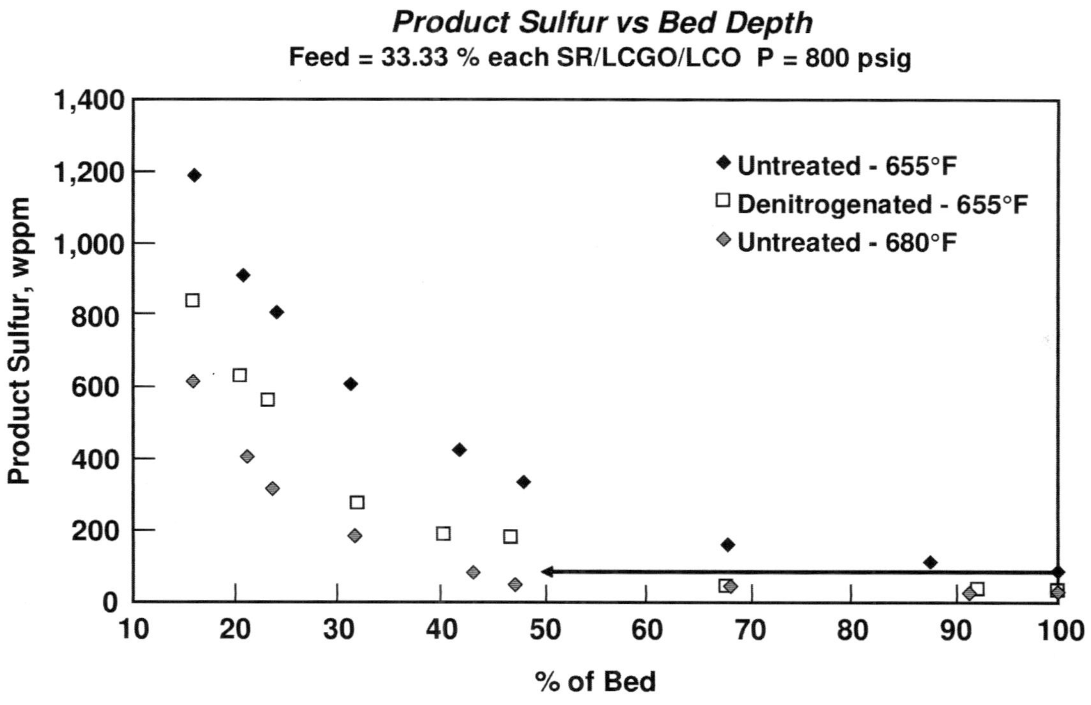


Nitrogen removal had also a significant impact on the nitrogen profile (Figure 7). Treating the feed with the IL1 allowed for ~ 70 % of the catalyst bed to operate in an essentially N free environment. As a result, the hydrodenitrogenation (HDN) activity was enhanced by the removal of the nitrogen compounds. Yang et al. [9] also observed that the removal of some of the nitrogen compounds from light cycle oil feeds enhanced the HDN rate of the remaining nitrogen compounds, and the greater the initial concentration of total nitrogen compounds, the lower the HDN rate, implying a self-inhibition of HDN by the nitrogen compounds.

**Figure 7. Product Nitrogen vs. Catalyst Bed Depth.**

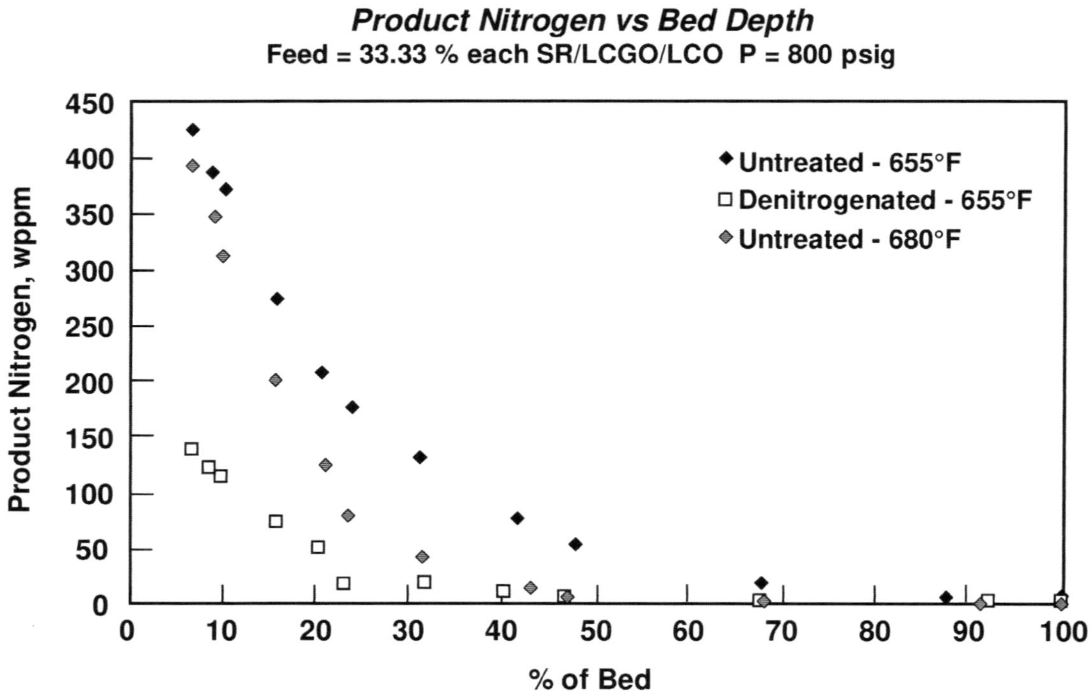


This improvement in catalyst activity that is achieved by incorporation of an ionic liquids extraction process can result in significantly improved economic performance for sulfur removal. A comparative economic analysis was completed to quantify the benefits of employing ionic liquids extraction. Three cases were evaluated with the process conditions summarized in Table 3:

Base Case: 30 MBPD, 600 psi Diesel Hydrotreating Unit (DHT) producing Sulfur Diesel (LSD) with hydrotreating catalyst “A”. ULSD is not produced in this case;

Alternate Case #1: Revamp the DHT unit to produce Ultra Low Sulfur Diesel (ULSD). This is a major revamp since it would require a new reactor, loading with modern hydrotreating catalyst “B”, and a new recycle gas compressor;

Alternate Case #2: Install an IL extraction unit upstream of the existing DHT unit and reload the reactor with modern hydrotreating catalyst “B” to make ULSD. No other revamp is required.

**Table 3. DHT Process Conditions Assumed for IL Extraction Case Study.**

| | *Base Case* | *ULSD Revamp* | *Ionic Liquid* |
|---|---|---|---|
| ***Catalyst*** | A | B | B |
| ***LHSV, $h^{-1}$*** | 3 | 1 | 3 |
| ***Life, months*** | 30 | 18 | 24 |
| ***Pressure, psi*** | 600 | 600 | 600 |
| ***Recycle Gas, SCF/bbl*** | 1500 | 2000 | 1500 |
| ***Makeup $H_2$, SCF/bbl*** | 450 | 600 | 600 |

Product pricing consistent with $40/bbl Brent crude oil was assumed for the analysis. The cost of the DHT Revamp in Alternate Case 1 was estimated to be $30MM and the cost of the IL extraction unit in Alternate Case 2 was assumed to be $20MM. The economics of the three options is compared in Table 4.

**Table 4. Ionic Liquids Extraction vs. DHT Revamp Economics.**

| *Case* | *Base Case LSD HDT* | *Revamp ULSD DHT* | *Ionic Liquid ULSD DHT* |
|---|---|---|---|
| ***Payback, years (after tax)*** | 0.0 | 5.4 | 3.2 |
| ***IRR, % (10 years, after tax)*** | n/a | 13.31 | 28.74 |
| ***NPV, $MM (10 years, after tax)*** | 7.44 | 8.87 | 24.62 |

It was clear from this economic analysis that incorporating IL extraction as a feed pretreatment process can have significantly better economics than a conventional DHT revamp to produce ULSD. The IL extraction process cost can also be substantial relative to the conventional revamp and still be attractive.

## Conclusions

A high-nitrogen containing diesel blend consisting of straight run diesel, light cycle oil and light coker gas oil was effectively denitrogenated via liquid-liquid extraction with an ionic liquid prior to hydrodesulfurization. The ionic liquid was highly selective towards the nitrogen compounds and it did not cross-contaminate the diesel blend. The ionic liquid solvent was shown to be regenerable.

Pilot plant runs aimed at quantifying the impact of removing the nitrogen compounds prior to desulfurization indicate that it is possible to use significantly less catalyst for a treated feed vs. non-treated feed to achieve the same product desulfurization level. This translates into significantly improved economic performance for sulfur removal through a combination of lower catalyst requirement, longer cycle length, and reduced utility consumption. An economic analysis showed that the IL extraction unit cost can be substantial relative to a conventional revamp and still be attractive.

## Acknowledgments

The authors gratefully acknowledge UOP's analytical laboratory staff for characterizing and analyzing the diesel feeds, and UOP's Riverside pilot plant staff for operating the hydrotreating unit.


## References


1. Schulz, H., et al. (1999), *Catal. Today*, 49, 87-97.
2. Kwak, C. et al. (2001), *Applied Cat. B: Env.*, 35, 59-68.
3. Yang, H., et al. (2004), *Fuel Processing Tech.*, 85, 1415.
4. Kim, J.H., et al. (2004), *ACS, Div.Pet.Chem.*, 49(1), 44.
5. Kim, H., et al. (2004), *Applied Cat. B: Env.*, 50, 17.
6. Ho, T.C. (2003), *J. Catal.*, 219, 442.
7. Ho, T.C., and Nguyen, D. (2004), *J. Catal.*, 222, 450.
8. Kasztelan, S. (1984), et al., *Appl. Catal.*, 13, 127.
9. Yang, H., et al. (2005), *Catal. Today*, 109, 16.

# Deep Desulfurization of Fuels Using Ionic Liquid Based on $FeCl_3$

*Yan Wang[1], Huaming Li[1], Wenshuai Zhu[1], Xue Jiang[1], Jidong Lu[2], Jiexiang Xia[1] and Yongsheng Yan[1]*
*[1] College of Chemistry and Chemical Engineering, Jiangsu University, Zhenjiang, China*
*[2] State Key Laboratory of Coal Combustion, Huazhong University of Science and Technology, Wuhan, China*

## Introduction

Sulfur in fuels is always a major source of $SO_x$ that contributes to acid rain and air pollution. Therefore, the threshold limit for sulfur in gasoline and diesel is expected to be regulated, and this limit becomes increasingly stringent. Therefore, Lower sulfur content even free-sulfur will be a trend for the development of quality in fules[1-2].

In industry, desulfurization of diesel is carried out by a traditional catalytic hydrodesulfurization(HDS) technology, which requires both high temperatures and high pressures of hydrogen gas, and high performance catalyst is also required. It is difficult to deal with BT and DBT, especially 4,6-alkyl-substituted DBTs for the sterically hindered adsorption of them on the catalyst surface[3]. To overcome this difficulty, many alternative processes such as selective oxidation, adsorption, extraction and bioprocesses have been studied recently[4].

Among these alternatives, extractive desulfurization (EDS) is an attractive technology, for its process is applicable under ambient temperature and pressure, and without the special need for catalysts and hydrogen. And the key to an EDS process is to find a good extractant featuring the following attributes[5], eg. (1) good extractive ability for the S-compounds; (2) free of contamination to the fuels; (3) nontoxicity and environmental benignity and stability for repetitive use. Ionic liquids (ILs), as a novel green solvent, have been extensively employed in green chemistry instead of organic solvent because of low melting point, wide liquid range, negligible vapor pressure and good solubility characteristics, etc. The new approach for desulfurization of diesel fuels by extraction with ILs has been described[6].

In this paper, we have synthesized ten types of Lewis acid ionic liquid, [bmim]Cl/$FeCl_3$ (bmim: 1-butyl-3-methylimidazolium), [omim]Cl/$FeCl_3$ (omim: 1-octyl-3-methylimidazolium), $T_8Cl/FeCl_3$ ($T_8Cl$: Tri-octyl methyl ammonium chloride), $T_4Cl/FeCl_3$ ($T_4Cl$: Tri-butyl methyl chloride), $D_{10}Cl/FeCl_3$($D_{10}Cl$: didecyl dimethyl ammonium chloride), $Et_3NHCl/FeCl_3$ ($Et_3NHCl$: hydrochloric acid triethylamine), which are utilized as extractants in extractive desulfurization of model oil.

## Experimental

### Preparation of Ionic Liquids

[bmim]Cl/$FeCl_3$ ionic liquid was synthesized as mentioned in literature procedure.[7] The procedure for synthesis of [omim]Cl/$FeCl_3$ ionic liquid was the same as that of [bmim]Cl/$FeCl_3$.

[bmim]Cl/$ZnCl_2$, [bmim]Cl/$SnCl_2$, [bmim]Cl/$MnCl_2$ and [bmim]Cl/$CoCl_2$ ionic liquids were prepared by adding the desired amount of chloride to the imidazolium salt. The reaction was left for stirring for 2 hours in flask. The synthesis of $T_8Cl/FeCl_3$, $T_4Cl/FeCl_3$, $D_{10}Cl/FeCl_3$ and $Et_3NHCl/FeCl_3$ were follwed by the similar procedure.

### Extractive desulfurization of the model oil

Model oil was prepared by dissolving DBT in n-octane to give solutions with sulfur contents of 1000 ppm. The desulfurization experiments were carried out in a homemade 40 ml two-necked flask. Extraction experiment was performed by mixing 3.5ml of model oil and some amount IL,(the mass ratio between them is 3:1) stirred vigorously at 30℃ for 30 minutes. When the reaction finished, the upper phase (model oil) was withdrawn and analyzed by gas chromatography (internal standard method nitrobenzene) coupled with a flame ionization detector (GC-FID). A 15 *m* x 0.32 *mm* inner diameter x 1.0 *um* film thickness SE-54 capillary column was used for separation.

## Results and Discussion

### Extraction capacity of the different ionic liquids for model fuel

The extraction capacities of [bmim]Cl/$FeCl_3$, [omim]Cl/$FeCl_3$, $T_8Cl/FeCl_3$, $T_4Cl/FeCl_3$, $D_{10}Cl/FeCl_3$, $Et_3NHCl/FeCl_3$, [bmim]Cl/$ZnCl_2$, [bmim]Cl/$SnCl_2$, [bmim]Cl/$MnCl_2$, and [bmim]Cl/$CoCl_2$ for model fuel are shown in Table1. The ionic liquids based on $FeCl_3$ showed the better extraction capacity than other ionic liquids. It was found that the extraction process went on quickly, and it could reach extraction equilibrium in little time.

**Table 1.** Competitive extraction for model fuel with different ionic liquids

| Entry | Ionic liquids | Sulfur removal/% |
|---|---|---|
| 1 | [bmim]Cl/$FeCl_3$ | 67.6 |
| 2 | [omim]Cl/$FeCl_3$ | 71.5 |
| 3 | $T_8Cl/FeCl_3$ | 70.1 |
| 4[a] | $T_4Cl/FeCl_3$ | 43.5 |
| 5 | $D_{10}Cl/FeCl_3$ | 51.3 |
| 6[b] | $Et_3NHCl$ /$FeCl_3$ | 32.3 |
| 7 | [bmim]Cl/$ZnCl_2$ | 40.1 |
| 8 | [bmim]Cl/$SnCl_2$ | 48.0 |
| 9 | [bmim]Cl/$MnCl_2$ | 35.8 |
| 10 | [bmim]Cl/$CoCl_2$ | 31.3 |

Conditions: mass ratio model oil/IL=3/1, t=30 min, T= 30℃, a: T= 70℃, b: T=50℃, initial sulfur content: 1000 ppm.

## Conclusion

In conclusion, [bmim]Cl/$FeCl_3$ ionic liquid can be used as extractant for desulfurization of model oil, and it exhibits a better extractive performance for DBT. The sulfur removal of DBT-containing model oil can reach 67.6% with single extraction under mild conditions. In addition, Lewis acid ionic liquid with long carbon chains can obtain 71.5% with simple extraction. It was shown that Fe-IL performanced a promising extraction power for DBT of model oil. Furthermore, Fe-IL system shows considerable promise for providing a technology to meet future needs for low sulfur diesel.

## Acknowledgments

This work was financially supported by the National Nature Science Foundation of China (No. 20676057, 20777029) and Jiangsu University Scientific Research Funding (No. 04JDG044).

## References

1. Shafi, R and Hutchings G. J. (2000), Hydrodesulfurization of Hindered Dibenzothiophenes: An Overview, *Catal. Today*, pp. 423-442
2. Eβer, J.; Wasserscheid P and Jess A. (2004), Deep Desulfurization of Oil Refinery Streams by Extraction with Ionic Liquids, *Green Chem*, June, pp. 316-322.
3. Lu, H. Y; Gao, J. B; Jiang, Z. X; Jing, F; Yang, Y. X; Wang, G and Li, C. (2006), Ultra-deep Desulfurization of Diesel by Selective Oxidation with $[C_{18}H_{37}N(CH_3)_3]_4[H_2NaPW_{10}O_{36}]$ Catalyst Assembled in Emulsion Droplets, *J. Catal,* August, pp. 369-375
4. Song, C. S. (2003), An Overview of New Approaches to Deep Desulfurization for Ultra-clean Gasoline, Diesel Fuel and Jet Fuel, *Catal. Today*, June, pp. 211-263
5. Jiang, X. C.; Nie, Y.; Li, C. X and Wang, Z. H. (2008), Imidazolium-based Alkylphosphate Ionic Liquids -A Potential Solvent for Extractive Desulfurization of Fuel, *Fuel*, March, pp. 79-84
6. Bosmann, A.; Datsevich, L.; Jess, A.; Lauter, A.; Schmitz, C and Wasserscheid, P. (2001), Deep Desulfurization of Diesel Fuel by Extraction with Ionic Liquids, *Chem. Commun*, November, pp. 2494-2495.
7. Sun, X. W and Zhao, S. Q. (2006), [bmim]Cl/$FeCl_3$ Ionic Liquid as Catalyst for Alkylation of Benzene with 1-Octadecene, *Chinese J. of Chem. Eng*, August, pp. 289-293

# High Quality Base Oil Production via the HyLube™ Process

***T.N. Kalnes and Mark VanWees**, UOP LLC, 25 E. Algonquin Rd., Des Plaines, IL 60017*
***Andreas Schuppel**, Puralube GmbH, Hauptstrasse 30, Gebaude 27, Elsteraue / OT Alttröglitz, D-06729, Germany*

## ABSTRACT

The successful commercialization of the HyLube™ process by Puralube GmbH, a subsidiary of Puralube Global was presented at the 2007 AICHE meeting[1]. This paper will expand upon the previous paper by providing more detailed analytical characterizations of the re-refined base oil products and a more in depth discussion of how the process manages the wide range of contaminants present in used motor oil.

Unlike most other re-refining processes, the HyLube process produces a very consistent high quality base oil product. In addition, byproducts are managed to minimize environmental impact while maximizing energy recovery. Key process features include (1) continuous processing of the entire used oil charge stock in a hydrogen-rich environment without use of intermediate oil storage, (2) efficient catalytic removal of contaminants such as chlorinated, sulfurous, and oxygenated organic compounds and polyaromatic hydrocarbons, (3) management of all sulfurous and odorous compounds to eliminate malodorous and toxic emissions and (4) production of deeply desulfurized and saturated base oil and distillate fuel products.

Conventional base oil production methods are energy intensive, consume a diminishing fossil fuel resource, and place a large burden on the environment. The current trend toward increasing percentages of ethylene-based synthetic base oils in the blended lubricant product further raises the overall life cycle burden of the finished product. Used oil containing high percentages of high viscosity index and low pour point base oil represents a valuable resource and its proper management should be given the most attention.

## INTRODUCTION

Tougher base oil standards have been established to facilitate auto manufacturers' requirement to improve engine performance from an environmental impact perspective. For example, General Motors' 2005 model year vehicles equipped with Engine Oil Life Systems averaged 8,500 miles between oil changes, compared to the 5,000 mile drain interval typical of GM's competitors. However, creating these longer-drain engine oils and fill-for-life transmission fluids and axle lubes requires Group II, II-plus or Group III base oils.[2]

The trend toward smaller engines that operate at higher engine revolutions per minute dictates the need for better motor oil quality. These quality requirements have been met by a variety of investments to produce enhanced quality base oils but also by the judicious application of performance related additive packages. The recent quality transition from International Lubricant Standard Approval Committee (ILSAC) GF3 designation (in 2000) to ILSAC GF4 (in 2005) and ILSAC GF5 (expected in 2009-2010) is a good example of this quality trend. Among other factors, these trends have sought to increase the level of saturates and decrease the volatility (in the base oil) both of which lead to superior quality finished oils that support extended oil drain intervals.[3]

The growing need for higher quality base oils provides an opportunity for a process that can rejuvenate used oil to these higher product standards. An example of such a process is the HyLube process commercialized by Puralube GmbH. This paper provides detailed analytical characterizations of the re-refined base oil products produced in the commercial HyLube process. It also provides an in depth discussion of how the process manages the wide range of contaminants present in used motor oil.

## HIGH QUALITY BASE OILS BY RE-REFINING

Conventional base oil production methods are energy intensive, consume a diminishing fossil fuel resource, and place a large burden on the environment. The current trend of including higher percentages of hydrocracked base oils and/or ethylene-based synthetics in the lubricant blend further increases the overall life cycle burden of the finished product. Spent lube oils containing large quantities of high viscosity index and low pour point base oil represent a valuable resource which should be properly managed. While simple blending of used oils with low quality fuels will recover the energetic value of this material, the latent value of an engineered material containing very low aromatics and waxes is lost.

Puralube is an environmental technology company engaged in the global commercialization of HyLube re-refining technology. The HyLube process uses a unique feed preparation concept known as direct contact hydrogenation, which eliminates the need for much of the equipment used in other lube re-refining processes. The used lube oil feed is heated to reaction temperature by direct contact with a hot circulating hydrogen stream. As a result, the feed stream is maintained in a hydrogen environment which inhibits the formation of the polymeric and carbonaceous by-products which can cause equipment and catalyst fouling.

Figure 1 shows a simplified block diagram of the process. The first part of the process involves separation of the lube range and lighter components of the feed from the non-distillable residue portion. The HyLube process reactions are carried out at elevated temperatures and pressures in a hydrogen atmosphere using UOP proprietary catalysts. Contaminants are removed and the quality of the lube base oil is rejuvenated and enhanced. In addition to converting hetero-atoms such as sulfur and nitrogen, the HyLube catalyst increase viscosity index via saturation of multi-ring aromatic compounds. After exiting the reactor, the purified products are separated from the reaction by-products and excess recycle gas. Acid gases such as $H_2S$ and HCl are present in the reactor effluent. To prevent these components from becoming corrosive, mild neutralization chemicals are injected into the lube separator vapor. The hydrogen rich vapor from the cold separator is scrubbed, compressed, reheated and returned to the mixer.

The hydrocarbon liquids collected in the separators are sent to the product fractionation section where the products are separated into various cuts to meet the desired lube oil viscosity grades. Unlike other re-refining processes, the lighter co-products are also severely hydrotreated. A photo of the Puralube unit, highlighting the lube fractionation column, is shown as Figure 2. Reference 1 provides more detailed description of the HyLube process flow scheme and product yields.

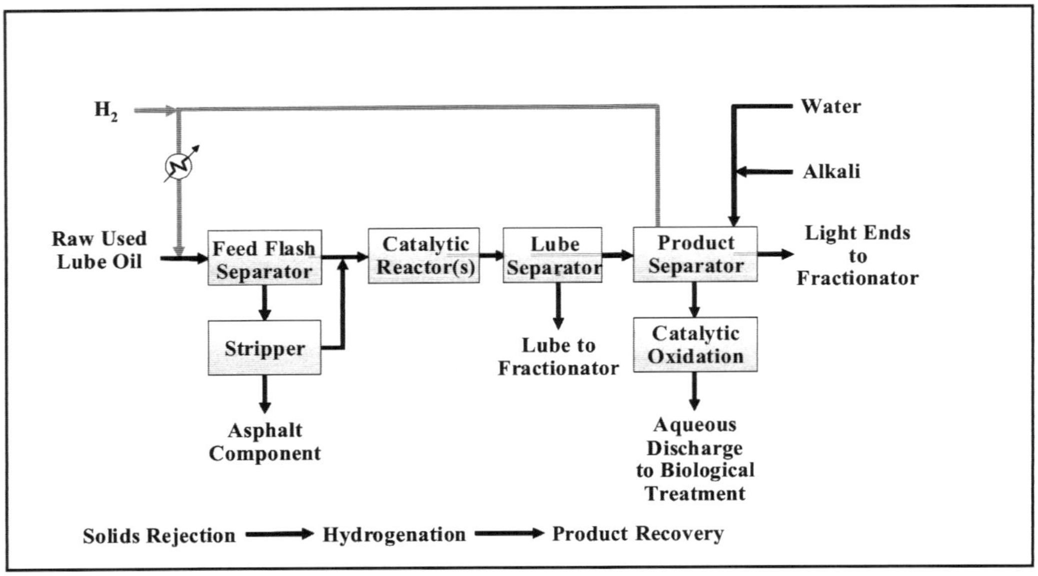


**Figure 1: HyLube Block Flow Diagram**

**Figure 2: Puralube GmbH HyLube Unit Photo**

Modern re-refined base oils meet the requirements of API Group II and are characterized by their high viscosity index, low sulfur content and water-white color.

Other characteristics are:

- ✓ Odourless
- ✓ High Oxidation stability
- ✓ Low evaporation loss
- ✓ Low content of aromatics
- ✓ Olefin free
- ✓ Halogen free
- ✓ No ash residues

The quality of the base oil products produced by Puralube is consistently very high (Group II, II-plus). This is due to both the improving quality of virgin base oils and to the attributes of the HyLube process which enables both rejuvenation and upgrading of the used oils. Tables 1-2 highlight the physical and chemical properties of the primary products; P-75, P-160, and P-300.

| Properties | Standard | Typical Data | | | Test Method |
|---|---|---|---|---|---|
| | | pur75 | pur160 | pur300 | |
| Color | | < L 0.5 | < L 0.5 | < L 1.5 | DIN ISO 2049 / ASTM D-1500 |
| Density @ 15 °C | kg / m³ | 850 | 855 | 860 | DIN 51757 / ASTM D-1298 |
| Flash point COC | °C | 190 | 215 | 228 | DIN ISO 2592 / ASTM D-92 |
| Pour point | °C | - 12 | - 12 | - 12 | DIN ISO 3016 / ASTM D-97 |
| Noack evaporation loss | % wt | 1.5 (150°C) | 9 | 3 | DIN 51581 / ASTM D-5800 |
| Kin. Viscosity | | | | | |
| @ 40 °C | mm²/s | 13.5 | 29.5 | 58 | DIN 51562-1 / ASTM D-445 |
| @ 100 °C | mm²/s | 3.19 | 5.2 | 8.4 | DIN 51562-1 / ASTM D-445 |
| @ 100 °F | SUS | 72 | 140 | 268 | ASTM D-2161 |
| @ 210 °F | SUS | 35 | 43 | 53 | ASTM D-2161 |
| Viscosity index | | 100 | 115 | 116 | DIN ISO 2909 / ASTM D-2270 |
| Sulfur content | ppm | 100 | 100 | 100 | EN ISO 8754 / ASTM D-4294 |
| Saturates | % wt | >> 90 | >> 90 | >> 90 | ASTM D-2007 |

**Table 1: Typical properties of Puralube base oil products**

| Properties | Standard | Typical Data | | | Test Method |
|---|---|---|---|---|---|
| | | pur75 | pur160 | pur300 | |
| Total Acid No. | mg KOH/g | < 0.03 | < 0.02 | < 0.03 | ASTM D-974/DIN 51558-1,2,3 |
| Copper Strip Corrosion | | 1a | 1a | 1a | ASTM D-130DIN ISO 2160 |
| Aniline Point | °C | 98.3 | 108 | 118 | ASTM D-611 |
| Refractive Index | | 1.4708 | 1.4725 | 1.4736 | ASTM D-1218/DIN 51423-1 |
| Sulfated Ash | wt.-% | <0.001 | <0.001 | <0.01 | ASTM D-874/DIN 51575 |
| Metals | ppm | < 1 | < 1 | < 1 | ICP |
| Chlorine | ppm | 0 | 0 | 0 | DIN 51408 P1 |
| Demulsibility | min | 4 | 5 | 12 | ASTM D-1401 |
| Air Release | min | 1 | 1 | 2 | DIN 51381 |

**Table 2: Additional properties of Puralube base oil products**

To be successful in a competitive marketplace, it is imperative that quality of the base oil products be consistent day in and day out. This is especially true for re-refiners. Figure 3 provides a one year history of the key properties of the Puralube P-160 product (29.5 cSt @ $40^0$C) demonstrating that Type II-plus quality can be consistently achieved. Because feed quality to a re-refinery can at times be difficult to control, the re-refining process needs to be robust in order to ensure product quality maintenance. Figure 4 illustrates the high conversion of aromatic oxygenates, sulfurs, and polyaromatic compounds achieved in the HyLube process as measured by high resolution mass spectrometry (HRMS) of feed and product.

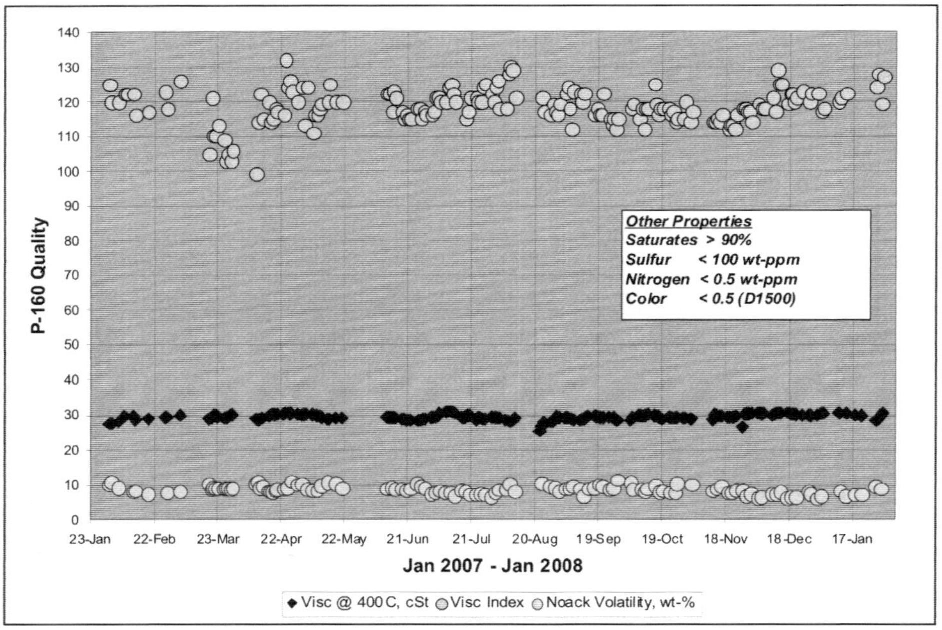


**Figure 3: Consistent product quality demonstrated**

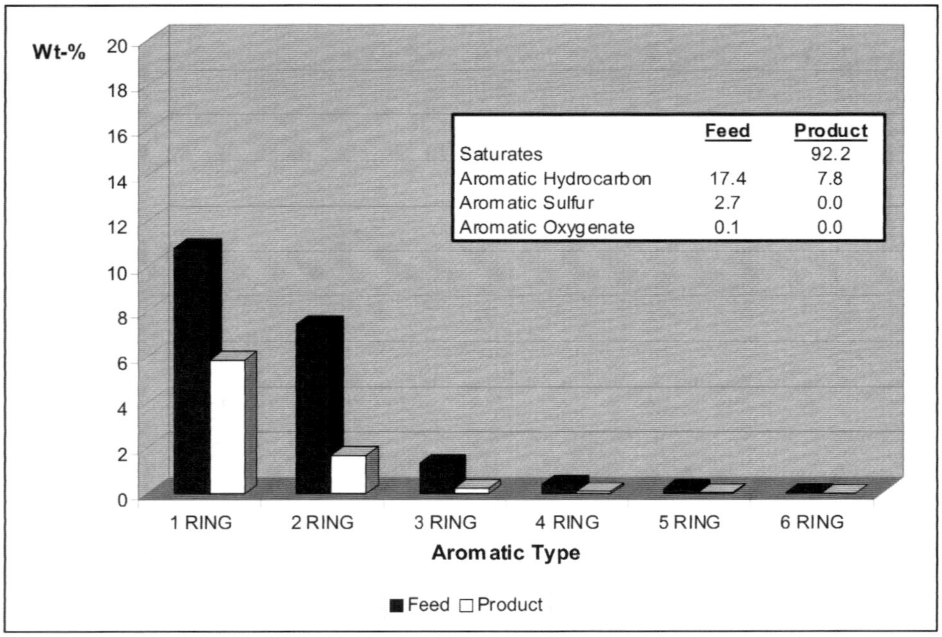


**Figure 4: HRMS data showing aromatic conversion**

As shown in Figure 5, the Puralube re-refined base oils passed all of the ecotoxicity tests with excellent grades. This is made possible by the high degree of polyaromatic saturation achieved in the HyLube process.

| IP 346 DMSO Refractive Index Method | Modified Ames Test (ASTM E 1687) MI | BUS Skin Irritation Test |
|---|---|---|
| pur75 < 0.76 % | pur75 < 0.11 | pur75 |
| pur160 < 0.20 % | pur160 < 0 | pur160 << 3.0 |
| pur300 < 0.10 % | pur300 < 0 | pur300 |
| (no labelling < 3 wt%) | (MI < 1.0 oils are considered no mutagenic) | (< 3.0 no skin irritation) |

**Benzo[a]pyrene:**
Grimmer Method PAH – 0397
(industrial fluids should contain < 1 ppm benzo[a]pyrene)

| SN 150 | pur75 | pur160 | pur300 | HC |
|---|---|---|---|---|
| < 0.1 | < 0.01 | < 0.2 | < 0.02 | < 0.01 |

[ppm]

**Figure 5: Ecotoxicity data for Puralube base oils**

Table 3 compares the Puralube base oils to key specifications for Group II and Group III oils. From this comparison it can be seen that this new generation of re-refined base oils produced in the HyLube process falls into a Group II-plus category similar to severely hydroprocessed base oils. Due to the high saturates, high VI and low volatility, Puralube base oils are able to outperform conventional Group I base oil products[4].

| | Group I | Group II | puralube® pur160 | Group III | Group IV | Group V |
|---|---|---|---|---|---|---|
| Saturates | < 90 % | > 90 % | > 95 % | > 90 % | PAOs | All base stocks not in Groups I, II, III and IV |
| | and/or | and | and | and | | |
| Sulphur | > 0.03 % | < 0.03 % | < 0.01 % | < 0.03 % | | |
| | and | and | and | and | | |
| VI | ≥80 - <120 | ≥80 - <120 | 115 | >120 | | |

**Table 3: Puralube P-160 product exceeds Group II requirements**

Significant advantages can be shown for lubricant formulations based on Group II or Group III base oils compared to Group I. These advantages include:

- the additive requirements to meet specifications such as ACEA E5-99 and API CH-4 are significantly lower for highly saturated, low sulfur base stocks
- Improved oxidation stability is observed in finished lubricants using Group II or III base stocks. Advantage can be taken of this by either the use of a more cost optimized inhibition system or by improved overall lubricant stability.
- Higher viscosity index base oils such as Group II+ or Group III (or Group IV) allow the blending of lower viscosity oils while continuing to meet viscometric and volatility requirements.

Recent literature indicates that severely hydroprocessed base oil can actually outperform a Group IV base oil (PAO) in several areas important to lubricants, such as additive solubility, lubricity and antiwear performance[5]. So even though there has been a worldwide excess base oil supply for some time, stocks in the Group II, Group II+ and Group III category should nevertheless be preferred for their advantages in engine oils, the biggest of all lubricant segments. This, plus the operating cost benefits of re-refining plants should position these stocks relatively well in an industry where worldwide future growth is expected to be modest.

## CONTAMINANT MANAGEMENT USING THE HyLUBE PROCESS

Re-refining used oil is complicated by the fact that the oil is difficult to characterize. Contaminant levels are very dependent upon the source of the used oil, the types of used oil, how it was collected, and the intended end-use of the oil[6]. The highly variable quality of used oils drives the complexity of the facilities necessary to treat it for re-use. Used oil is made up of base oils and various additives that impart the desired quality characteristics to the finished product. High tech motor oil, for example, contains as much as 20–25 percent additive components in the finished product. This includes additives like viscosity improvers, demulsifiers, detergents, and antiwear additives, to name a few.

The additive content of lubricants and motor oils in particular, gives rise to the environmental concerns involved with combustion of used oils[3]. It is common that up to 20 to 25 percent of typical motor oil blends are made up of additives that are used to improve the quality, stability, and longevity of motor oils in combustion engine applications. Although industrial oils are also components of used oils, they do not contain the same level of additization and thus they tend to dilute the impact of motor additives. Table 4 details some of the common additives and their chemical makeup that are present in typical used oils.

High concentrations of a wide variety of synthetic additives also represent one of the major challenges to re-refiners that use hydroprocessing technology to rejuvenate used oils. Furthermore, contaminants such as soot, water, organic acids, and solvents that collect in the oil during normal service and collection make used oil a challenging feedstock that requires re-refining technology that is more robust than traditional oil refining processes.

Feed to the Puralube HyLube unit (Table 5) includes a mixture of recovered motor oils, hydraulic oils, gear oils, and other paraffin-rich industrial oils. In addition to lube base oil and assorted additives, the feed contaminants include gasoline, diesel, combustion byproducts, and water. Varying concentrations of particulate matter and corrosion metals are also present. Day-to-day variations in feed composition are buffered by blending in a primary feed tank which holds approximately seven days of inventory.

| | |
|---|---|
| Antiwear | Zinc dithiophosphates, acid phosphates, organic sulfur and chlorine compounds. Sulphurized fats, sulfides and disulfides |
| Detergent | Metallo-organic compounds of sodium, Calcium and magnesium phenolates. Phosphonates and sulphonates |
| Anticorrosion | Zinc dithiophosphates, metal phenolates, fatty acids and amines |
| Dispersant | Alkylsuccinimides, alkylsuccinic esters |
| Friction Modifier | Organic fatty acids. Lard oil. Phosphorous based compounds |
| Pour point depressant | Alkylated naphthalene and phenolic polymers, polymethacrylates |
| Seal Swell Agent | Organic phosphates aromatic hydrocarbons |
| Viscosity Modifier | Polymers of olefins, methacrylates, di-enes or alkylated styrenes |
| Antifoamant | Silicone polymers, organic copolymers |
| Antioxidant | Zinc dithiophosphates, hindered phenols. Aromatic amines, sulphurized phenols |
| Metal Deactivator | Organic complexes containing nitrogen and sulfur amines, sulphides and phosphates |

**Table 4: Listing of Common Additives in Used Oils[7]**

The patented HyLube approach to contaminant management is based on the continuous direct contact hydrogenation of the entire used oil feedstock. Volatile feedstock components are vaporized at reactor pressure by direct contact with a hot hydrogen rich recycle gas stream and then immediately hydrogenated without intermediate storage. Lighter than lube co-products are also severely hydrotreated. The non-volatile components and ash are concentrated and stabilized in a heavy asphalt product using a steam-stripped vacuum column.

| Property | Units | Min | Max | Average |
|---|---|---|---|---|
| Viscosity @ 40C | cSt | 46.8 | 70.6 | 56.2 |
| Density | Kg/m3 | 881 | 907 | 888 |
| Flash Point | °C | 174 | 310 | 202 |
| Total acid number | mg KOH/g | 2.8 | 7.0 | 4.3 |
| Water | wt% | 3.1 | 10.7 | 6.0 |
| Ash | wt% | 0.33 | 1.11 | 0.72 |
| Chloride | wt% | 0.01 | 0.216 | 0.05 |
| Sulfur | wt% | 0.35 | 0.92 | 0.61 |
| Nitrogen | wt% | 0.00 | 1.60 | 0.12 |
| Lead | wt ppm | 6.9 | 72.7 | 29.4 |
| Arsenic | wt ppm | 0.02 | 3.8 | 0.81 |
| Chromium | wt ppm | 1 | 798 | 62 |
| PCBs | wt ppm | 0.1 | 1.2 | 0.3 |

**Table-5: HyLube Process Unit Feed Characterization**

The fate of the various feed contaminants is illustrated in Figure 6. Toxic organics are destroyed by severe catalytic hydrogenation which converts polyaromatic and heteroatomic compounds to useful products. Metals are stabilized in the asphalt product.

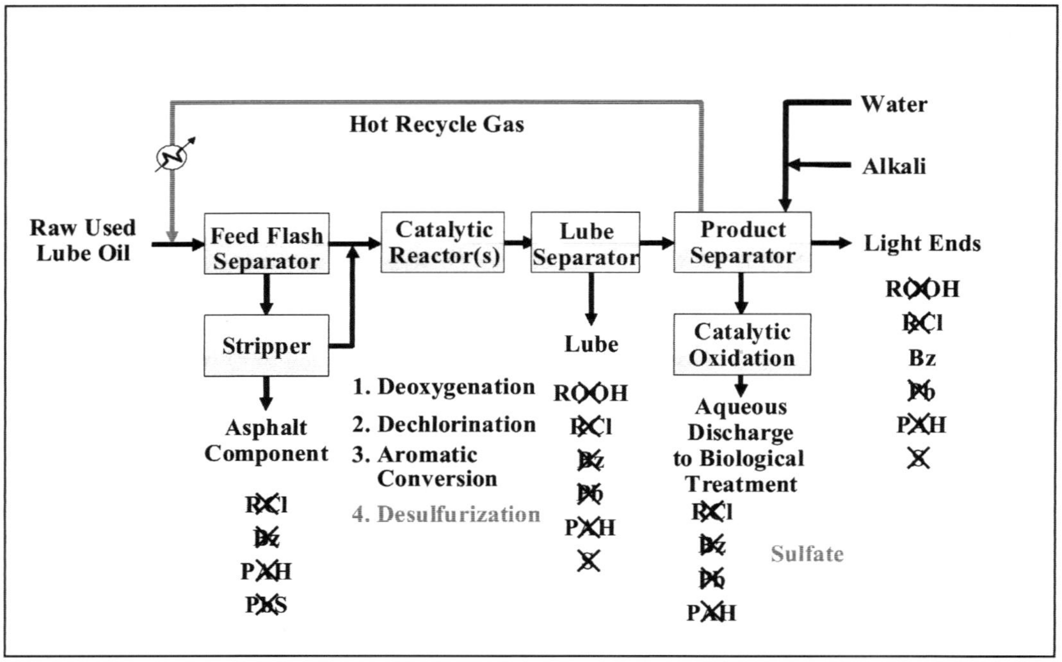


**Figure 6: Fate of Contaminants in HyLube Process**

Because the Puralube HyLube operation is a stand-alone facility, all of the sulfur, chloride, nitrogen, and oxygen removed from used oil feed must be managed in an environmentally acceptable manner. The byproduct of acid gas neutralization is an aqueous stream containing sulfide and chloride salts as well as small amounts of ammonia and water produced in hydrotreating reactions. All of the aqueous streams leaving the HyLube unit are collected in the main column overhead receiver and then routed to the Sulfide Oxidation section of the unit.

The Sulfide Oxidation process is a patented extension of the widely used UOP Merox™ process. Here the sulfides are catalytically oxidized to sulfate salts with a resultant reduction in chemical oxygen demand. After sulfide oxidation, the process water is sent to a two-stage biological treatment facility prior to discharge.

## ENVIRONMENTAL CONSIDERATIONS

The waste management hierarchy detailed in Table 6 is expressed in the waste management policy sections of the Resource Conservation and Recovery Act and Pollution Prevention Act. The "Considered Action Step" is the logical application of this hierarchy to used oil from the most preferred environmental option to the least preferred environmental option[3].

| Waste Management Ranking (from Environmental Perspective) | Option | Considered Action Step |
|---|---|---|
| Most Preferable | Prevent the waste in the first place | Source reduction (e.g. extended oil drain intervals) |
| | Re-use and reclaim the product | Re-refine used oil |
| | Recover energy by burning | Combust used oil for heating value recovery* |
| Least Preferable | Dispose of the waste by land filling or incineration | Recover and collect used oil for proper disposal |

**Table 6: RCRA Policy applied to Used Oil[3]**

Several recent life cycle assessments (LCAs) have concluded that re-refining of used lube oils is a preferable form of recycling when compared to combustion for energy recovery from an environmental perspective [8, 9, 10, 11]. One of the most recent LCA studies modeled the effects of re-refining used oil generated in the European market in the HyLube process[12]. These effects were then compared to the impacts of burning the same oil in cement kilns, as a substitute for the primary energy sources of coal, heavy fuel oil (HFO) or natural gas (Nat Gas). ISO 14040 standards were followed in the study and SimaPro6.0 software was used for calculations.

Figure 7 presents a comparison of environmental impacts for the four alternatives. For the categories of acidification / eutrophication and fossil fuel consumption, re-refining of used oil (UO) is the best alternative[12]. The greatest benefit for climate change (primarily $CO_2$ emissions) is for the case of UO combustion with coal displacement. This is logical because coal has the highest emission of $CO_2$ per unit of fuel energy for any of the fossil fuels in this study, and therefore when coal is displaced by a less carbon intensive fuel, lower greenhouse emissions occur. The next best case is re-refining, followed by UO combustion with fuel oil displacement. The least desirable alternative for climate change is UO combustion with natural gas displacement. When UO displaces natural gas in cement kiln combustion, a higher rate of $CO_2$ release occurs, but this increase is almost exactly compensated for by avoiding emissions from natural gas production.

An analysis of the environmental fate of the key contaminants present in recycled lube oils was conducted by the DOE in 1995. It highlights the end disposition of the key contaminants depending on whether the used oil was combusted or re-refined. The key conclusion in the case of re-refining is that the noxious compounds are solidified and stabilized due to their disposition as a solid and pose minimal environmental risks. On the other hand, combustion of used oil results in air emissions, the magnitude dependent on the quality of the air pollution control equipment utilized.[3]

A peer reviewed study conducted by the California EPA and University of California (Berkeley) compared the life cycle environmental impacts of combusting used oil and re-refining used oils to produce base oils. The results showed that heavy metal emissions from uncontrolled combustion of used oil fuels could be hundreds of times higher than the comparable re-refining process or from the combustion of virgin fuel oil. The heavy metals emissions may lead to 150 times the ecotoxicity impacts compared to re-refining if air pollution control technology is not used.[8]

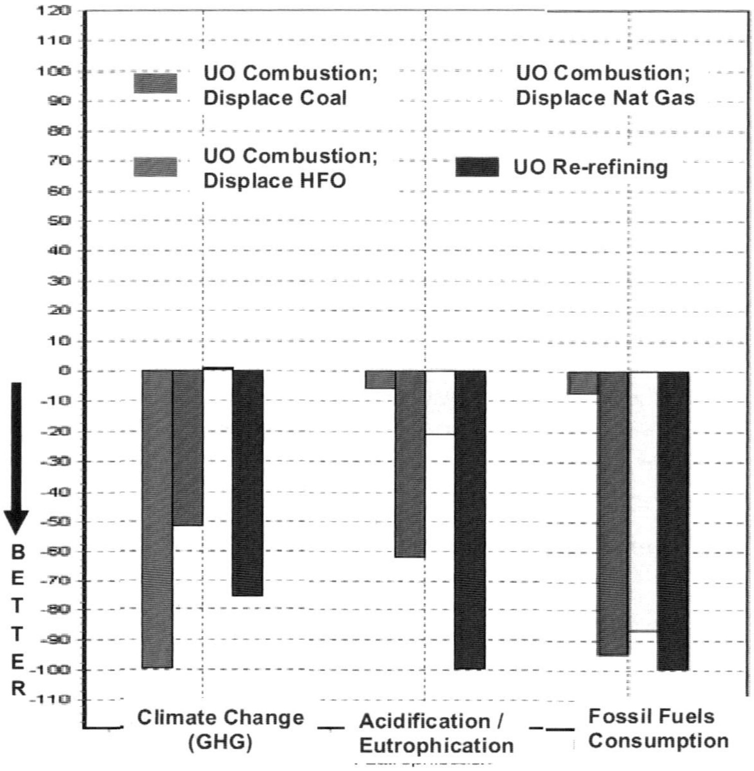


**Figure 7: Environmental Impact Comparison (UO=Used Oil)** [12]

## SUMMARY

Spent lube oils containing large quantities of high viscosity index and low pour point base oil represent a valuable resource which should be properly managed. The HyLube approach to re-refining maximizes the recovery of high value lube products while destroying toxic organics and stabilizing metals. Sulfur is also managed in an environmentally friendly manner. Several recent life cycle assessments (LCAs) have concluded that re-refining of used lube oils is a preferable form of recycling when compared to combustion for energy recovery from an environmental perspective. As a result, RCRA guidelines for used oil management rank re-refining ahead of combustion for heating value recovery.

As automotive engine oil standards become more stringent and crude oil quality declines, production costs for base oil will increase. This trend coupled with more stringent environmental standards creates an opportunity to expand used oil recycling efforts worldwide. The successful commercialization of the first HyLube unit by Puralube GmbH represents an important step forward in the quest for a sustainable lube oil re-refining process.

The quality of the base oil products produced by Puralube is consistently very high (Group II, II-plus). Commercially re-refined base oils pass all of the ecotoxicity tests with excellent grades. This is made possible by the high degree of polyaromatic saturation achieved in the HyLube process. This new generation of re-refined base oils falls into a Group II plus category similar to severely hydroprocessed base oils. Based on high saturates content, high VI and low volatility these base oils are preferred for use in engine oils, the biggest of all lubricant segments. High quality coupled with the operating cost benefits associated with HyLube re-refining plants position these products relatively well in an industry where worldwide future growth is expected to be modest.

Puralube has moved forward with a project to build a second HyLube unit in Germany for startup in October 2008. Their longer term vision is to become the pre-eminent refiner of used oils in the world based on the UOP HyLube Process with a target to install 1.5 mm mta of processing capacity. Activities are already underway to expand the waste oil collection side of the business.

## REFERENCES

1. Kalnes, T., Kadlec, L., Schuppel, A., "Hylube™ Process Commercialization: Recovering Value from Used Motor Oil", Spring 2007 AICHE Meeting, Houston, Texas

2. Lube Report (ISSN 1547-3392), "Lubes'n'Greases Magazine and Lubricants Industry Sourcebook"

3. "Used Oil Re-refining Study to Address Energy Policy Act of 2005 Section 1838", Office of Oil and Natural Gas Office of Fossil Energy, U.S. Department of Energy, July 2006

4. The 3rd Pan-American Base Oil & Lubricants Conference, "Re-refining Update, Puralube presentation", November 29, 2007, New York, C. Hartmann

5. David C. Kramer, Brent K. Lok, Russ R. Krug and John M. Rosenbaum, ChevronTexaco Global Lubricants, "Performance of Base Oils and Future Trends - The Evolution of Base Oil Technology - Part 3". *Machinery Lubrication* Magazine. September 2003

6. "End Uses for Used Oil, A Market Perspective", by Don Kress Par Excellence Developments Inc., Sudbury Canada, http://www.ped.vianet.ca/article1.pdf.

7. David Fitzsimons, Nicholas Morley, Peter Lee, Oakdene Hollins LTD, "UK Waste Oils Market 2001", page 8; http://www.defra.gov.uk/environment/waste/topics/oils/pdf/wasteoils.pdf

8. Bob Boughton, “Environmental Assessment of Used Oil Management Methods”, Arpad Horvath Environmental Science and Technology, Vol. 38 No. 2, page 353, 2004, http://pubs.acs.org/cgibin/download.pl?es034236p/D2gn

9. Frankl, P., Fullana, P., Baitz, M., 2005, Europe Life Cycle Considerations on Waste Oils and Implications for Public Policy - Ecobilancio Italia, ESCI

10. Fehrenbach, H. 2005, Ecological and energetic assessment of re-refining used oils to base oils: Substitution of primarily produced base oils including semi-synthetic and synthetic compounds, Institut für Energie- und Umweltforschung GmbH (IFEU), a study commissioned by GEIR - Groupement Européen de l’Industrie de la Régénération.

11. Hartmann, C., 2005, Clean, Competitive, and Sustainable: Why the Recycling of Waste Oils Must Remain an EU Policy Priority, GEIR (Groupement Européen de l’Industrie de la Régénération), position paper, Square Marie Lousie 49 – 1000 Brussels High Performance Automotive Lubricants, 1995, Chem Systems, May 1995, 93S4

12. Kalnes, T., Shonnard D., Schuppel, A., 2006, LCA of a Spent Lube Oil Re-refining Process, 9th International Symposium on Process Systems Engineering, (Marquardt et al editors) © 2006 Elsevier B.V./Ltd.

## 67b Unravelling the Detailed Characterization of Hydroprocessing Feeds and Products

**Cyril Dartiguelongue**[1], Damien Hudebine[2], Fabrice Bertoncini[3], Frédérick Adam[1], Marion Courtiade[1], Didier Espinat[1], and Thierry Chapus[2]. (1) Physics and Analysis Division, IFP, BP3, Vernaison, 69390, France, (2) Process Development and Engineering Division, IFP, BP3, Vernaison, 69390, France, (3) Catalysis and Separation Division, IFP, BP3, Vernaison, 69390, France

Clean fuel legislation in Europe, America and Asia is driving the demand for ultra-low sulfur diesel (ULSD) fuel, requiring a total sulfur content around 10 ppm after 2008. Further improvement in diesel hydrotreatment (HDT) will depend on the accurate understanding of the phenomena involved at the molecular scale. Therefore, it is of major importance for refiners to get an accurate characterization of HDT feeds and products, for both hydrocarbons, sulfur and nitrogen species. Due to the complexity of diesel cuts, conventional gas phase chromatography is unable to provide a detailed description of diesel samples.

The recent development of comprehensive two-dimensional gas phase chromatography (GC*GC) over the last years has been a turning point in the molecular characterization of gas-oils. The presentation will describe the advances obtained by IFP in the characterization of hydrocarbons, sulfur and nitrogen compounds.

GC*GC hyphenated with a flame ionisation detector allows a detailed quantification of hydrocarbons both by families and by numbers of carbon atoms. This technique can especially be applied to the quantification of mono, di and tri-aromatics, as well as mono, di and tri-naphtenes.

The use of specific detectors based on chemiluminescence enables to separate and quantify sulfur and nitrogen species. Through the sulfur speciation method, alkylated benzothiophenes and dibenzothiophenes from diesel feeds are separated and quantified by numbers of carbon atoms. On the other way, the nitrogen speciation method developed in IFP allows an unambiguous separation of basic and non basic compounds into well defined families (pyridines, anilines, quinolines, acridines, indoles, carbazoles). Both speciation techniques have recently been applied to diesel HDT products, allowing the identification and quantification of the most refractory compounds towards hydrodesulfurization and hydrodenitrogenation.

Therefore, GC*GC offers an interesting perspective in the monitoring of reactions involved in hydrotreatment processes.

# LCO UPGRADING

## A NOVEL APPROACH FOR GREATER ADDED VALUE AND IMPROVED RETURNS

Vasant P. Thakkar, Suheil F. Abdo, Visnja A. Gembicki, James F. Mc Gehee, and Bart Dziabala
UOP LLC, a Honeywell Company
Des Plaines, Illinois, USA

## INTRODUCTION

### LCO IN THE CLEAN FUELS REFINERY

As refiners plan to meet regulations for clean fuels, one of the many considerations they face is the disposition of light cycle oil (LCO). It is a poor diesel fuel blending component due to its poor engine ignition performance and its high sulfur. Beyond middle distillate blending, LCO has also historically been used as a blend-stock into heavy fuel oil for viscosity adjustment. This opportunity is also becoming constrained by declining demand for heavy fuel oil. In the overall context of a high conversion, clean-fuels refinery, light cycle oil is an issue, both in terms of product blending and product-value maximization. In addition to the use of conventional hydrotreating, high pressure hydrocracking units have historically been used to crack LCO into naphtha and lighter products. These units are relatively high in capital cost, consume large quantities of hydrogen, and the naphtha product requires reforming before blending to gasoline. This paper will address an alternative LCO processing solution to the growing demand for clean fuels and product slate flexibility with much lower capital investment.

## UOP LCO Unicracking™ Process Technology

UOP recently introduced and commercialized the UOP LCO Unicracking process. The technology solves the problem of LCO disposition by economically converting it to ULSD and high octane gasoline suitable for direct blending without need for further reforming. UOP's recently introduced UOP HC™ 190 catalyst, coupled with process innovations, enables an LCO Unicracking unit to operate at much lower pressure relative to conventional hydrocracking unit. The key feature of the technology is selective hydrogen addition, focusing on desulfurization and hydrogen addition to the diesel product, while minimizing naphtha aromatic saturation. The low pressure flow scheme, coupled with efficient hydrogen utilization, provides superior economics compared to a conventional hydrotreating solution.

## World Wide Refining Trends

World oil demand is projected to continue increasing, at a rate of about 1.8% per year, with increased growth of transportation fuels and a decrease in fuel oil. Worldwide the diesel demand now exceeds that of gasoline, with the ratio of diesel to gasoline exceeding 1.0 in some areas, such as Europe. In the US most of the 400,000 barrel annual increase in crude consumption will be used for transportation fuels, with gasoline representing about 45% of the total petroleum consumption.

Coupled with this demand pattern is the changing regulatory picture for transportation fuels. By 2006, the US on-road diesel pool specification will be set at a maximum of 15 ppm sulfur and minimum 40 cetane number (or maximum aromatics level of 35%), with gasoline sulfur specified at a maximum of 30 ppm, and total aromatics at maximum 35%. The gasoline benzene specification for 2011 is set for the maximum average gasoline pool level of 0.62 lv-%. Beyond 2006, there are pending regulations calling for a gradual merging and harmonization of sulfur specifications for on road and off road diesel. This second wave of regulations is expected to be complete by the end of decade.

How will these trends affect LCO as a distillate pool component? LCO comprises about 15% of the total US distillate pool, and by 2006, it will mainly be suitable as a blend component for lower quality distillate products including off-road diesels, heating oil and marine fuels, or as a cutter stock for heavy fuel oils. It will become increasingly difficult to blend LCO as the specifications for the on-road and off-road diesel pools are harmonized to the same ultra low sulfur levels.

## Characterization of LCO

The total worldwide installed FCC capacity is approximately 14 million BPSD, with an overall production of about 3 million BPSD of LCO. The majority of FCC capacity is in North America, followed by Europe and Asia. Operating severity varies depending upon the market. For example, North American refiners operate their FCC units at high severity to maximize gasoline production with LCO yield of less than 20%, whereas European refiners operate a lower severity for greater LCO production.

Cracked products such as LCO and coker distillates have a considerably lower cetane value compared to straight run distillates derived from most of the world's crude sources. LCO cetane ranges from 15-25, compared to 40-60 for the straight run distillates produced from the same crude.

The sulfur content in average light cycle oils can range from 0.2 to 2.5 wt-%. A detailed sulfur speciation of LCO shows that a significant portion of the sulfur is found in alkyldibenzothiophenes (DBT), which are relatively difficult to desulfurize by hydrotreating. The aromatics content of LCO from FCC units in a normal gasoline-oriented operation can be as high as 80 wt-%. The organic nitrogen is almost entirely composed of non-basic aromatic compounds, such as carbazoles, with a concentration range of 100-750 ppm. The components of LCO boil in the diesel range with a 95% point of 360ºC or higher, representing thermally stable cracked hydrocarbons that are not further reacted in the FCC process. Over 70% of the aromatic hydrocarbons present in LCO have two rings, while the remainder is typically evenly split between single ring and 3-plus ring aromatics. Two and 3+ ring aromatics combust poorly in the diesel engine. They have very low cetane values and are the root cause of the low blending quality of LCO.

It is necessary to saturate and open the di-aromatic rings to increase the fuel value of products from LCO upgrading. These reactions are a fundamental pathway in hydrocracking reaction chemistry and thus this process is ideal for converting LCO to a higher quality diesel product. Single ring aromatics boiling in the gasoline range are excellent high octane components in the gasoline pool. Aromatic ring manipulation is the key to producing higher value gasoline and diesel.

## Options for Upgrading LCO

As discussed earlier, pending diesel sulfur specifications effectively eliminate the option to directly blend the LCO into the on-road diesel pool. Further, harmonization of on-road and off-road diesel specifications will mean that blending LCO into off-road diesel will also no longer be possible. Blending raw LCO into the fuel oil and heating oil pool will remain an option but those products will experience market demand shrinkage in the future. Thus, cost-effective upgrading of LCO is becoming increasingly important.

The two main processes for upgrading LCO are hydrotreating and hydrocracking. UOP licenses the UOP Unionfining™ process and the UOP Unicracking process, respectively, for these applications.

## UNIONFINING PROCESS

Table 1 shows an inspection of a typical LCO sample containing 0.7 wt-% sulfur. Approximately 300 ppm of the sulfur is contained in molecules that are very difficult to desulfurize by hydrotreating. As a result, hydrotreating LCO to improve its diesel pool blending characteristics is a challenge not only from the standpoint of its high aromatics content but also due to the nature of its sulfur species. Hydrotreating LCO to reduce its sulfur content to an ultra-low level requires high severity operation. However, high pressure hydrotreating results in over saturation of aromatics and inefficient use of hydrogen for only a modest cetane improvement. Some tri- and higher ring aromatics are converted to lower aromatics but the total aromatic content of the LCO still remains high and relatively little cetane improvement is realized. Figures 1 and 2 show that even at low pressure there is significant conversion of polyaromatics to monoaromatics and most of the cetane boost is accomplished at low hydrotreating pressures. As pressure is increased to promote deep desulfurization, little further increase in cetane takes place.

Another important aspect of LCO upgrading is that in the absence of significant conversion of aromatics through cracking reactions, the saturation reactions proceed only to their equilibrium limit. A typical temperature and pressure range of most hydrotreaters places a ceiling on how far the cetane index of the LCO can be improved via simple partial saturation of aromatic rings.

In summary, hydrotreating of LCO for deep sulfur removal requires relatively high pressure and hydrogen consumption, yet achieves only limited improvement in cetane number, total aromatics, and density.

**Table 1**

***Sulfur Speciation of Typical LCO***

| | |
|---|---|
| **API** | **15.4** |
| **IBP – EP, °F** | **410-680** |
| **Cetane Index** | **23.1** |
| **Total Sulfur, wt-%** | **0.7** |
| **4,6 dimethyldibenzothiophene, ppm** | **56** |
| **4-methyldibenzothiopene, ppm** | **231** |
| **Other sulfur species, ppm** | **702** |
| **Total Nitrogen, ppm** | **257** |

**Figure 1**

*Hydrogenation of Aromatics as a Function of Pressure*

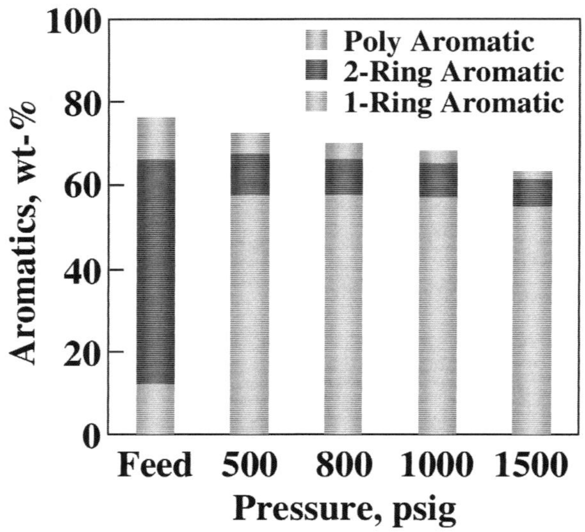


**Figure 2**

*Cetane Number as a Function of Pressure*

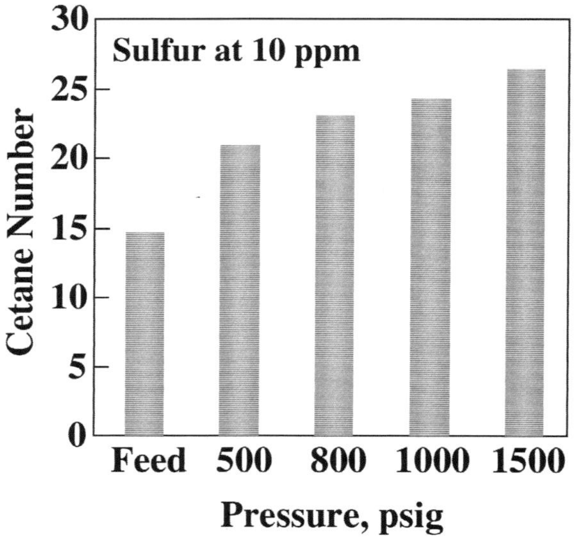


## FULL CONVERSION UNICRACKING PROCESS

As an alternative to conventional hydrotreating, the hydroprocessing severity can be expanded to include hydrocracking reactions. The process goal in this case is to saturate and open the multi ring aromatic compounds to increase the blending value of products. These reactions are a fundamental part of hydrocracking reaction chemistry, and full conversion hydrocracking of LCO to naphtha is a well established commercial process. However, unlike the high octane gasoline made in the LCO Unicracking process, the gasoline boiling-range product from full conversion hydrocracking is a highly naphthenic, low octane naphtha that must be reformed to produce the octane required for product blending, incurring additional operating expense.

## LCO UNICRACKING PROCESS TECHNOLOGY FROM UOP

UOP's recently commercialized ***LCO Unicracking*** technology enables substantially reduced capital and operating costs while producing gasoline and diesel streams for direct ULSD and ULSG pool blending. It is highly selective in controlling the hydrocracking reactions in a partial conversion operating mode to achieve good hydrogenation of the diesel fraction, while preserving aromatics in the gasoline range product. Depending on a refiner's product needs and product quality targets, aromatic ring manipulation is the key to producing high-value gasoline and diesel from LCO. In order to accomplish these parallel objectives, UOP has combined process innovations with HC 190 catalyst, a high-activity catalyst for effectively upgrading LCO in a partial conversion hydrocracking process configuration.

## HC 190 CATALYST

HC 190 catalyst was shown to be well suited to accomplish the desired chemistry for partial conversion LCO hydrocracking. It is designed to provide significantly higher activity than the current naphtha hydrocracking catalysts. For example, its performance on a cracked stock blend, containing light cycle oil, shows about 15°F higher activity and higher naphtha yield at equivalent hydrogen consumption compared to current commercial catalysts. It has demonstrated superior performance in pilot plant tests for partial conversion of LCO by maximizing retention of single ring aromatics, thus producing a higher octane naphtha product. It also produces diesel with very low sulfur levels.

## LCO UNICRACKING PROCESS

The LCO Unicracking process uses partial conversion hydrocracking to produce high quality gasoline and diesel stocks in a simple once-through flow scheme. The basic scheme is shown in Figure 3.

**Figure 3**

***LCO Unicracking Process***

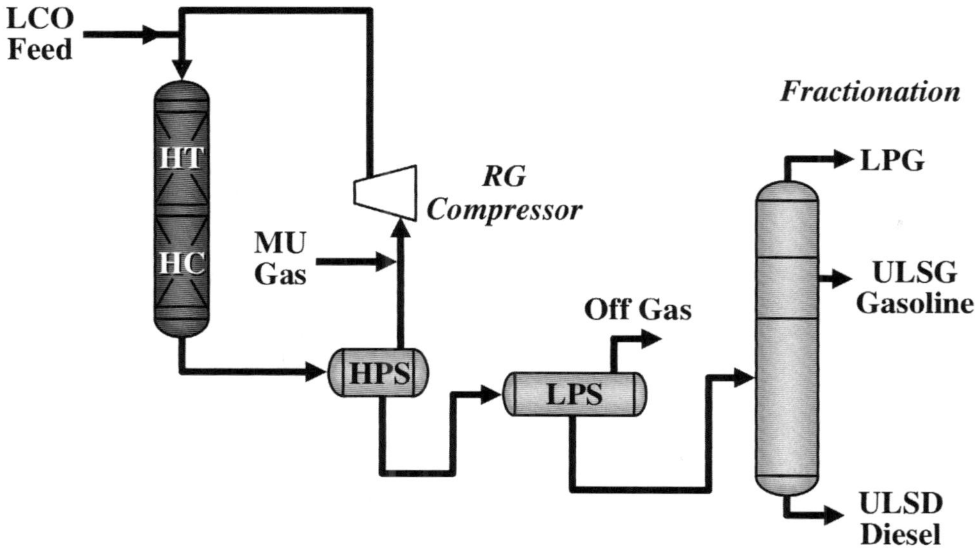


The feedstock is processed over a pretreatment catalyst and then hydrocracked in the same stage. The products are subsequently separated without the need for liquid recycle. The advantage of the LCO Unicracking process is that it can be designed for lower pressure operation. The pressure requirement will be somewhat higher than high severity hydrotreating but significantly lower than a conventional partial conversion and full conversion hydrocracking unit design. The upgraded middle distillate product makes a suitable ultra-low sulfur diesel (ULSD) blending component. The naphtha product from low-pressure hydrocracking of LCO has ultra-low sulfur and high octane and can be directly blended into the ultra-low sulfur gasoline (ULSG) pool. Thus, the LCO Unicracking process offers excellent flexibility to increase the value of LCO by upgrading its products to ULSG and ULSD blending targets.

## ANALYTICAL TECHNIQUES

Recent advances in hydrocarbon analysis have made it possible to speciate distillate fuels at the molecular level. UOP has developed a technique, which uses comprehensive gas chromatography, also called GCxGC, that enables resolution of hydrocarbon species in a three-dimensional map. One axis distinguishes by boiling point, the second axis distinguishes by polarity, and the vertical axis measures the relative concentration (Figures 4 and 5). This type of graphical representation significantly enhances the understanding of complex hydrocarbon mixtures like LCO. Advanced procedures to analyze and interpret this data are being developed at UOP and have been crucial in understanding the fundamental chemistry and pathways associated with different catalysts and reaction conditions. It will allow more precise feedstock characterization and facilitate better understanding and control of the desired molecular transformations

**Figure 4**

***Comprehensive GC Speciation of LCO***

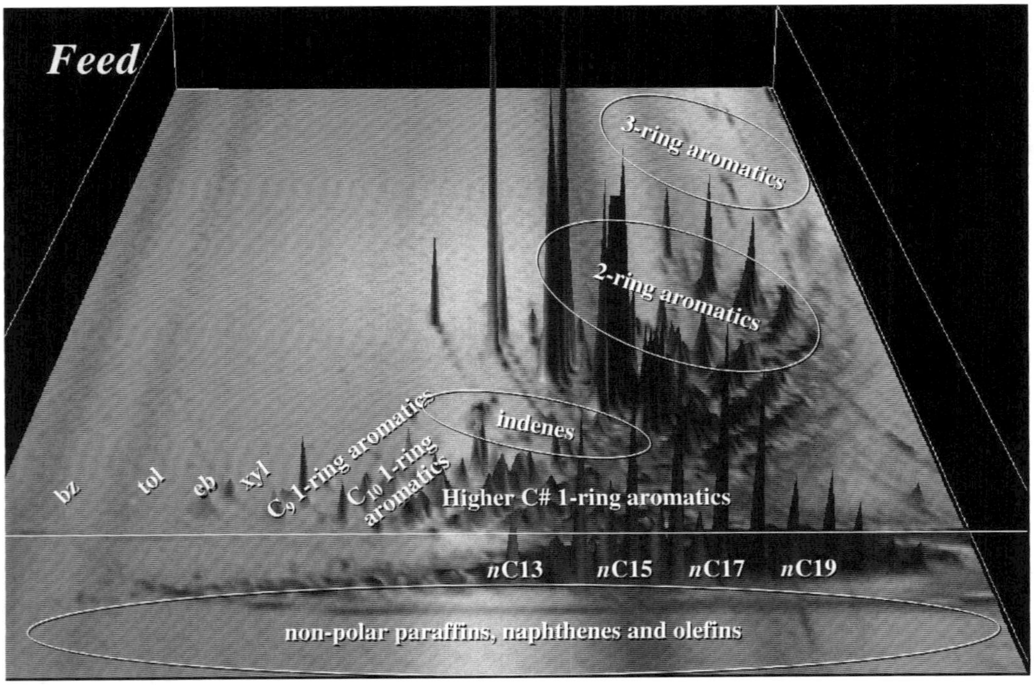

**Figure 5**

*Comprehensive GC Speciation of Hydrocracked LCO*

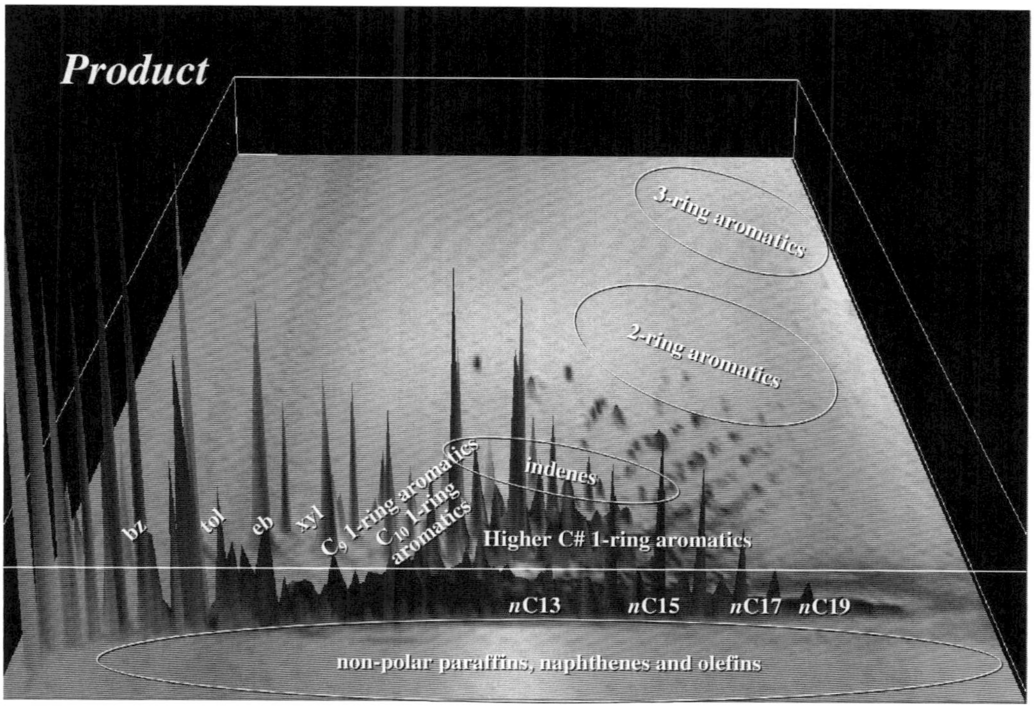


## PILOT PLANT EVALUATION

In partial conversion hydrocracking of LCO, the severity must be optimized so that the naphtha product has high octane for direct blending to gasoline without the need for reforming. At the same time, the diesel product must be hydrogenated to produce a higher cetane, ultra-low sulfur blendstock. Finally, ring saturation and ring opening need to be highly selective to make efficient use of hydrogen. These objectives require unique catalyst chemistry and operating conditions to maximize the yields and qualities of the desired products.

An example of a preferred reaction is shown for a typical two-ring aromatic hydrocarbon, 1,3-dialkylnaphthalene, in Figure 6. The desired reactions shown in the figure are those to saturate and open one ring to yield a mixture of alkylaromatics in the naphtha boiling range.

Pilot plant tests were conducted on representative commercial LCO feedstocks. The key properties of these feeds covered wide ranges of aromatic content and distribution, sulfur and nitrogen content and distillation characteristics. Their properties are listed in Table 2. The tests were all conducted in partial conversion hydrocracking mode at relatively mild conditions.

**Figure 6**

***Preferred Reaction Pathway***

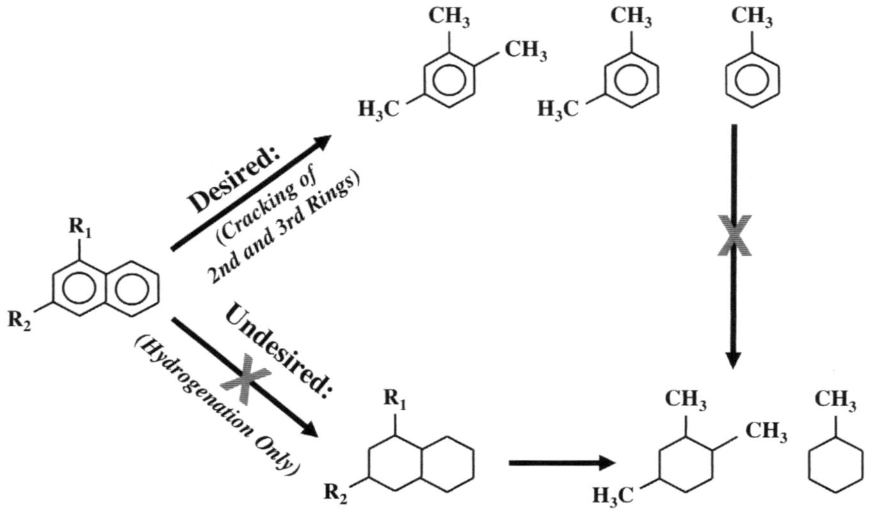


Typical cracked product characteristics using HC 190 catalyst are listed in Table 2. It can be seen that the high activity and unique cracking characteristic of HC 190 catalyst enable good

**Table 2**

***LCO Unicracking Process Pilot Plant Evaluation***

| | |
|---|---|
| **Feedstocks:** | |
| **Origin** | **Commercially derived LCOs** |
| **API Gravity** | **15.1 – 19.0** |
| **Sulfur, ppm** | **2290 – 7350** |
| **Nitrogen, ppm** | **255 – 605** |
| **Aromatics, IP391, wt-%** | |
| **1 – ring** | **12 – 21** |
| **2 – ring** | **40 – 55** |
| **3+ rings** | **8 – 14** |
| **Distillation, ASTM D-2887, °F** | |
| **95%** | **660 – 710** |
| **EP** | **725 – 790** |
| **Cetane Index, ASTM D-4737** | **22 – 25** |
| **Operating Mode:** | **Partial Conversion** |
| **Products:** | |
| **Light Naphtha Yield, wt-%** | **10.5 – 13.5** |
| **Heavy Naphtha** | |
| **Yield, wt-%** | **35 – 37** |
| **RONC** | **90 – 95** |
| **Sulfur, ppm** | **< 10** |
| **Diesel** | |
| **Yield, wt-%** | **46 – 51** |
| **Cetane Index Improvement** | **+6 to +8** |
| **Sulfur, ppm** | **<10** |

aromatic retention in the gasoline range product as demonstrated by the high research octane number. The diesel range product contains less then 10 ppm sulfur and low aromatic content, well within ULSD fuel specification limits. Its cetane index ranges from 6 to 8 numbers higher than the LCO feed.

Extended stability tests of more than six months of continuous operation at low pressure have confirmed excellent catalyst stability while maintaining the cracking conversion and product qualities. Stable yields of products were obtained over the entire period, with slight improvement observed in the gasoline octane with time on stream.

## REFINERY APPLICATIONS

The potential impact of LCO Unicracking technology employed in a typical US refinery was examined by comparing it against the alternative of deep LCO desulfurization by high pressure hydrotreating.

## BLENDING OF TRANSPORTATION FUEL POOLS

In order to assess the potential impact of upgrading these two options, a "typical" US coker/FCC-based refinery was considered. The refinery capacity was set at 300,000 BPSD of crude oil. The refinery and gasoline pools had the following products available for blending:

### *Gasoline Pool*

Treated FCC gasoline
Hydrotreated light straight run naphtha
Reformate
Alkylate (to premium)
Purchased MTBE (to meet oxygen specification for RFG)
Butanes for RVP adjustment
LCO Unicracking unit gasoline

### *Diesel Pool*

Jet Fuel blending components
- Kerosene
- Hydrotreated kerosene

Ultra-low sulfur diesel components
- Hydrotreated kerosene
- Hydrotreated straight run diesel
- LCO Unicracking unit diesel or hydrotreated LCO (case study)

All process units were sized to be consistent with the design crude oil capacity, and assumed to operate at "typical" conditions for a US refinery. The new LCO Unicracking unit or the hydrotreating unit cases were sized at a 20,000 bpsd capacity. The gasoline and diesel pools were blended to meet the 2006 US quality regulations. In order to reflect the next wave of diesel regulations, the diesel pool blend was based on harmonized on-road and off-road diesel specifications, i.e. the entire pool must attain an ultra low sulfur level of 15 ppm or less.

The estimated quantities and qualities of the blended gasoline and diesel products are detailed in Table 3. The results show that both solutions, i.e. the LCO Unicracking process or high pressure hydrotreating will enable production of acceptable ultra-low sulfur diesel which can be entirely blended into the diesel pool. Due to selective chemistry of the LCO Unicracking process however, this route produces 3-5 points higher cetane index than the hydrotreating route. Both routes produced a cetane index in excess of the 2006 specification. By employing the LCO Unicracking solution, the refiner will be better prepared for expected future upward changes in the cetane specifications. Additionally, LCO Unicracking process yields a high octane and ultra low sulfur gasoline, which can be blended directly into the gasoline pool without jeopardizing the pool specifications for octane, sulfur, or aromatics. The small amount of naphtha produced in the hydrotreating case is sent to a reformer prior to blending into the pool. Due to the high octane of the gasoline produced via the LCO Unicracking process, the overall refinery pool octane is not changed by direct blending of this product into the pool.

The installation of an LCO Unicracking unit increased gasoline production by about 10% while decreasing the diesel production by 13% compared to the hydrotreating unit option. The addition of an LCO Unicracking unit has the benefit of varying diesel/gasoline slate to meet the refiner's seasonal demand.

**Table 3**

***Refinery Gasoline and Diesel Pool Yields and Qualities As a Function of the LCO Upgrading Option***

| *Type of LCO Upgrading* | *Hydrotreating* | *LCO Unicracking Process* |
|---|---|---|
| **Gasoline pool:** | | |
| **Volume, BPSD** | **134,616** | **148,242** |
| **RONC / MONC** | **92.2/83.7** | **92.1/83.8** |
| **Sulfur, ppm** | **<30** | **<30** |
| **ULSD Pool:** | | |
| **Volume, BPSD** | **104,763** | **92,871** |
| **Cetane Index** | **45** | **48** |
| **Sulfur, ppm** | **<10** | **< 10** |

## ECONOMICS OF LCO UPGRADING OPTIONS

The economics of LCO upgrading are a function of many factors, such as product pricing, and also the value assigned to LCO feedstock, which is in turn a function of its end-product disposition. The refiner has generally three options:

- Invest in a hydrotreating unit, which requires lower capital investment and operating cost, but results in inflexible product slate, i.e. mainly diesel product.
- Invest in an LCO Unicracking unit, which requires higher capital and operating cost than the hydrotreating unit. This solution enables a flexible yield of both ULSD and high octane gasoline, ready for the pool blending.
- In a special case where the refiner needs to add new unit capacity for straight run or coker distillate hydrotreating, in addition to investing in an LCO Unicracking unit, an integrated plant solution is a very attractive option. The synergy resulting from parallel reaction system for hydrocracking LCO and hydrotreating the other refinery distillate streams within the shared recycle gas system and common downstream separation, can provide as much as 30-40% capital savings compared to building two stand alone units. Furthermore, with this strategy a refiner's ULSD project and capital investment can be consolidated into a single project versus multiple unit revamps.

## ECONOMIC PROJECTION CASE STUDY

The following study compares the relative economics for upgrading light cycle oil using an LCO Unicracking process versus an LCO hydrotreating process. The scenario is that of a nominal 300,000 bbl/d FCC-coker based US refinery. The refiner wishes to upgrade LCO to produce both ULSD and ULSG products. Prices for this study are based on market averages from early 2004 period.[2,3] The value of LCO was derived from its alternate use as a fuel oil blending component. Additional hydrogen need was met with purchased hydrogen, while the estimated ISBL erected costs are based on US Gulf Coast construction for the 4th quarter of 2004.

The purpose of the case study was to:

- Determine the investment payback on an LCO Unicracking unit designed to produce a flexible product slate of both ultra-low sulfur diesel and high octane gasoline blendstocks, and compare the economics to hydrotreating only.
- Determine the sensitivity of payback and net present value to the price differentials between the primary products, gasoline and ULSD, and to the differential between LCO and ULSD value.

## PERFORMANCE AND ECONOMICS

Table 4 summarizes economic projections for the addition of either a hydrotreating unit or an LCO Unicracking unit solution for LCO upgrading in the refinery study.

**Table 4**

***Process Economics***

| *Type of LCO Upgrading* | *Hydrotreating* | *LCOUnicracking* |
|---|---|---|
| **Unit Capacity, BPSD** | **20,000** | **20,000** |
| **Yield, Lt and Hvy Naphtha, wt-%** | **1.6** | **57** |
| **Yield, ULSD, wt%** | **98.9** | **41.6** |
| **Estimated ISBL Investment Cost, $M** | **36.4** | **61.4** |
| **10 yr NPV, $M** | **(1)** | **203** |
| **Simple Payout, yrs** | **6.5** | **2.1** |

The LCO Unicracking project produces a very attractive positive 10 yr NPV of $203M and results in a simple payout of two years. LCO hydrotreating option has a negative NPV and a payout time of six years, i.e. it is a 'stay-in-business' solution imposed by the regulatory requirements.

The sensitivity analysis of the two solutions was done using two comparative parameters; price differential between ULSD and LCO, and a price differential between gasoline and ULSD. The results are depicted in Figures 7 and 8. It is shown that the LCO Unicracking solution retains the same advantage over the hydrotreating solution for a range of price differentials. As expected, the Unicracking solution advantage increases as the price differential between gasoline and diesel increases. It should be noted that this analysis is specific to market where gasoline is a higher priced product, as for example, in the US. A somewhat different case study was described recently in another publication which was focused on the European refinery setting and market pricing [(1)].

**Figure 7**

*NPV Sensitivity to ULSD-LCO Differential*

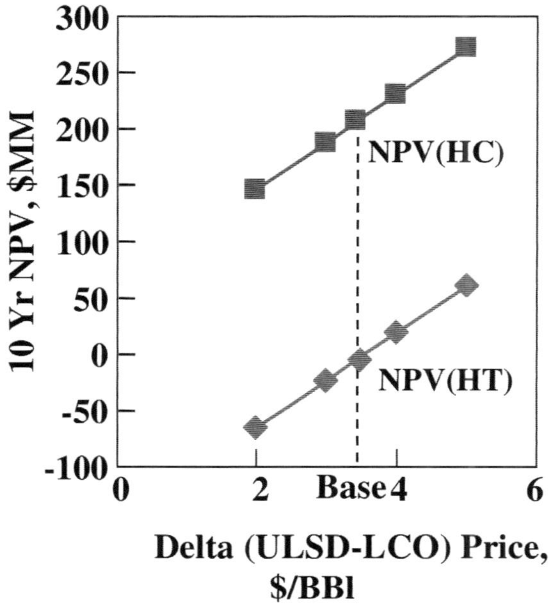


**Figure 8**

*NPV Sensitivity to Gasoline-ULSD Differential*

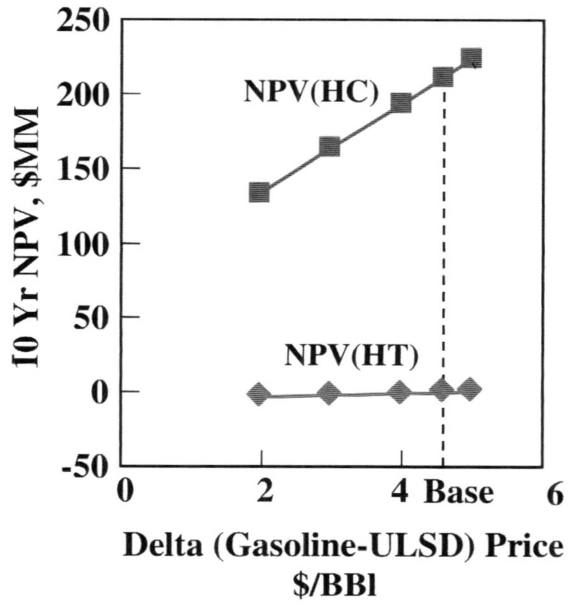


## CONCLUSIONS

LCO is a distressed refinery stream with limited future disposition options without further significant upgrading. Refiners will have to consider investments in technology to upgrade the LCO to higher value transportation fuels as they prepare for clean fuels production and increased market demand in the future.

In this paper we have outlined the benefits of our recently commercialized, LCO Unicracking process, ideally suited for production of ULSD and ULSG blendstocks from LCO at relatively mild conditions. The unique selectivity characteristics of the HC 190 catalyst are also an important enabler in this process. The LCO Unicracking process makes efficient use of hydrogen by employing advanced catalytic know-how to selectively increase the hydrogen content of the diesel range product while minimizing saturation of aromatics in the naphtha range.

The process economics for a US based refinery have shown that the investment in an LCO Unicracking unit yields superior NPV and provides a good payout compared to a conventional hydrotreating solution.

# REFERENCES

1. V. P. Thakkar, V.A. Gembicki, D. Kocher-Cowan, S. Simpson, "LCO Unicracking Technology – A Novel Approach for Greater Added Value and Improved Returns," ERTC, 2004

2. Crude Oil Prices, EIA petroleum marketing Monthly, www.eia.gov

3. Product prices obtained from historical North American data of Platts, and CMAI, www.platts.com, and www.cmaiglobal.com

A Honeywell Company

**UOP LLC**
25 East Algonquin Road
Des Plaines, IL 60017-5017


UOP 4399B

## Chemical Tracers Find Elusive Leak in Hydrotreater Feed/Effluent Exchangers

*By Troy Witherill*
*Refinery Process Engineer*

*Dave Ferguson*
*Tracerco Business Development Manager*

### Introduction

High sulfur content in the effluent of a hydrotreater stream is a signal that the unit has a problem with either the reactor catalyst bed or the feed/effluent exchangers. Plant resources can often isolate the offending vessel by analyzing samples on-line or pressure testing exchangers off-line. These approaches take time and can be expensive when they involve a shutdown to search for the problem. At times, conventional efforts fail to locate the offending vessel. On-line techniques, such as radioisotope tracers, have been used for many years to leak test these exchangers. These techniques avoid the maintenance and lost production costs of a search during a shutdown. But, every technique has its limit and some leaks are too small for the previously available techniques. A new service using stable chemical tracers has been employed to find very small leaks.

### Statement of the Problem

A Mid-West Refinery in the U.S. operates a Naphtha Hydrotreater with a bank of four feed/effluent exchangers. See Figure 1.

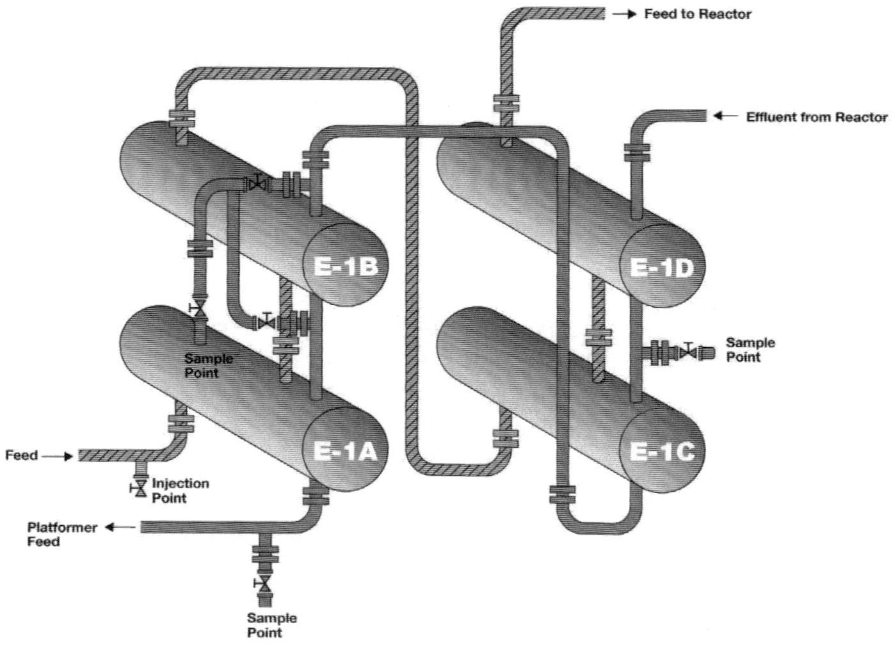


**Figure 1 – Four Feed/Effluent Exchangers in the Naphtha Hydrotreating Unit**

The flow though the Hydrotreater varies from 5,000 to 10,000 bbls/day. The effluent from the reactor is cooled by the exchangers and then sent as feed to the Platforming Unit. Results of Platformer Feed sample analysis showed that sulfur levels, which had been staying between 0 to 2 ppm, were now fluctuating between 2 and 10 ppm. Figure 2 shows a distinct trend to higher sulfur levels. Plant personnel had tried to determine which exchanger was leaking by sample analysis, but had been unsuccessful. They contacted Tracerco to gain assistance in identifying the leaking exchanger before taking an outage to fix the problem.

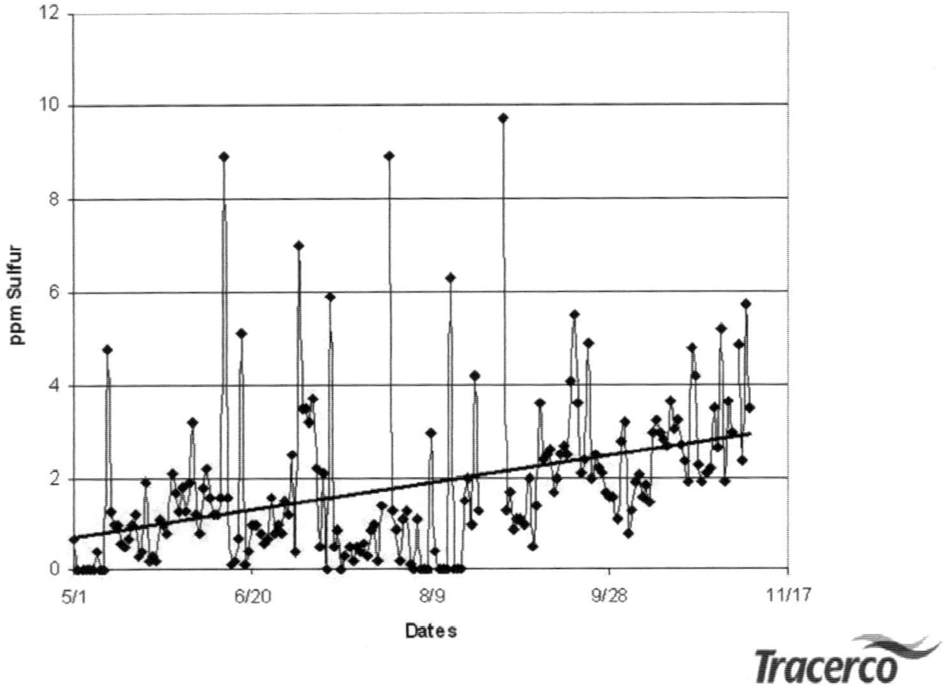


**Figure 2 Platformer Feed Sulfur Levels**

## The Solution

Tracerco has a long history of performing on-line heat exchanger leak tests using radioactive materials as tracers. The standard technique is to mount radiation detectors on the feed and effluent lines of each exchanger and inject a radioisotope tracer that is compatible with the feed stream. As the tracer flows through the exchangers with the higher-pressure feed stream, the detectors on the feed piping will respond to the passage of the tracer, producing bell-shaped Gaussian response curves on the computer. If one of the exchangers has a leak, the leak to the effluent side will be at the same concentration of tracer as its percentage in the feed. Though diluted by the effluent, the detectors on the effluent exit lines will respond, producing a response curve on the computer. By comparing the area under the leak detector response curve to the area under the feed detector response curve, the percentage leak can be approximated.

The limitation of this technique is in its sensitivity. This technique is the easiest and quickest way to perform an on-line leak test if 0.5 percent or more of the feed is leaking into the effluent. Sometimes, the sensitivity can be improved to 0.2 or even 0.1, but the more typical limit is 0.5 percent.

Smaller leaks can be found with radioisotope tracers by a sampling technique. The tracer material and the injection procedure are exactly the same as before, but instead of mounting detectors on the feed and effluent lines, sample points on the effluent lines have to be established. Sample points generally consist of a sample cooler where hot effluent is condensed/cooled and collected in sample containers. The samples are counted with a very sensitive radiation detector. The each sample can be counted for several minutes. Compared to the on-line tracer technique where the tracer material flashes past the detector in a few seconds, much smaller amounts of radiotracer can be measured in the samples. This technique has been used to find leaks as low as 100 ppm. Unfortunately, finding smaller leaks than this would require the injection of so much radiotracer that disposal of the radiotracer becomes an issue, as well as the safety of the crew performing the leak test.

To address the need to find leaks smaller than 100 ppm, Tracerco researched chemical tracer techniques. Tracerco identified a range of compounds, some of which are compatible with organic and some with aqueous streams, that are chemically and thermally stable, and can be readily recognized in samples by specialized gas chromatography. These chemicals can be found in hydrocarbon samples at concentrations as low as 1 part per billion (ppb).

The four feed/effluent exchangers each had a sample point, which meant the tracers could be injected into a single injection point. See Figure 1. Each exchanger was tested with a separate tracer. This allowed a two man crew to test the four exchangers. One tracer could have been used, but all four exchangers would have to have been sampled at the same time, requiring a five man crew. It was more cost-effective to use two people and four tracers.

Because the effluent stream contained hydrogen sulfide ($H_2S$), the decision was made to collect the samples in stainless steel sample cylinders instead of using a sample cooler. Hydrogen and $H_2S$ would have been continuously released to the atmosphere while collecting the samples with a sample cooler. Ten sample cylinders were mounted to a manifold at the first sample point. As the tracer was injected, the first sample cylinder was opened and closed in 10 seconds. The other nine sample cylinders were opened and closed at 10 second intervals. The sample cylinders were then removed from the manifold and immersed in an ice bath to condense the naphtha and tracer. Each cylinder was depressurized through a tube containing activated charcoal and then the naphtha in the cylinder was collected in plastic bottles.

The activated charcoal tubes and the sample bottles were sent to the Tracerco lab for analysis. The results showed that three of the sets of samples were negative. The set of samples for the E-1B exchanger showed tracer in samples 5 through 10. See figure 3.

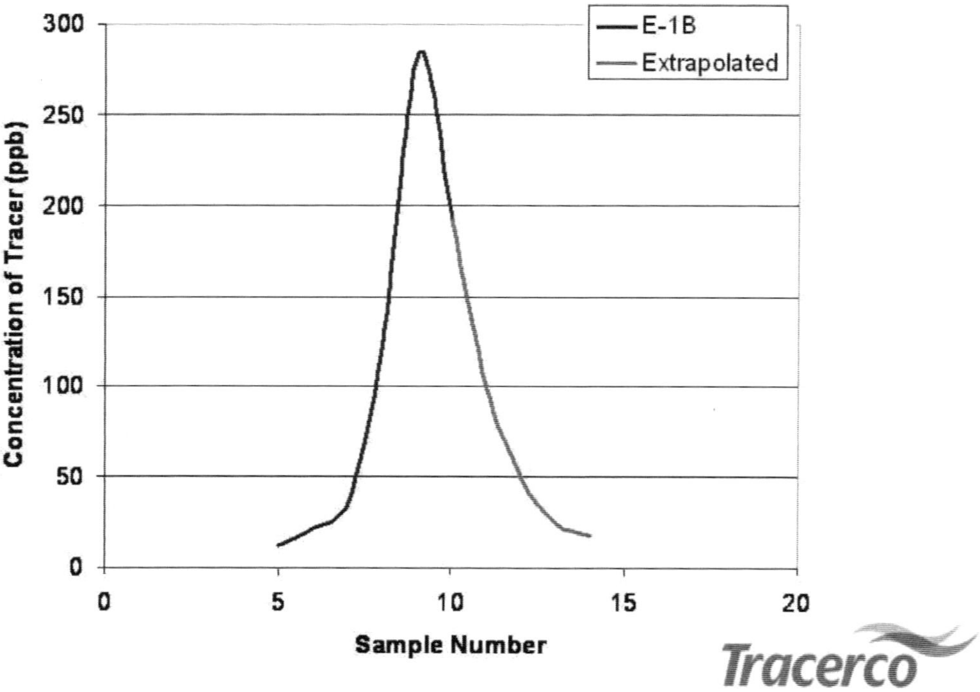


**Figure 3 Chemical Tracer Response Curve**

Unfortunately, the tracer was slower coming through the exchangers than had been expected. The ninth sample had the highest reading and the tenth sample was the first sample on the trailing edge of the response curve. If another 10 samples had been caught, there would have been tracer in samples 11 through 13, at least, and probably in a few more. The trailing edge of a response curve is usually more drawn out than the leading edge.

For this analysis, however, the conservative approach was applied and a short trailing edge was extrapolated. The average concentration for the samples was determined and the grams of chemical tracer that passed the sample point while the samples were being collected was calculated. When this result was divided by the total grams of chemical tracer that was injected, the resulting percentage leak was revealed. The results indicated that the rate of leakage of the feed into the effluent was 0.0044 percent or 44 ppm at the time of the test.

At a convenient time for the refinery, the unit was brought down and the E-1B exchanger was pressure tested with water. One tube had an obvious leak and four other tubes had small leaks. The tubes were plugged and the unit was brought back on line. The sulfur levels returned to the 0 to 2 ppm level.

## Conclusion

The plant personnel were able to avoid the lost production from a longer shutdown and the maintenance costs of opening and hydrotesting the other three exchangers. This project demonstrated the capabilities of the new approach to leak testing exchangers. Ultra Low Sulfur Diesel Hydrotreaters operate at such low sulfur specifications that very small leaks in the feed/effluent exchangers can result in off-specification effluent. With the new chemical tracers that are now available, leak rates as low as 5 ppm can be detected.

## Resumes

Troy Witherill was graduated from the University Of North Dakota with a B.S. in Chemical Engineering in 1992. He has worked for Ashland Petroleum Company, Basis Petroleum Inc., Murphy Oil USA, Inc., and Countrymark Cooperative, LLP in various technical and process engineering and supervisory roles.

Dave Ferguson was graduated from Texas A&M University with a B.S. in Chemical Engineering in 1980. He has worked for Olin Corporation, Dow Chemical, Iris Power Engineering, Tru-Tec Services, and Tracerco in various research, engineering, commercial, and management roles.

# 92a Framework for Operability Assessment of Production Facilities: An Application to a Primary Unit of a Crude Oil Refinery

**Sampath Yela**, Omar Galan, and Jose Romagnoli. Cain Department of Chemical Engineering, Louisiana State University, 110 Chemical Engineering, South Stadium Road, Baton Rouge, LA 70803

Nowadays due to environmental concerns, the high cost of energy and the tight product specifications the existing production facilities have been pushed to their capacity design limits. However, in order to enhance the performance of any process with some complexity involves a large number of process variables that must be modified in a coordinated fashion. Production engineers have improved several processes after a great deal of effort and trial and error experimentation. It is costly in terms of engineering hours and production lost, but it can also compromise operating personnel safety and equipment integrity. Therefore, there is a need of methodologies and prediction tools to assist the production engineers in their tasks.

Although, the maximization of production has been traditionally the most important target and concern in the day to day operation of many production facilities, recently, some other issues related with the overall impact of waste streams and emissions are considered more seriously due to environmental regulations. It is common to find production facilities operating in an economically viable fashion, conversely, these processes still struggle to embrace all the restrictions imposed for the up to date environmental requirements. Also, a growing competitive international market is demanding the enhancement of existing plants and the construction of innovative facilities with the state-of-the-art technology. This work focuses on the development of a methodology for the optimization, control and operability of existing production facilities. Nevertheless, the ideas and concepts presented in this work can be extended to the design and plant synthesis. Existing production facilities present an interesting challenge from the operability point of view. The main constraint is the fact that many production managers are not willing to change the plant layouts. Modifications in an existing plant are usually costly in time down and capital expenditure. However, there is room for improvement when choosing the set points optimally. The optimal operating conditions are not unique and they are subject to the constant fluctuations in the raw material quality and sustained unknown disturbances on the process. Secondly, in existing plants there are a priori limitations related with the capacity and performance of equipment units involved in the process. Considering the above limitations related with the equipment and environmental regulations, it is possible to put forward a methodology that embraces the optimization and control of an existing production facility reducing its environmental impact.

In this work, the methodologies of Life Cycle Assessment (LCA) and Environmental Damage Assessment (EDA) are applied within the optimization problem. The LCA evaluates the environmental impact of a process from the raw material to a final product. The EDA can supply the necessary information about the damage caused by the process to the environment. Basically we are looking at damages involving human health, ecosystem quality and resources usage.

The model-based methodology proposed in this work, starts with the implementation of first principle models for the process units on consideration. Secondly; steady-state and dynamic simulations are performed for model validation purposes. It is followed by rigorously posing the optimization problem, that is, objective function and constraints. In this assessment, besides the common raw materials and energy costs penalties; the so-called triple-bottom-line constraints are incorporated in the objective function. The triple-bottom-line constraints are related with sustainable and environmental costs respectively. The solution to the optimization problem gives the optimal set points for the process variables under steady-state conditions. At this stage, the influences of exogenous disturbances are not taken into account since these are mathematically feasible solutions only. Subsequent to the solution of the optimization problem, it is the critical importance an assessment of process controllability in view of the fact that optimal set points may be difficult to maintain under sustained disturbances or process variability. The process controllability feature relays on a back-off strategy where the constraints are restrained avoiding process excursions that violate a hard constraint compromising the plant stable operation when subject to disturbances. Another equally critical concern addressed in this work is related with the transition task, that is, the trajectory followed by a process when changing from an operating regime to another one. Optimal trajectories are calculated using a model predictive control strategy (MPC) which can handle constraints and presents good robustness features against model mismatch and perturbations.

The present framework is implemented in HYSYS® and a user friendly front end in MS Excel® where the-state-of-art optimizer is implemented. An application to a primary unit of a crude oil refinery is used for benchmarking the methodology above mentioned. Although the primary unit does not have waste streams neither emissions, the optimization of this section has a remarkable effect on the up and down streams units reducing the overall environmental impact of the process. The operability assessment of primary unit presents several challenges due to that the process is extremely interactive due to a high degree of energy integration. It is an example where steady-state optimization was performed and controllability issues were no part of the design. Optimal set points are calculated and implemented in a closed loop fashion driving the process along an optimal trajectory without violating any of the constraints. The assessment compares the impact of the primary unit operating at the current conditions and at the optimal conditions. When implementing the optimal set points, there is a reduction in the amount energy usage and it implies a reduction in the production of NOx, CO2 and SO2 to the atmosphere. The presented methodology is a promising approach for the optimization of existing production facilities and synthesis of future chemical plants

## 92b Fidelity, Dynamics, Integrity – the Real Needs for Automation System Testing and Training

**Martin Berutti**, MYNAH Technologies, 504 Trade Center Blvd, Chesterfield, MO 63005

The value of simulators for testing automation system application software and training operators in the hydrocarbon and petrochemical industry is proven. However, users often make mistakes when specifying or implementing these systems that reduce their effectiveness. In this paper we will discuss practical, proven approaches to addressing the following simulation issues:

· Fidelity – what is it and how much does the user need.

· Dynamics – what makes a simulation dynamic and how does the user achieve good dynamics

· Integrity – how does the user ensure simulation system integrity and why is it important

· Testing – unique requirements for simulators used for application software testing

· Training – proven approaches to ensure good operator training results

# A Modified Extended Recursive Least-Squares Method for Identification of FIR Type Models from Closed-Loop Data

Christopher Betts and James Riggs[†]
Department of Chemical Engineering
Texas Tech University, Lubbock, TX 79409

## Abstract

Performing a plant test under closed-loop conditions can be desirable for model identification. Production loss, resulting in loss of profits, and safety problems may result when control loops are opened for plant testing. However, identification of models from closed-loop data is more difficult compared to models that are identified from open-loop data due to the correlation between the colored noise and the process inputs created by the feedback. This correlation creates a bias in the estimated process model parameters when the normal least-squares estimator is used. The batch solution for the least-squares estimate assumes the prediction error to be zero mean white noise. This assumption fails for closed-loop data. A new method for identifying FIR type models using closed-loop data is proposed. A time-varying bias term with a moving average process is introduced into the model structure and identification is performed using a modified extended recursive least-squares algorithm which eliminates the bias from the process parameter estimates. Once this bias model is introduced, the open-loop step response model (SRM) can be estimated from the closed-loop data which can be used for control purposes. Simulation case studies on a fluidized catalytic cracking (FCC) unit with DMCPlus© as the controller were conducted to support the proposed method.

Figure 1 shows a typical FCC unit that is used in this work which was modeled in the FORTRAN programming language. This simulator exhibits nonlinearity and coupled dynamics due to the regulatory control loops within the system. These features are considered the "real-world" aspects of the FCC simulator.

---

† To whom correspondence should be addressed. Tel: (806)742-1765. Email: jim.riggs@ttu.edu.

The fundamental problem with closed-loop identification is that the noise is correlated with the process inputs due to the feedback of the closed-loop system. This correlation creates a bias in the process parameter estimates. Therefore, a bias model is introduced into the identification procedure which eliminates most of the bias, which is shown below.

$$C(q^{-1})e(t) + \mu(t)$$

$C(q^{-1})e(t)$ is a moving average process and $\mu(t)$ is the bias parameter which forces the residual (e.g. prediction error) to a zero mean. The bias parameter models the "colored" components of the noise while the moving average process models the "white" components of the noise. A recursive least-squares algorithm is then used to identify the process model.

Closed-loop data is generated by introducing setpoint excitation into the DMC controller. The proposed method is then used to identify a step response model (SRM) for the FCC unit which is shown in figure 2 with moderate levels of noise. This model is compared to the "true" model which is shown by the dashed line. CV3 is the stack gas oxygen concentration and CV1 is the riser outlet temperature. MV5 is the slide valve position on the regenerated catalyst transport line and MV4 is the air blower suction valve position. The SRM identified from the closed-loop data using the proposed method does compare to the "true" model very well. Therefore, the SRM identified from the closed-loop data can be used for control purposes due to the high model quality.

**Key Words**: Closed-loop identification, finite impulse response model, recursive identification algorithms, bias modeling

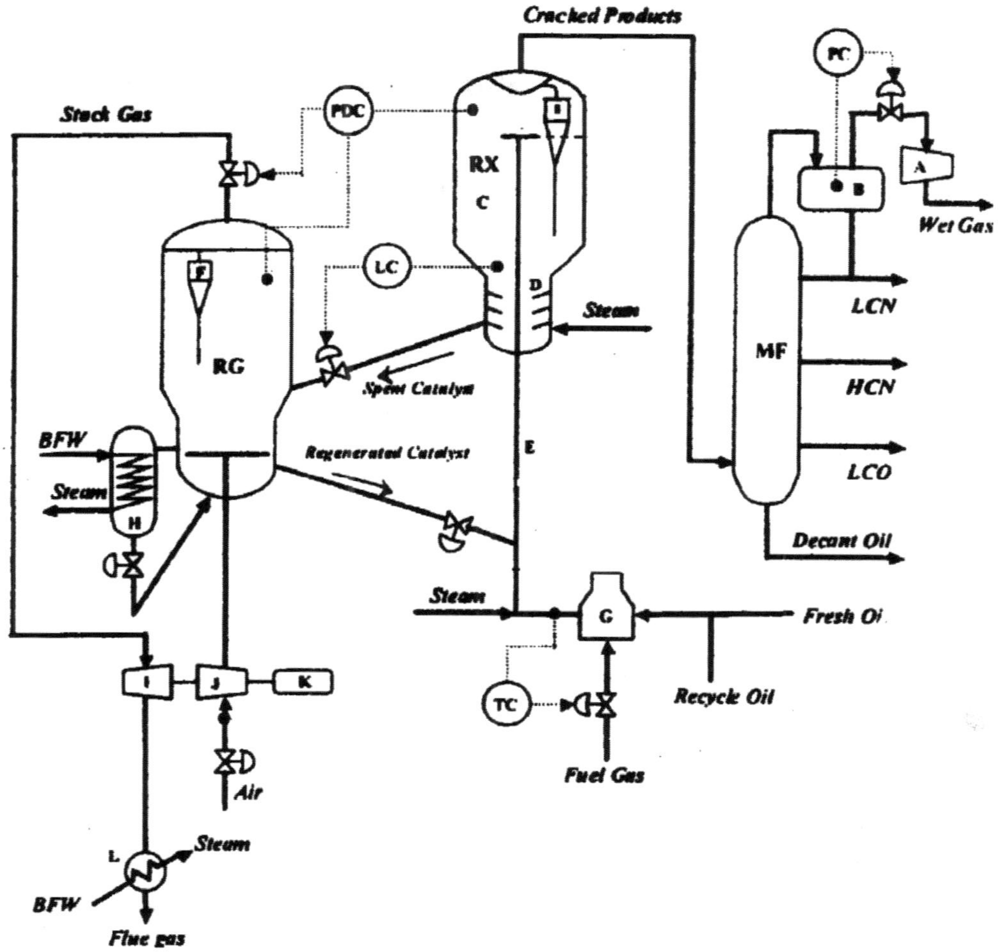


Figure 1: Diagram of a typical FCC process: (RX) reactor; (RG) regenerator; (MF) main-fractionator; (A) wet gas compressor; (B) overhead drum; (C) disengager; (D) stripper; (E) reactor riser; (F) cyclones; (G) feed pre-heater; (H) catalyst cooler; (I) stack gas expander; (J) air blower; (K) motor; (L) CO boiler

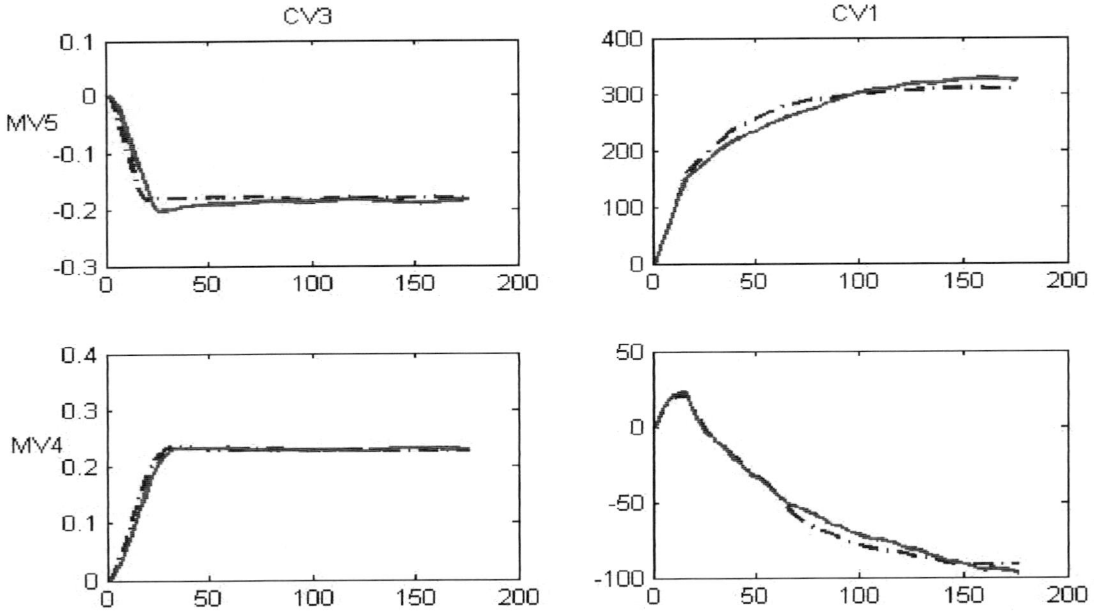


Figure 2: Identified SRM from closed-loop data using the MX-FIR method with colored noise added. (±1x$10^{-3}$ to CV3 and ±0.1 to CV1)

# 92d Experimental Test Design for Mimo Identification Subject to Integral Controllability

**Mark Darby**, Department of Chemical and Biomolecular Engineering, University of Houston, 4800 Calhoun Ave., Houston, TX 77204-4004 and Michael Nikolaou, Chemical and Biomolecular Engineering, University of Houston, 4800 Calhoun Ave., Hoston, TX 77204-4004.

Model-based multivariable control requires good-quality process models which, in addition to standard specifications for model quality, also satisfy the integral controllability condition. To identify such models it is necessary to design appropriate identification experiments. In recent work we developed a rigorous methodology for the design of identification experiments that produce models satisfying the integral controllability condition. This methodology encompasses well known experiment designs but also suggests many new designs of practical importance. In this presentation we focus on what is a general challenge for experiment design methods, namely the dependence of the design on the very model to be identified. We demonstrate that a simple adaptive approach can produce reasonable results. We illustrate the proposed approach through simulations on a fluid catalytic crackling unit (FCCU) example and compare it to known experiment design schemes.

# THE LP AS UTILITY SYSTEMS DESIGN TOOL

Thomas Phillips (**thomas.phillips@aramcoservices.com**) & Ahmed Sirag Khogeer (khogeer@hotmail.com ). Saudi Aramco. Houston, TX.

## ABSTRACT

Petroleum refining is a capital-intensive business. The cost of building a new large complex refinery can far exceed $10 billion in which utilities and offsites compromise more than a third of that. Typically, an LP model of a grassroots refinery project would only be used in the initial configuration studies. After that was done, the LP would be quietly retired early in the Front End Engineering Design (FEED) and only occasionally resurrected to answer the inevitable 'what if' questions. Once FEED was well underway, overall refinery material and utility balances would be performed not with an LP, but with spreadsheets. This resulted in only a limited number of balances for select points in the operating envelope.

Although this approach was common in the past, increased computing power and enhanced LP software has allowed the LP to keep pace with and even lead the FEED work. By continuously updating the LP model with the latest unit yields, blending properties and utility consumption/production figures as developed by the licensors and the FEED contractor, the LP can now quickly produce unlimited balances for the overall refinery as well as for individual units. This allows a complete assessment of the operating envelope which in turn leads to enhanced knowledge of and better designs for OSBL utility systems, including steam, power, water and hydrogen. Simulation runs were carried out using ASPEN PIMS to solve the optimization problem. The optimization was done using FEED-quality data developed by the process licensors and the FEED contractor.

## I INTRODUCTION AND LITERATURE REVIEW

### I.1. BACKGROUND

Petroleum is one of the most important sources of power today. In addition, it is the only source of many petroleum-based chemicals known as petrochemicals.

The refining industry has suffered considerably during the past two decades. The last US grass roots refinery was built during early 1980s I[3]. This is due to increasingly low refining margins and strict environmental regulations. Many small refineries were forced to quit business since upgrading small and old refineries is usually not economically feasible. In 8 years, the US refining industry has decreased by 40 refineries, from 192 to 152 I[1]. In particular, smaller refineries with zero conversion have closed. Crude capacity has grown by 8% and conversion capacity has grown by 5%. Although 5% seems small, it represents a significant addition of 460 KBPD of new conversion capacity. In such tight situations, a new grass roots refinery should be a large throughput refinery, usually greater than 300 KBPD with a high complexity factor, or even a full conversion refinery, having the ability to produce high quality products for different markets, and preferably being integrated with a petrochemical complex.

## I.2. LP IN REFINERY OPTIMIZATION & DESIGN

Dantiz introduced Linear Programming (LP) in 1947 I[2]. In LP, all constraints and objective functions are linear. One of first principal uses of LP was to solve refinery optimization problems. In 1956, Symons used LP to solve a simplified gasoline-blending problem. Since then, LP has found wide acceptance in refining industry.

In planning and scheduling, the constraints are linear (supply and demand). The objective function is also linear, and if not it can be linearized with acceptable accuracy for simple supply chain problems I[4]. On the other hand, for refining industry process modeling, the relations describing the conversion and treatment processes are nonlinear by nature.

The scope of refinery optimization LP does not cover heat integration, and usually does not cover steam and hydrogen management. An exception is cases in which the optimization of such items is related to a change in operating conditions or parameters such as flow, operating temperature, type of feed, diverting one or more intermediate products from one unit to another, or in a refinery-wide optimization model (such as utilization of hydrogen from the CCR unit instead of producing fresh hydrogen in the PSA unit). The model that was the inspiration for this paper is unique in that it contains extensive modeling for all major utilities as well as hydrogen distribution and management.

Heat integration requires hardware management, such as changing the heat exchanger train configuration. The change of the cooling fluid type or waste heat recovery (such as recovering heat from an FCC unit) is usually a design problem requiring mechanical work. However, this is usually expensive and is beyond the scope of regular optimization models which optimize existing hardware for optimum goals, which is what most refiners are seeking.

The Linear Program or LP as it is commonly called is the quintessential tool for understanding overall refinery operations and the complex and numerous interactions between units. In the proper hands, an LP exposes the anatomy of a processing facility and greatly enhances the LP operators understanding of 'how it works' and 'what makes it tick'. When skillfully driven, an LP will bring to light the multi-dimensional operating topography complete with peaks, valleys and the 'edges of the operating envelope'. The LP operator is the pilot who can successfully steer a complex operation through this terrain, whether the facility is up and running or in design.

Most refinery engineers are familiar with how an LP is typically used; it sorts through a collection of thousands of refinery variables and finds the combination of these variables that provides the highest margin. In order to do this, the LP must perform precise material and utility balances. The ability to do this quickly and accurately throughout the operating landscape makes the LP a wonderful tool for ensuring that interconnecting piping and utility systems are properly sized and configured.

# II THIS WORK

## II.1. CONTRIBUTION

The contribution of this work to the field of chemical engineering in general and to the fields of petroleum refining and refinery design is that:

It links the utilities system design to the process system LP, thus the engineer can see immediate results for each operating point in the feasible solution envelope.

It discusses how LP capabilities can be used to expand the role of the LP model in the design of a grassroots refinery project. Given the insight and enhanced understanding made possible with a good LP model it makes sense to use the LP in a project oversight role from initial configuration studies all the way through to startup and hand-off to operations.

It utilizes user-friendly industry standard software in order to have this work accepted and used by the refining industry. Many refining engineers do not use academic research results for actual plant operation.

## II.2. MODEL FORMULATION

### II.2.1. Objective Function

Utilities are introduced to the objective functions as costs. Thus the LP will try do decrease the cost of utilities as much as possible. The price structure affects the refinery in relation to both feed cost and product revenue. For the feed cost, cheaper crudes may require more processing and treatment to produce salable products and may also be more expensive to ship. The operation of refineries varies with the seasons, and the price structure is affected by the market demand.

The operating cost includes catalyst, additives, fuel, power, and cooling water. Most design or optimization models lump all these costs in one category called utilities since the operating costs are smaller in comparison to other items in the objective function. However this simplification can result in misrepresentations of refinery operations and economics and make the LP unsuitable for use in the latter stages of FEED. If sufficient utility detail is included in the LP model, it can be a unique and insightful tool for offsite system design.

The objective function can be expressed mathematically as follows for a given optimal refinery operating case that is for a fixed feed and products streams:

$$\text{Minimize } \Phi = Pt \qquad (2)$$

Where

$\Phi$ is the cost -based objective function,

Pt is the total cost of utilities given by the following equation:

$$P_t = \sum (\text{Prod}_j . \text{Price}_j), \qquad (3)$$

where

j = 1, ..., 7 is the utility for the refinery, such as LPG, gas, HP steam, etc.,

$\text{Prod}_j$ is the consumption (or production rate) of utility $_j$, units can be different for each utility such as KWH for electricity, Tons per hour for steam,

$\text{Price}_j$ is the price of a unit quantity of product j, for example, \$/Ton steam.

### II.2.2. Constraints

In addition to the refinery constraints are usually capacity limits, process variable constraints and product specifications.

Table 1: Typical Refinery Constraints:

| | Unit | Constraint Name | Lower limit | Upper limit |
|---|---|---|---|---|
| Capacity constraints | Steam | HP steam | 160 | 250 |
| | | MP steam | 20 | 50 |
| | | *LP steam* | 50 | 100 |
| | Water | cooling | | |
| | | fresh | | |
| | | BFW | | |
| Process constraints | Steam balance | | | |
| | Water balance | | | |

These constraints can be written as:
Steam total = $\sum$ (Steam $_k$ X wt $_f$)
MP Steam $\geq 0.18$ Steam Total ; MP Steam $\leq 0.26$ Steam Total etc...

### II.2.3. Termination Criterion

The termination criterion is

$$\left| \Phi_{n+1} - \Phi_n \right| < \varepsilon, \qquad (6)$$

Where $\Phi$ denotes the objective function value, the subscript indicates the iteration number and $\varepsilon$ represents the absolute difference in the objective function in two successive iterations.

## II.3. A DESCRIPTION OF THE REFINERY

The model was based on a 400 KBPD full conversion refinery in the Middle East with the ability to export products worldwide. The refinery has a coker, hydrocracker, CCR, distillate hydrotreater and no FCC.

## II.4. SOFTWARE SELECTION AND FEATURES

Simulation runs were carried out using ASPEN PIMS. It was preferred to continue the runs using PIMS since it is the primary operations optimization tool for many refineries and its use is more convenient for process engineer.

PIMS uses several tools to perform the optimization, including a third party optimization engine known as XPRESS from the Dash Optimization software company. In addition, PIMS uses the same basic chemical engineering equations needed for mass and material balance. This software has the capability of using various techniques to solve the nonlinear parts of the model. These include Successive Linear Programming, Recursion and Distributive Recursion for the global model case.

# III IMPLEMENTATION

## III.1. THE LP TEAM – ONE RECIPE THAT WORKS

Building and maintaining an LP is a tedious job. Usually, this is done by a dedicated individual working alone. We have found however that the process can be more effective when done by teams of two or more people. Experience has shown that an effective two person team consists of an LP architect and an LP operator.

The LP architect knows all the tricks and traps of the LP modeling system. He has the knowledge, patience and perseverance to build and debug LP structure. Ideally, he is also quite knowledgeable of refinery processes and possesses a rich data bank of refinery process data. The LP operator ideally has years of refinery engineering, operations and planning experience and understands the LP structure well enough to make minor structural changes. He thoroughly understands refining, product blending, refinery economics and the capabilities and power of the LP but does not necessarily have the patience or in depth LP knowledge of the LP architect. Ideally, the LP operator has extensive experience actually using an LP to plan and optimize refinery operations.

Having such a two person team with these diverse skill sets reduces the tedium of LP model construction and helps create an atmosphere conducive to creating a truly inspired product. The operator and the architect together ensure that the model truly and accurately represents the capabilities of the process facility. Together they can accomplish more than could be done by either one alone.

It works like this: After making multiple runs, the operator determines that the LP needs more structure and functionality in order to perform a specific task or better represent a specific operation. Following detailed discussions with the operator, the architect adds the rather complex structure required and after a few debugging runs turns the model back to the operator. The operator then extensively tests the model under various constraint scenarios to determine if this new structure works properly and truly represents what his experience tells him will actually happen in the refinery.

Extensive communication between the architect and the operator is required to efficiently 'grow' an LP. As in most work environments, this communication is ideally facilitated by a high degree of camaraderie. This camaraderie in turn is fostered in an environment where the LP is viewed as entertainment rather than work, i.e., a fun 'video game simulator' with the goal of truly understanding the essential process and economic nature of the refinery. When this understanding is achieved it becomes a relatively simple matter to optimize the design of a new refinery or the operation of an existing one.

## III.2. THE OLD WAY OF SIZING OFFSITES

Normally a project LP is shelved shortly after the early configuration work is done. During this configuration phase of the project, generic yield and utility consumption data are used in the LP for the process units. For high level configuration decisions, this degree of accuracy is normally all that is required.

As the project moves into FEED and data is received from the licensors, static spreadsheet tools would then be used by the FEED contractor to perform the material and utility balances in each individual unit as well as across the entire refinery. Because of the complex interactions between refinery units and the engineering labor required to generate the balances, these spreadsheet balances would be performed for a few cases, i.e., normal operations and a handful of off-design extremes. Offsites facilities would then be designed against these few cases.

## III.3. A BETTER WAY OF DOING BUSINESS

In order to optimize the design of the offsite systems, the operation of the refinery should be accurately simulated at 20 or 30 points inside and on the edge of the operating envelope. A team representing both operations and project design engineers should determine these points and then the material and utility balances should be done for each identified case. Ordinary spreadsheets are not up to this task. In addition to their rigidity and labor intensiveness, they cannot accurately deal with product blending and the multiple modes of operation possible in many of the ISBL units. Typically, only one mode is modeled for each unit when in reality a continuum of modes is possible. Consequently a more capable tool is required if a more in-depth understanding is desired. That tool is the up-to-date and accurate project LP.

After the early configuration work is done, work should continue on the development of the project LP model. As soon as it is available, yield and utility data from the licensors should be incorporated into the LP. Stream blending data and utility summaries should be requested from the licensors for all base and rated cases. This data must then be entered into the LP. In addition, structure must be added as required to properly simulate the utility generation options under consideration.

Some ISBL units have multiple modes of operation with distinct utility profiles for each mode. For example, suppose the CCR is designed to operate both with and without benzene precursors and at 100 and 102 RONc. This would result in four distinct modes of operation with four distinct sets of utility balances and stream blending data, all of which should be obtained from the licensor and incorporated into the LP model. Once data for the four modes is available, others modes can be interpolated or extrapolated from the data. For example, a 96 RONc mode could be extrapolated for both precursor modes resulting in a total of 6 CCR modes in the LP model.

By comparison, a typical or conventional utility balance spreadsheet used for offsites design would contain only the base CCR design case (precursors to the CCR, 102 RONc). During the multiple LP runs made to characterize the operation of the refinery, it may become clear, for example, that 102 RONc is not the 'normal' mode of operation. Maybe it's closer to 100 RONc if this mode maximizes refinery margins while making on-test motor gasoline product. This lower severity mode in turn changes steam production in the hydrogen plant and steam consumption across the gasoline block.

Steady state operation is but one point on the curve, and it is often not the case that sets the design conditions for the offsite facilities. What are these extreme design cases? This can be determined by carefully examining and comparing the utility balances produced by the LP for the many operating cases mentioned above. These cases can include partial and total shutdown of various process units during various motor gasoline blending seasons for both expected and guaranteed hydrogen production/consumption across the CCR and hydroprocessing units.

## III.4. UTILITY DATA INCLUDED IN THE LP

In order to use an LP for utility design, accurate utility data must be entered and structure added as required to properly simulate the utility generation options under consideration. As a minimum the LP must include the following utility data for each ISBL unit:

- Electrical power consumption
- Boiler feed water consumption
- Steam production/consumption for each level of steam used in the refinery. Typically, this would include separate entries for each ISBL unit for high, medium and low pressure steam
- Condensate production/consumption
- Cooling water consumption
- Fuel consumption/production

For the offsites, the following should be modeled in the LP:

- Steam generation system
- Power generation
- Steam letdown stations
- Steam condensing stations, if any
- Boiler feed water de-aeration
- Flash steam production
- Cooling water distribution
- All miscellaneous offsite steam and power consumers. These can be modeled as fixed rates or tied to appropriate process variables. Examples of these miscellaneous items include:
    - Tankage
    - Flare System
    - Waste Water Treatment & Drains
    - Fuel Gas System
    - ISBL De-superheating sprays
    - Flash Steam
    - Small IOU Steam Drivers
    - Misc. OSBL Allowances & Steam Losses
    - Water Treatment & Supply
    - Cooling Water System
    - Sea Water System
    - Air System
    - Nitrogen System
    - Fire Protection
    - Coke Transport & Storage – driven by coke production rate
    - Sulfur Pellet Loading –driven by sulfur production rate
    - Plant wide lighting, HVAC & Power Losses Allowance

## III.5. INTERPRETING THE DATA

Once the utility design cases are defined and run (there could be 30 or more) an MS Excel workbook can be prepared that uses LP produced spreadsheets as input. This workbook can then produce simplified, easy to view summaries for the fuel gas system (including composition and composite heating value data), the hydrogen plant feed system (including rate and properties of each feed stream), and all major utility systems (including boiler duty, break over station duty, LP steam condensing duty, fuel gas, cooling water, BFW, cooling water and electrical power used/consumed, imported fuel requirement, condensate flash steam produced, etc). All parameters required to close the major utility balances can be provided by the LP. A careful evaluation of all cases will then reveal the extreme design points for all major utility systems. Once these various extreme design cases are determined, the FEED contractor can then produce the traditional spreadsheet balances for these identified cases in order to verify the LP output and to document the IOU design decisions.

This approach provides a high level of confidence that the offsite systems will not be over or under designed and will be capable of handling a wide range of operating scenarios. This is a degree of confidence and design efficiency not possible with the traditional methods.

## III.6. CASE STUDY: HYDROGEN OPTIMIZATION AND DISTRIBUTION

The same material balance and utility balance strengths inherent in an LP that make it an ideal tool for understanding and designing utility systems also make it an ideal tool for optimizing and designing the hydrogen distribution system.

This can only be done if the LP is populated with accurate hydrogen production and consumption data for all hydrogen-involved units, at all levels of hydrogen purity throughout the refinery. Typically, this would include separate tracking of high purity PSA hydrogen, CCR net gas and any and all other low purity hydrogen streams. These lower purity streams would include:

- Hydrocracker cold low pressure separator gas
- Solution hydrogen in the Hydrocracker, hydrotreaters and CCR off gas
- Saturate gas plant off gas

Hydrogen distribution is a complex issue in a modern refinery. There are many sources and many destinations. When designing the interconnecting piping for hydrogen distribution, it is important to understand the operating circumstances and the flow rates and composition for each branch of the network. When accurately modeled, the LP can help the FEED engineers more completely understand the dynamics of the hydrogen distribution system.

For example, the primary destination of CCR net gas may be a diesel hydrotreater. But how should this stream be handled when the diesel hydrotreater is down? Options may include a PSA unit, the Hydrocracker, feed to the hydrogen plant, the fuel system, and finally the flare. Because of its powerful material and utility balance capabilities, the LP can quickly, accurately, and optimally re-distribute hydrogen under any possible operating scenario. This is a powerful design tool which again gives the system designers a high degree of confidence in the final system design.

As was done with the utilities systems, the LP spreadsheet output can be quickly copied and pasted into a workbook to produce tabular and graphic depictions of the hydrogen distribution system, an example of which is shown below:

**Figure 1: Hydrogen Distribution Diagram**

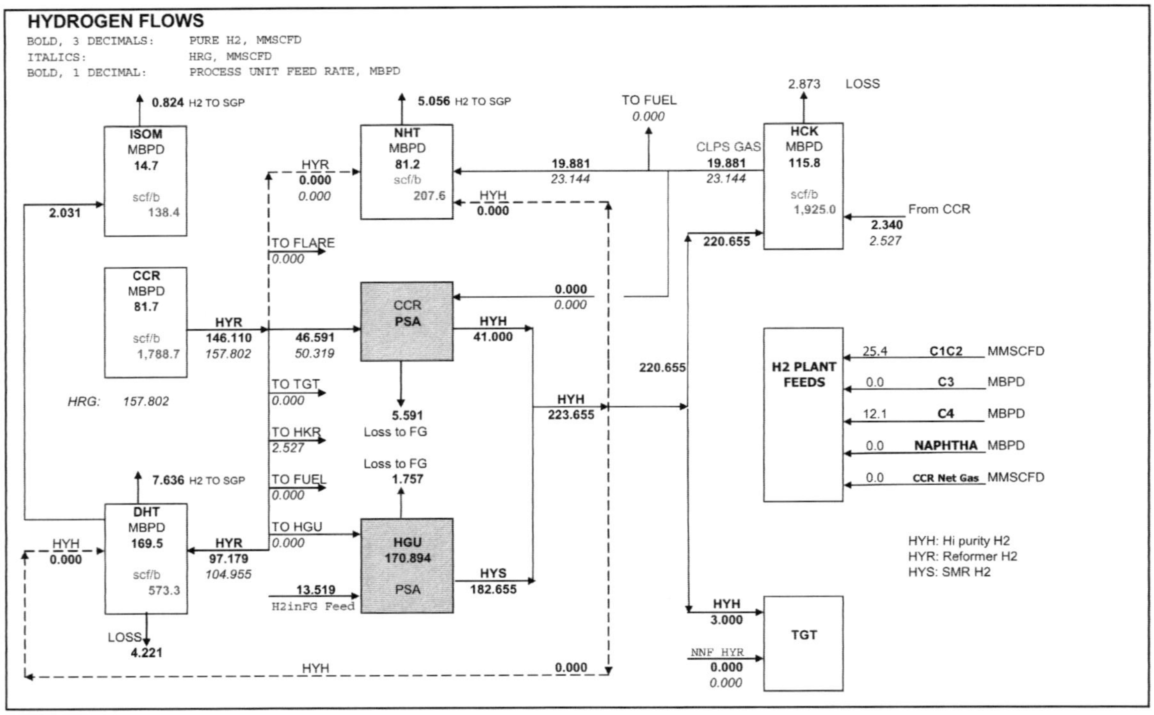


## III.7. THE NEW PARADIGM: MAINTAINING THE LP THROUGHOUT THE PROJECT LIFE CYCLE

An accurate LP, in the hands of a skilled and dedicated team, can quickly and accurately address a multitude of design, operations and economic questions that arise during FEED. In addition to addressing the inevitable 'what if' configuration questions that arise throughout the FEED process, the LP will also provide input into the project economic model, and as we saw above, assist with the design of the utilities and interconnecting piping. As such, it is a tool that should be maintained and utilized throughout FEED.

When it is all said and done, LP is the only tool capable of accurately and quickly assessing overall project performance. Maintaining the LP throughout FEED then provides project and company management with key information they need to make final investment decisions. Without a good LP in the hands of a skillful team, the picture is far less clear and less accurate.

As the project moves into detailed design, the LP should continue to be updated with design data as it is made available by the EPC contractors. During this EPC phase, all functionality required by Operations to start up and operate the facility should also be added. At some point during the EPC phase, Operations should assume ownership of the LP.

Smooth startup and operation of the new facility will be greatly enhanced by skillful use of the LP. If possible, the LP architect and LP operator from the FEED phase will continue to support the LP during startup. If this is not possible, they should at least spend as much time as possible 'turning over' the LP to Operations before they move on to new ventures. In order to enhance this process, the LP should be designed from the beginning with the understanding that it will eventually be handed over to operations. This means providing adequate operating flexibility (not just rigid, one mode unit models) and copious documentation to facilitate understanding of model structure. Normally, this documentation will take the form of comments and notes inside each unit submodel.

As is the case with most LP models, survival of the model is dependent in large part on the how effectively the model is handed over from one team to the next.

## IV CONCLUSION & RECOMMENDATIONS

a. The LP is much more than just an optimization tool. In addition to performing its legacy task of finding the optimum configuration, it is also an important design tool that can effectively be used throughout FEED for utilities and interconnecting piping design.

b. Maintaining the LP throughout FEED will improve overall project integrity and will improve the quality of project information transmitted to management.

c. Effectively transferring the LP to Operations during the EPC phase will facilitate a smooth project startup and maximize overall project profitability.

d. It is recommended that this work be extended by combining the LP with off-the-shelf scheduling software to optimize product tankage and blender designs.

e. It is the opinion of the authors that essentially all refinery and petrochemical project FEED efforts for both ISBL and OSBL systems can be enhanced by effectively utilizing an accurate LP throughout the entire project life cycle.

## V ABBREVIATIONS

| Abbreviation | Description |
|---|---|
| CCR | Continuous Catalyst Regeneration (Type of Catalytic Reforming) |
| CONST | Constraint |
| DHT | Distillate Hydrotreater unit used to reduce diesel sulfur content using hydrotreating process |
| FCC | Fluidized Catalytic Cracking unit |
| KBPD | Thousand barrels per day |
| LP | Linear Programming (also known as linear planning) |
| PIMS ( or ASPEN PIMS) | Process Industry Modeling System by Aspen-Tech |
| ISBL | Inside battery limits units that refers usually to process units |
| OSBL | Outside battery limits that includes storage, interconnections |
| FEED | Front end engineering design |

# VI REFERENCES


[1] DUNCAN, N. REFINERS CAN BOOST PROFITS USING THE CONVERSION INDEX. OIL & GAS JOURNAL. V98, 2000.

[2] LI, X. REFINERY-WIDE OPTIMIZATION. PHD DISSERTATION. TEXAS TECHNICAL UNIVERSITY, 2000

[3] SAUDI ARAMCO TECHNICAL COURSES SERIES ON PETROLEUM REFINING PROCESSES MANUALS, 2000.

[4] UOP FCC PROCESS TECHNOLOGY COURSE MANULAS. UOP. 1996.

[5] ZHANG, J, ZHU, X AND TOWLER, G. A SIMULTANEOUS OPTIMIZATION STRATEGY FOR OVERALL INTEGRATION IN REFINERY PLANNING. INDUSTRIAL AND ENGINEERING CHEMISTRY RESEARCH. V40, 2001.

# 131a Back to the Basics - Process Control Diagnostics Improves Refinery Performance

**James Beall**, Advanced Applied Technologies Team, Emerson Process Management, PO Box 630865, Nacogdoches, TX 75963

This presentation discusses process control diagnostic techniques. Many times the limit on process control performance is due to undetected, poor performing instrumentation and final control elements. The presentations shows techniques identify problems with instrumentation and final control elements that limit process performance. Another problem that is discussed is substandard control schemes and problems with proper signal tracking for bumpless transfers. In addition to just finding these problems, this presentation shows how to find solutions so the process performance can be improved.

## 131b Modern Loop Tuning Tutorial

Jim Dunbar, Advanced Applied Technologies, Emerson Process Management, 12301 Research Blvd., Research Park Plaza, Bldg. III, Austin, TX 78759 and **James Beall**, Advanced Applied Technologies Team, Emerson Process Management, PO Box 630865, Nacogdoches, TX 75963.

This presentation is a tutorial on tuning PID process controllers. The course will explain the operation of the PID controller, typical variations of the PID controller, common process responses and how to identify the dynamics of the process. The instructor will demonstrate how to calculate PID tuning parameters from the process dynamics and the desired controller response.

## 131c Troubleshooting of Oscillatory Control Loops

**Srinivas Karra** and M. Nazmul Karim. Chemical Engineering, Texas Tech University, Lubbock, TX 79409-3121

The presence of oscillations in control loop increases the variability of the process variables thus causing inferior products, larger rejection rates, increased energy consumption and reduced average throughput. Existing methodologies for oscillation diagnosis in control loops are suffered by many draw-backs such as; indecisive or wrong diagnostic regions in the spectrum of process dynamics, failure in case of multiple faults occurring simultaneously, incapability to extend to multi-variable control system diagnosis and, absence of quantification of root-cause. In this work, a novel methodology has been devised to bridge all these gaps in oscillation diagnostics. Power spectral density (PSD) based oscillation detection method followed by a model based approach for identifying and quantifying the root cause of the oscillations is proposed. Proposed technique is validated on first order with time delay (FOPTD) system and integrating system with variety of dynamics under various cases of stiction, poor controller tuning and external disturbances.

# PID CONTROLLER DESIGN FOR CONSTRAINED OPTIMAL CONTROL OF INVENTORY LOOP

*Moonyong Lee, Yeungnam University, Kyongsan, Korea*
*Joonho Shin, JongKu Lee, LG Chemical, Daejun, Korea*

## Introduction

Liquid level loops are commonly encountered in process industries. Since the desired production rates and inventories are achieved through the proper control of flows and levels, level control is quite important for the successful operation of most chemical plants. The industrial importance of level loops has led to extensive research interest to achieve the enhanced control performance of the level loop[1~8].

In a level control loop, the manipulated flow is often situated upstream of a critical unit. In this case, controlling the behavior of the outlet flow is as important as that of the liquid level itself, in order to avoid rapid variations with a significant magnitude. Therefore, the level controller is required to provide non-aggressive and smooth control action as well minimizing the deviation of the level. Furthermore, a level loop normally has two important requirements : (1) the rate of change of the outlet flow should be kept below a specified allowable limit, (2) the deviation of the level should also be within a specified allowable limit. For this reason, level control problems can be considered as a typical constrained optimal control problem. In spite of its industrial and economic importance, the constrained optimal control strategy has rarely been employed in practical level loops. One of the main reasons for this is that most industrial level loops use simple P-only or PI controllers, which are generally accepted as being too simple to implement any sophisticated control strategy. The main obstacle to achieving optimal control is that it normally requires an optimization package, which most practitioners are not so familiar with, to find the optimal control action. Using a sophisticated advanced controller such as a model predictive controller might be a solution for the constraint control of the level loop, but it cannot be considered to be a practical approach for many reasons.

In this study, we developed an analytical design method for an PI controller that enables the constrained optimal control of the liquid level loop. The constrained optimal level control problem is firstly formulated and then converted into a simple form with two independent variables by using a proper variable transformation. The Lagrangian multiplier method is applied to handle the constraint optimization problem and the optimal PI tuning rule is finally found from the analysis of the global optimum condition. The proposed method is shown to deal with the two major control specifications in level loops explicitly, while minimizing the optimal control performance measure.

## Constrained Optimal Control of Inventory Loop

The liquid level control system presented in Figure 1 is simply described as

$$H(s)=\frac{1}{As}Q_i(s)-\frac{1}{As}Q_o(s) \tag{1}$$

where $Q_o(s)=K_L(1+\frac{1}{\tau_I s})(H(s)-H^{set}(s))$ and $K_L=K_c\frac{Q_{o\max}}{\Delta H}$

The closed-loop transfer functions for the level control system are then

$$H(\mathrm{s})=\frac{\tau_{\mathrm{H}}\tau_I}{A}\frac{s}{\tau_{\mathrm{H}}\tau_I s^2+\tau_I s+1}Q_i(s)+\frac{\tau_I s+1}{\tau_{\mathrm{H}}\tau_I s^2+\tau_I s+1}H^{set}(s) \tag{2}$$

$$Q_o(s) = \frac{\tau_I s + 1}{\tau_H \tau_I s^2 + \tau_I s + 1} Q_i(s) - \frac{K_L \tau_H s(\tau_I s + 1)}{\tau_H \tau_I s^2 + \tau_I s + 1} H^{set}(s) \tag{3}$$

Where $\tau_H = \frac{A}{K_L} = \frac{\tau_V}{K_c}$ and $\tau_V = \frac{(\Delta H)A}{Q_{o\max}}$

The damping factor of the above closed-loop characteristic equation is expressed as

$$\zeta = \frac{1}{2}\sqrt{\frac{\tau_I}{\tau_H}} \tag{4}$$

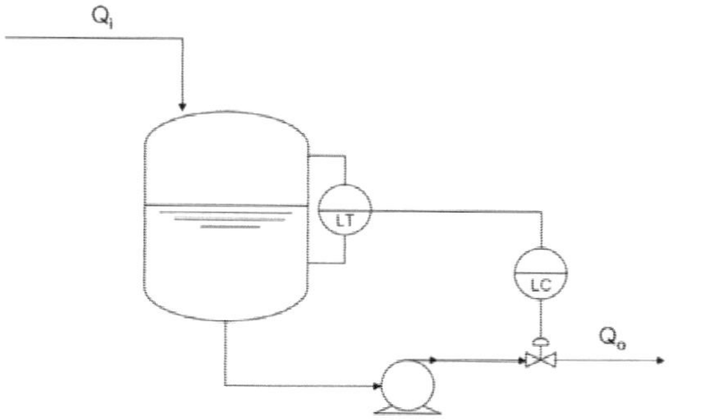


Figure. 1. Schematic of a level control loop featuring manipulation of the outlet stream

Regulatory control against a load change is the major concern in level control. The main disturbance in the level loop is the variation of the inlet flow. The control objective is to minimize both the rate of change of the outlet flow and the deviation of the level, while keeping (1) the rate of change of the outlet flow within a specified allowable limit and (2) the deviation of the level within a specified allowable limit for a given load variation. Therefore, the constrained optimal control problem of the level loop can be defined as finding the controller parameters that minimize the performance measure in Equation (5a) for a given step change in $Q_i$, subject to the constraints in Equations (5b~c):

$$\min \Phi = \omega \int_0^\infty [\frac{H(t)}{\Delta H}]^2 dt + (1-\omega)\int_0^\infty [\frac{Q'_o(t)}{Q'_{omax}}]^2 dt \tag{5a}$$

subject to $|Q'_o(t)| \le Q'_{omax}$ and $|H(t)| \le H_{\max}$ (5b,c)

By performing certain mathematical manipulations, the constrained optimal control problem can be expressed in terms of $\tau_H$ and $\zeta$ as follows :

$$\min \Phi(\tau_H, \zeta) = \alpha \tau_H^3 \zeta^2 + \beta \cdot \frac{1}{\tau_H}\left(1 + \frac{1}{4\zeta^2}\right) \tag{6a}$$

subject to $\tau_H \ge \gamma_L h(\zeta)$ and $\tau_H \le \frac{\gamma_U}{g(\zeta)}$ (6b,c)

where

$$\alpha = 2\omega\left(\frac{\Delta Q_i}{A\Delta H}\right)^2,\ \beta = \left(\frac{1-\omega}{2}\right)\left(\frac{\Delta Q_i}{Q'_{o\max}}\right)^2,\ \gamma_U = \frac{AH_{\max}}{\Delta Q_i},\ \gamma_L = \frac{\Delta Q_i}{Q'_{o\max}},$$

$$h(\zeta)=\frac{\sqrt{1+x^2}}{2}\exp\left(-\frac{3\tan^{-1}x-\pi}{x}\right) \quad for \quad 0<\zeta<0.5 \quad ; \quad h(\zeta)=1 \quad for \quad 0.5\le\zeta$$

$$g(\zeta)=\frac{2}{\sqrt{1+x^2}}\exp\left(-\frac{\tan^{-1}x}{x}\right) \quad for \quad 0<\zeta<1 \quad ; g(\zeta)==\frac{2}{\sqrt{1-x^2}}\exp\left(-\frac{\tanh^{-1}x}{x}\right) \quad for \quad 1>\zeta$$

where $x=\frac{\sqrt{1-\zeta^2}}{\zeta} \quad for \quad 0<\zeta<1 \quad ; \quad x=\frac{\sqrt{\zeta^2-1}}{\zeta} \quad for \quad \zeta>1$

It is preferable for a level controller to allow smaller $Q'_{o\max}$ and $H_{\max}$ specifications. However, since the feasible region is surrounded by the two constraints, as smaller $Q'_{o\max}$ and/or $H_{\max}$ specifications are applied, the feasible region bounded by these constraints also becomes smaller and eventually disappears. For a given $Q'_{o\max}$ ( or $H_{\max}$ ) specification, the tightest $H_{\max}$ ( or $Q'_{o\max}$ ) specification occurs at $\zeta^t=0.4040$ and thus is calculated by

$$H_{\max}Q'_{o\max}=h(\zeta^t)g(\zeta^t)(\Delta Q_i^2/A)=0.5206(\Delta Q_i^2/A) \tag{7}$$

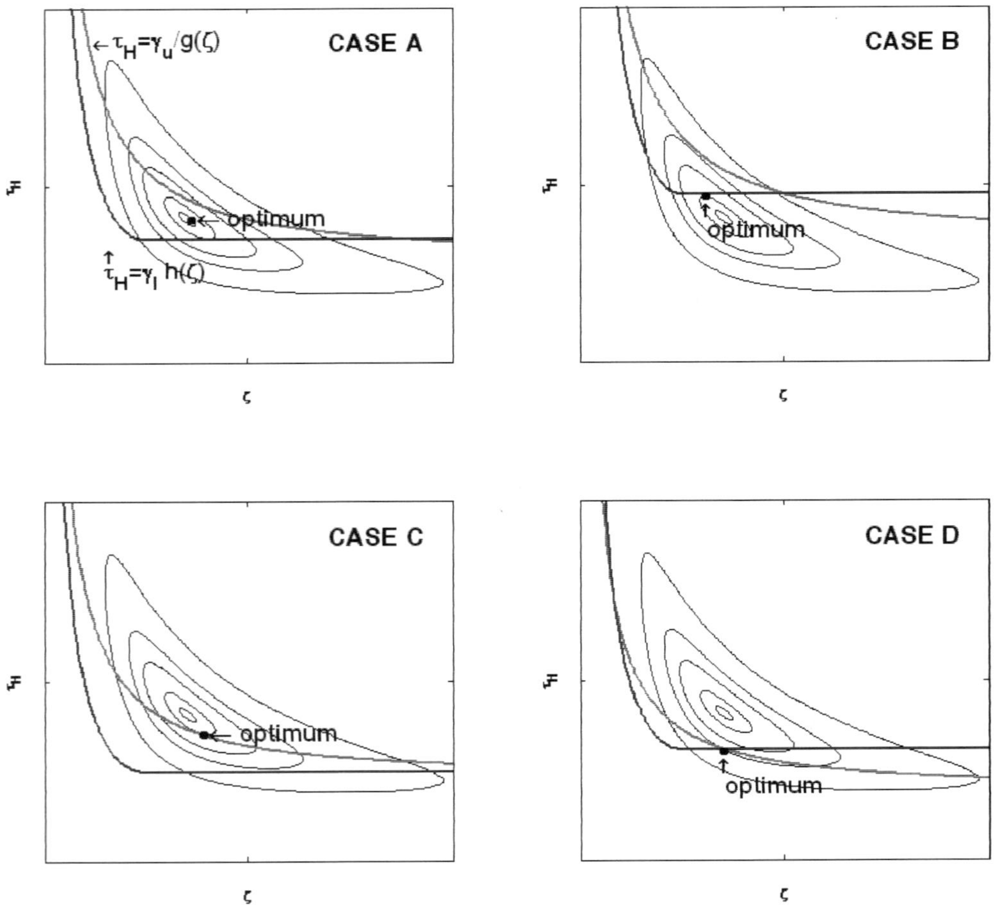


Figure 2. Typical contours and constraints with four possible cases of optimum location

Figure 2 shows several possible instances of a global optimum with the contours of the objective function and the constraints. Four cases are possible with respect to the global optimum location: (1) the global optimum is in the interior of the constraint set; (2) the global optimum is on the constraint $\tau_H = \gamma_{\mathrm{L}} h(\zeta)$; (3) the global optimum is on the constraint $\tau_H = \dfrac{\gamma_{\mathrm{U}}}{g(\zeta)}$; (4) the global optimum is located on the right vertex point formed by the two constraints.

Based on these characteristics of the optimum point location, the global optimums of $(\zeta, \tau_H)$ are found as follows if the feasible region is bounded (i.e., $\gamma_{\mathrm{L}} \geq \gamma_{\mathrm{U}}$) :

**Case A** . The extreme point $(\zeta^{\dagger}, \tau_H^{\dagger})$ is in the interior of the constraint set and the global optimum is :

$$\tau_H^{\dagger} = \left(\frac{\beta}{\alpha}\right)^{\frac{1}{4}} ;\ \zeta^{\dagger} = \frac{1}{\sqrt{2}} \tag{8}$$

**Case B** : The global optimum is located on the constraint $\tau_H = \gamma_{\mathrm{L}} h(\zeta)$ and denoted by $(\zeta^*, \tau_H^*)$:

$$\tau_H^* = \gamma_{\mathrm{L}} h(\zeta^*) \ ;\ \zeta^* = \frac{1}{\sqrt{2}} \frac{\tau_H^{\dagger}}{\gamma_{\mathrm{L}} h(\zeta^*)} \tag{9}$$

**Case C** : The global optimum is located on the constraint $\tau_H = \dfrac{\gamma_{\mathrm{U}}}{g(\zeta)}$ and denoted by $(\zeta^{**}, \tau_H^{**})$ :

$$\tau_H^{**} = \frac{\gamma_{\mathrm{U}}}{g(\zeta^{**})} \tag{10a}$$

and $\zeta^{**}$ can be calculated by a simple root-finding method from the following equation.

$$2\alpha\gamma_{\mathrm{U}}^3 g(\zeta^{**})\zeta^{**} - \frac{\beta}{2\gamma_{\mathrm{U}}} \frac{g(\zeta^{**})^5}{\zeta^{**3}} - \left\{3\alpha\gamma_{\mathrm{U}}^3 \zeta^{**2} - \frac{\beta}{\gamma_{\mathrm{U}}} g(\zeta^{**})^4 \left(1 + \frac{1}{4\zeta^{**2}}\right)\right\} g'(\zeta^{**}) = 0 \tag{10b}$$

**Case D.** The global optimum denoted by $(\zeta^{v}, \tau_H^{v})$ is located on the vertex point formed by $\tau_H = \dfrac{\gamma_{\mathrm{U}}}{g(\zeta)}$ and $\tau_H = \gamma_{\mathrm{L}} h(\zeta)$. The value of $\zeta$ at the global optimum point can be found by a simple root finding method from

$$g(\zeta^{v}) h(\zeta^{v}) = \frac{\gamma_{\mathrm{U}}}{\gamma_{\mathrm{L}}} \tag{11}$$

The value of $\tau_H$ at the global optimum point is

$$\tau_H^{v} = \gamma_{\mathrm{L}} h(\zeta^{v}) \tag{12}$$

In summary, the procedure for finding the global optimum is as follows :

(i) Calculate the unconstrained extreme point $(\zeta^{\dagger}, \tau_H^{\dagger})$.

If $\gamma_{\mathrm{L}} h(\zeta^{\dagger}) \leq \tau_H^{\dagger} \leq \dfrac{\gamma_{\mathrm{U}}}{g(\zeta^{\dagger})}$, then $(\zeta^{\dagger}, \tau_H^{\dagger})$ is the global optimum.

(ii) If $\tau_H^{\dagger} < \gamma_{\mathrm{L}} h(\zeta^{\dagger})$, then calculate $(\zeta^*, \tau_H^*)$ and $(\zeta^{v}, \tau_H^{v})$. If $\zeta^* \leq \zeta^{v}$, then $(\zeta^*, \tau_H^*)$ is the global optimum ; If $\zeta^* \geq \zeta^{v}$, then $(\zeta^{v}, \tau_H^{v})$ is the global optimum.

(iii) If $\tau_H^{\dagger} > \dfrac{\gamma_U}{g(\zeta^{\dagger})}$, then calculate $(\zeta^{**}, \tau_H^{**})$ and $(\zeta^{v}, \tau_H^{v})$. If $\zeta^{**} \le \zeta^{v}$, then $(\zeta^{**}, \tau_H^{**})$ is the global optimum ; If $\zeta^{**} \ge \zeta^{v}$, then $(\zeta^{v}, \tau_H^{v})$ is the global optimum.

If the feasible region is unbounded, no vertex point formed by the two constraints exists for $\zeta > 0.4040$, and accordingly case D does not exist. Also, the inequality conditions $\zeta^{*} \le \zeta^{v}$ and $\zeta^{**} \le \zeta^{v}$ do not need to be evaluated for cases B and C, respectively.

Once the global optimum is obtained in terms of $\zeta$ and $\tau_H$, the corresponding optimal PI parameters can be directly calculated as

$$K_C = \frac{(\Delta H)A}{Q_{o\max}\tau_H} \; ; \; \tau_I = 4\zeta^2\tau_H \tag{13}$$

## Illustrative Example

The liquid level of a drum with a cross section area of 1 $m^2$ and a working volume $A\Delta H$ of 2 $m^3$ is controlled by a PI controller. The maximum outlet flow rate $Q_{o\max}$ is 4 $m^3$/min. The initial steady-state level is 50% and the nominal flow rates of the inlet and outlet are both 1 $m^3$/min. The maximum expected change of the inlet flow rate $\Delta Q_i$ is 1 $m^3$/min. The weighting factor for optimal control is set to $\omega = 0.8$. Consider the following four cases :

**Case 1** ($Q'_{o\max}$ is 3 $m^3$/min$^2$ and $H_{\max}$ is 1.0 m) ; **Case 2** ($Q'_{o\max}$ is 1.2 $m^3$/min$^2$ and $H_{\max}$ is 1.0 m) ; **Case 3** ($Q'_{o\max}$ is 4 $m^3$/min$^2$ and $H_{\max}$ is 0.2 m) ; **Case 4** ($Q'_{o\max}$ is 1.5 $m^3$/min$^2$ and $H_{\max}$ is 0.4 m)

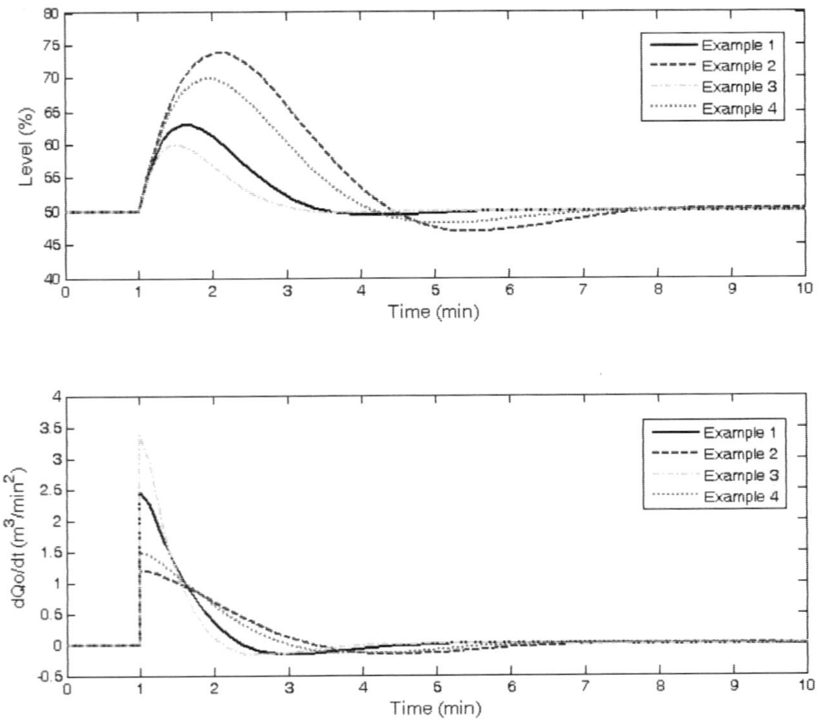


Figure 3. Responses of level and rate of change of outlet flow for cases 1,2,3, and 4

Figure 3 compares the responses for the liquid level and the rate of change of the outlet flow rate for each case. A step change of 1 $m^3$/min is made in the inlet flow at t=0 in the simulation. As seen in the figure, the PI controllers designed by the proposed method give the optimal responses while strictly satisfying the given $H_{max}$ and $Q'_{o\max}$ specifications.

## Acknowledgment
This research was supported by Yeungnam University research grants in 2007.

## References

1. Buckley, P. (1983). Recent Advances in averaging level control, in: *Productivity through Control Technology*, 18-21, Houston.
2. Cheung, T. and Luyben, W. (1979). Liquid-level control in single tanks and cascades of tanks with P-only and PI feedback controllers, *Ind. Eng. Chem. Fund.*, **1**, 15-21.
3. Dennis, J.B. (1959). *Mathematical Programs and Electron Networks*, Jhon Wiley & Sons, New York.
4. Luyben, W. and Buckley, P.S. (1977). A proportional-lag controller, *Instrum. Tech.*, **24**, 65-68.
5. MacDonald, K., McAvoy, T., and Tits, A. (1986). Optimal averaging level control, *AIChE J.*, **32**, 75-86.
6. Rivera, D.E., Morari, M., and Skogestad, S. (1986). Internal model control. 4. PID controller design, *Ind. Eng. Chem. Process Des. Dev.*, **25**, 252-265.
7. Seki, H. and Ogawa, M. (1998). Japan Patent # 2811041.
8. Wu, K., Yu, C., and Cheung, Y. (2001). A two degree of freedom level control, *J. of Process Control*, **11** 311-319.

## 165a Revamp of Heat-Integrated Crude Oil Distillation Systems for Energy Reduction and New Product Demands

**Lu Chen**, Megan Jobson, and Robin Smith. Centre for Process Integration, School of Chemical Engineering and Analytical Science, The University of Manchester, Manchester, United Kingdom

Heat-integrated crude oil distillation systems (i.e. the preheat train and crude oil distillation columns) are complex, energy intensive and expensive to modify. Strong interactions exist between the distillation system and the preheat train. Therefore, many degrees of freedom and many constraints need to be taken into account during process revamps.

An optimisation-based design methodology is proposed for the retrofit design and analysis of heat-integrated crude oil distillation systems. The approach considers the distillation column and preheat train in a single framework and aims to improve the overall performance of the system. In the approach, distillation units are simulated using simplified models that allow the specification of the products and the heat exchanger network is studied in terms of its detailed structure. The hydraulic constraints of the existing distillation column design are accounted for, and the existing numbers of stages in each section are considered. Moreover, exiting product qualities are constrained within acceptable range. Simulated Annealing has been employed as the optimisation algorithm because of its ability to optimise both structural and operational process variables. The number of structural modifications to the existing system can also be controlled to allow practical and simple retrofit solutions to be obtained.

Case studies are presented addressing the reduction of energy consumption in an existing heat-integrated atmospheric distillation column, and for improving profit by adjusting product yields according to market value.

PETROBRAS

# Heat Integration in a Crude Distillation Unit Using Pinch Analysis Concepts

(AIChE 2008 Spring Meeting – 165b)


## PETROBRAS R&D Center– CENPES

Antonio V. S. de Castro*, M.Sc.
Carlos Ney da Fonseca
Claudio L. M. Kuboski
Silvia Waintraub, M.Sc.
Washington de O. Geraldelli, Ph.D.

# Introduction

- Higher prices of energy and oil
- Crude Distillation Unit:
  - Energy-intensive Process
  - Heat Integration
  - Fractionation Constraints
- Pumparound Design
  - Number of Pumparound Sections
  - Location of Pumparound Sections
  - Pumparound Section Heat Duty

PETROBRAS

# Outline for Simulation Approach

- Design procedure:
  - Location of pumparounds (PA)
  - Analyse Pumparound Duty concerning the Fractionation constraints
  - Evaluate alternatives to improve Heat Recovery: global costs (Pinch Design Method)
    - Evaluate PA heat duty distribution at atmospheric tower (vacuum constant)
    - Evaluate changing pinch stream possibilities by process modifications (modify vacuum tower configuration, considering atmospheric tower best result fixed)
    - Evaluate modifying pinch stream return temperature (if PA)

PETROBRAS

## Pumparound Section

- Heat Recovery at higher temperature
- Maximum heat recoverable – Heat of vaporization of the liquid from the tray above the pumparound section
- Trade-off:
  - ↑ Pumparound Duty
  - ↓ Fractionation above the pumparound
- Fractionation Quality:
  - Internal reflux
  - Gap and Overlap

PETROBRAS

# Pumparound Section

- Max Heat Duty at PA:
  - Zero Internal Liquid Reflux above PA return.
- By Simulation:
  - Internal Liquid Reflux above PA return Enthalpy Difference at bubble and dew point;
  - Simulate the tower specifying near Zero Internal Reflux above PA, varying PA duty.
- In all studies, products specification were a target. However, stripping steam optimization was not part of this present work.

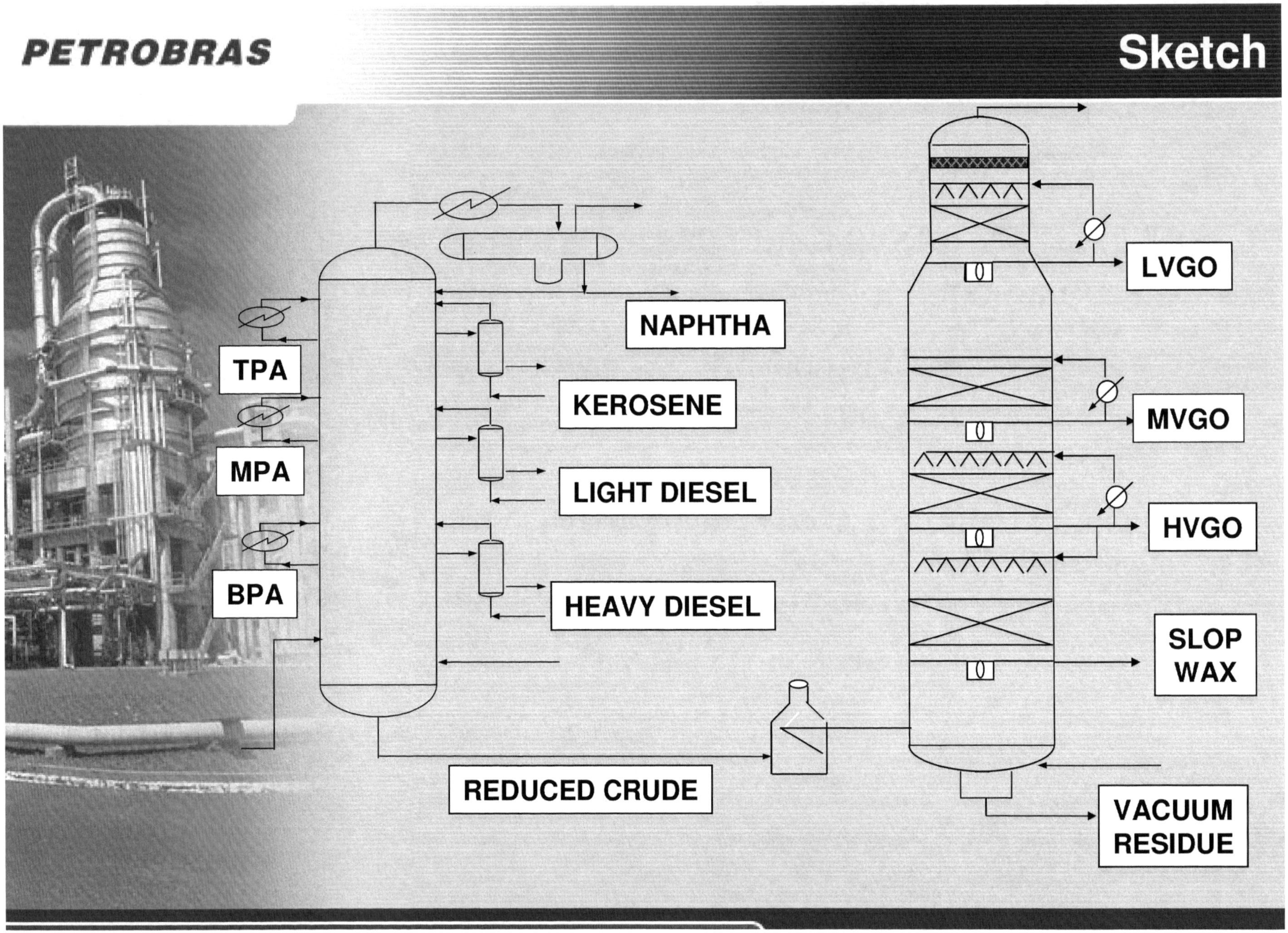
PETROBRAS
Sketch
TPA
MPA
BPA
NAPHTHA
KEROSENE
LIGHT DIESEL
HEAVY DIESEL
REDUCED CRUDE
LVGO
MVGO
HVGO
SLOP WAX
VACUUM RESIDUE

PETROBRAS

# Pumparound Section - Example

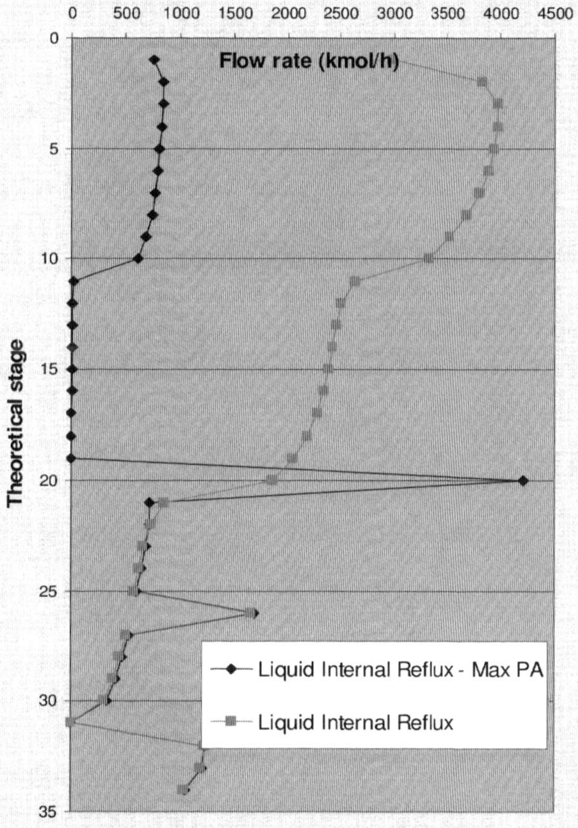


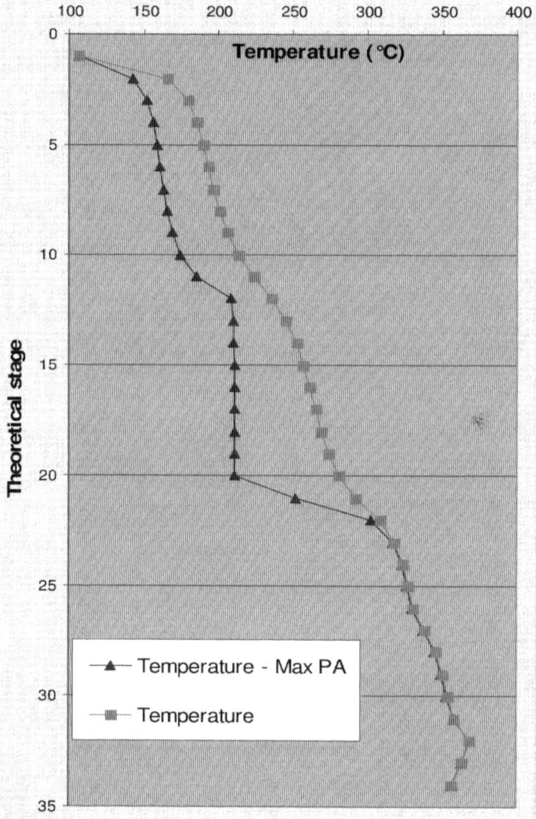


These graphics compare both liquid internal reflux and temperature profile at atmospheric column, considering BPA is already defined. Data refering to Max PA are at near zero liquid reflux, while the other data refer to maximum liquid internal reflux above Mid PA section.

PETROBRAS

# Simulation Basis

- 19° API Brazilian Crude
- Kept Constant:
  - Atm Furnace Outlet Temperature
  - Vacuum Furnace Outlet Temperature
  - Atm Ovhd Drum Temperature
  - Overflash Rate
  - Number of stages
- HVGO / LVGO ~ 1
- Pumparound Withdraw at Product Drawoff Pans
- Fractionation Constraints:
  - Naphtha – Kerosene: 0 °C min gap
  - Kerosene – Light Diesel: 5 °C min gap
  - Light Diesel – Heavy Diesel: 30 °C max overlap
- Cost basis:
  - Brent: US$ 30.00 / bbl
  - Fuel oil: US$ 20.60 / $10^6$ kcal
  - Cooling water: US$ 0.066 / m3
  - Equipment Cycle Life: 10 years

**PETROBRAS**

# Fractionation vs Heat Recovery

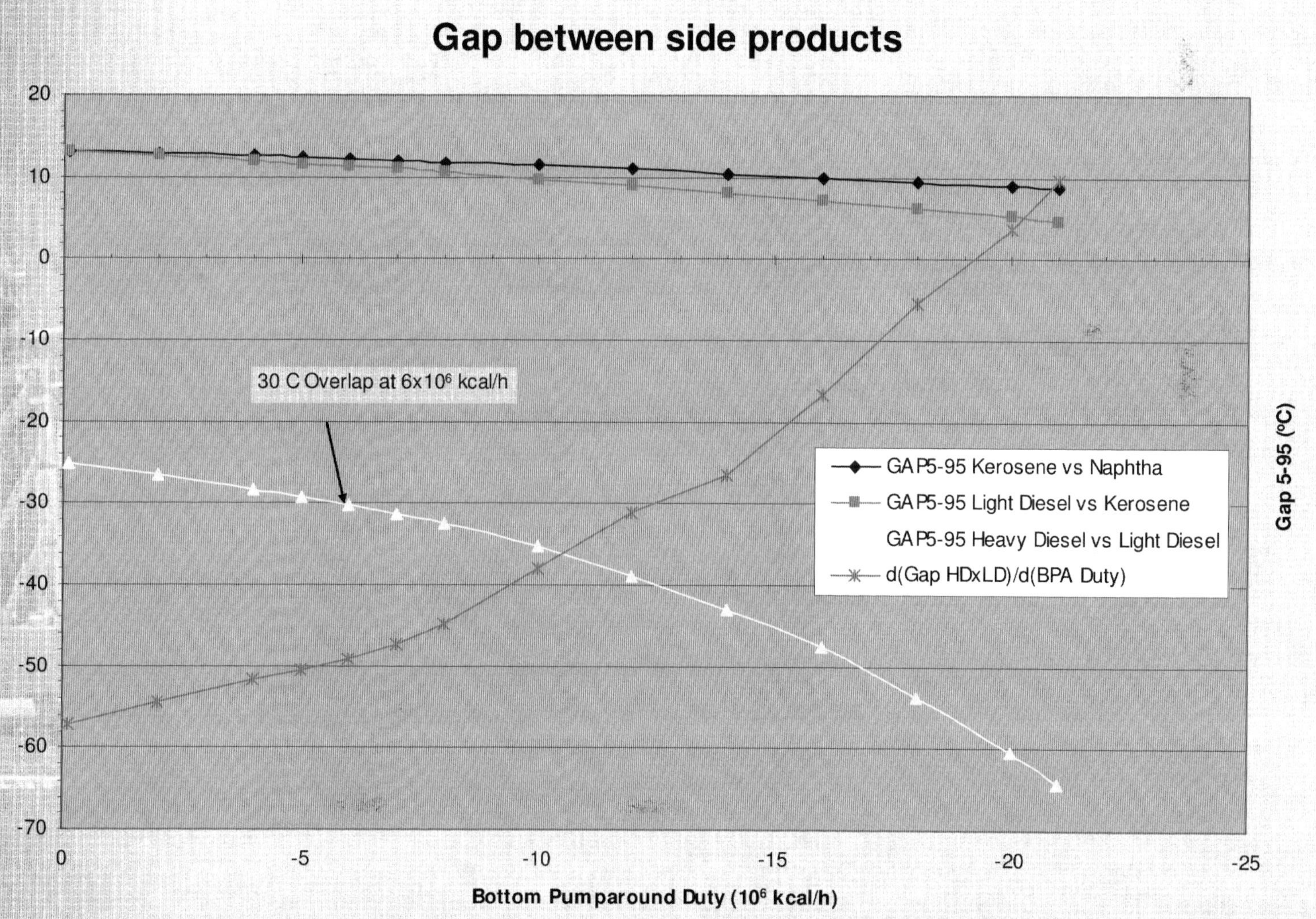

PETROBRAS

# Fractionation vs Heat Recovery

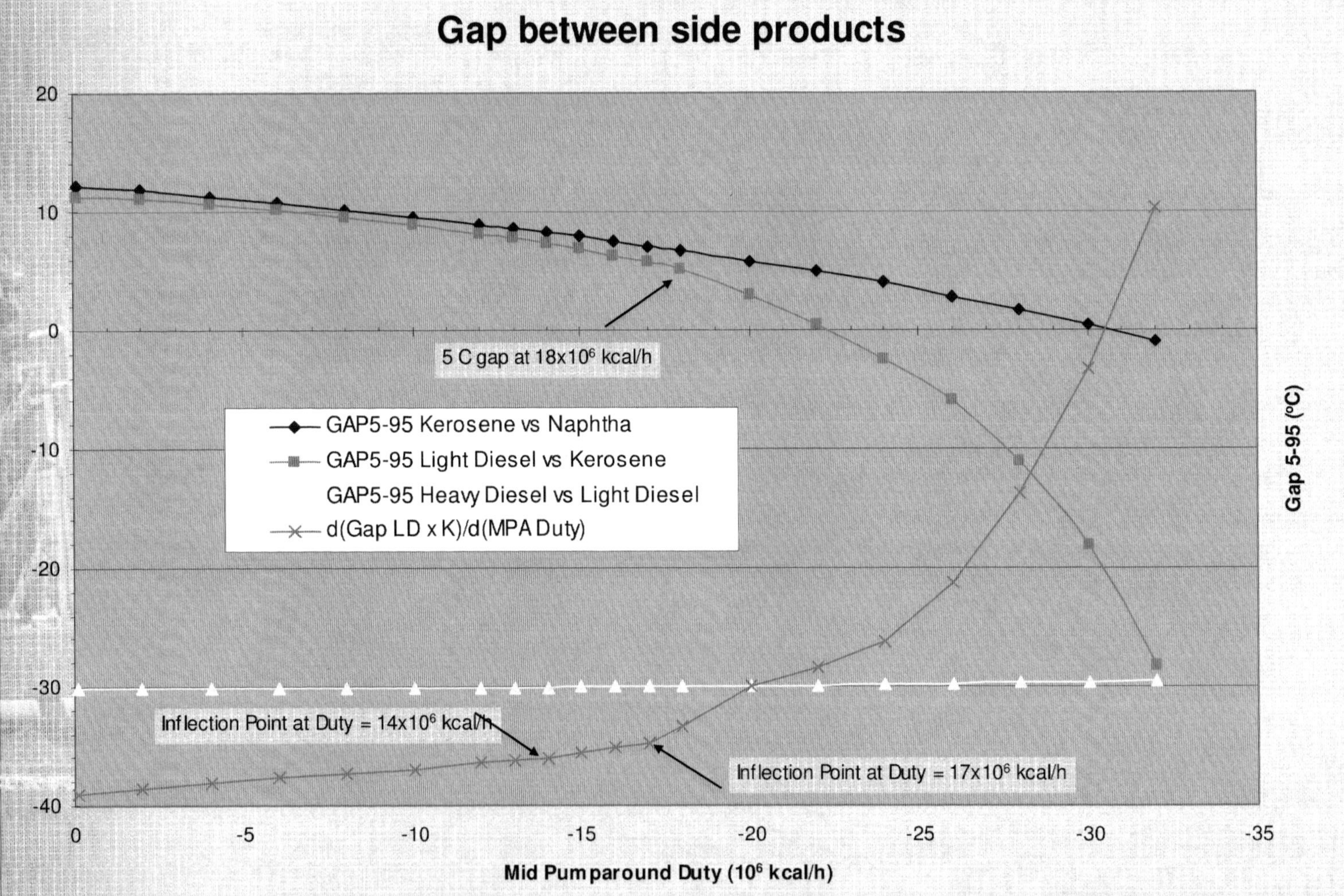

PETROBRAS

# Fractionation vs Heat Recovery

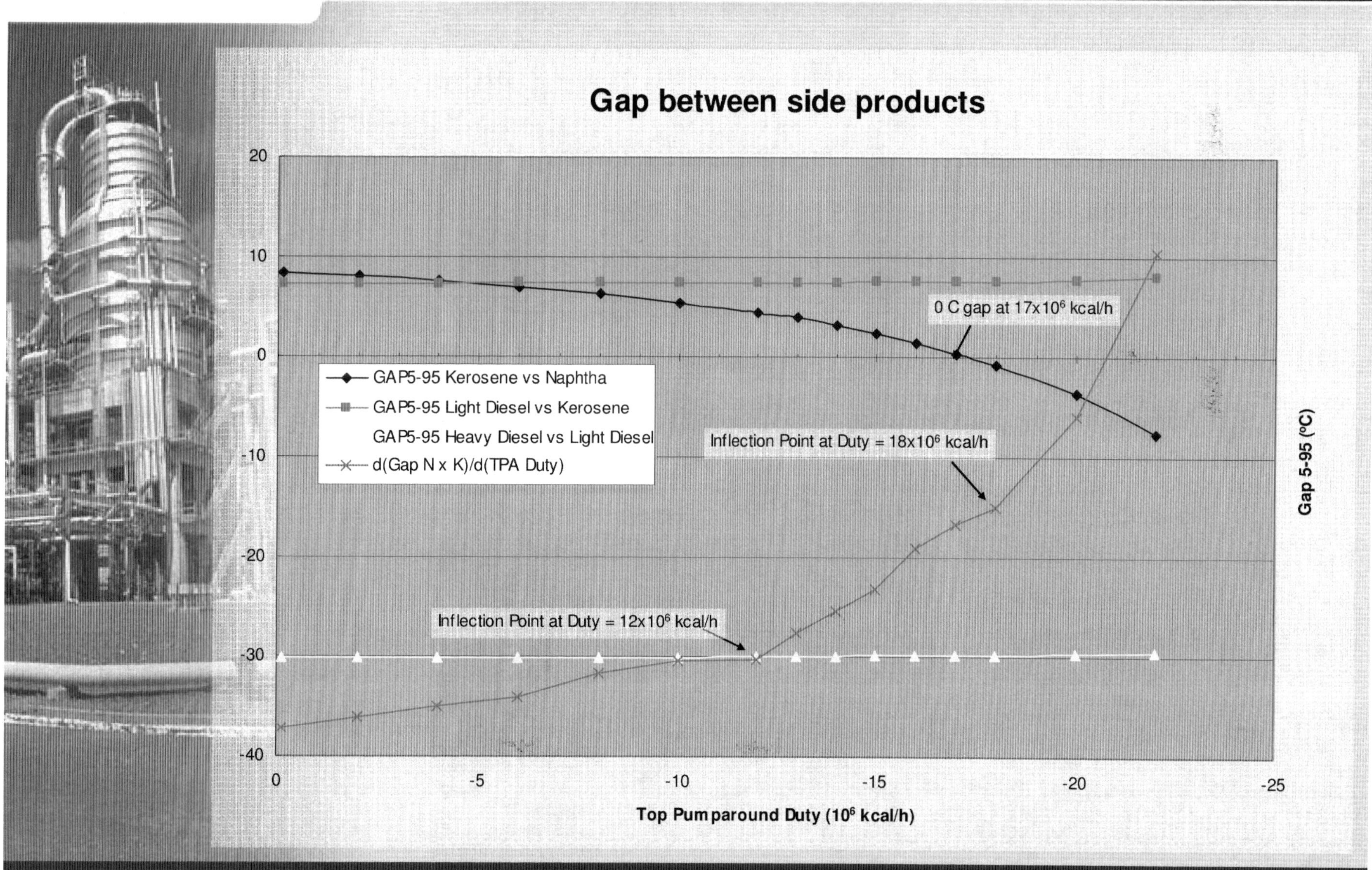

PETROBRAS

# Case Study 1 – max heat recovery

Evaluate PA heat duty distribution in atmospheric tower (vacuum configuration constant)

**Duties in $10^6$ kcal/h**

| | **Base Case** | **16,20,0** | **16,14,6** |
|---|---|---|---|
| **Ovhd Condenser** | 58.8 | 22.7 | 23.0 |
| **Top PA** | 0 | 16 | 16 |
| **Mid PA** | 0 | 20 | 14 |
| **Bottom PA** | 0 | 0 | 6 |
| **Overall** | 58.8 | 58.7 | 59.0 |

**PETROBRAS**

# Results – Case Study 1

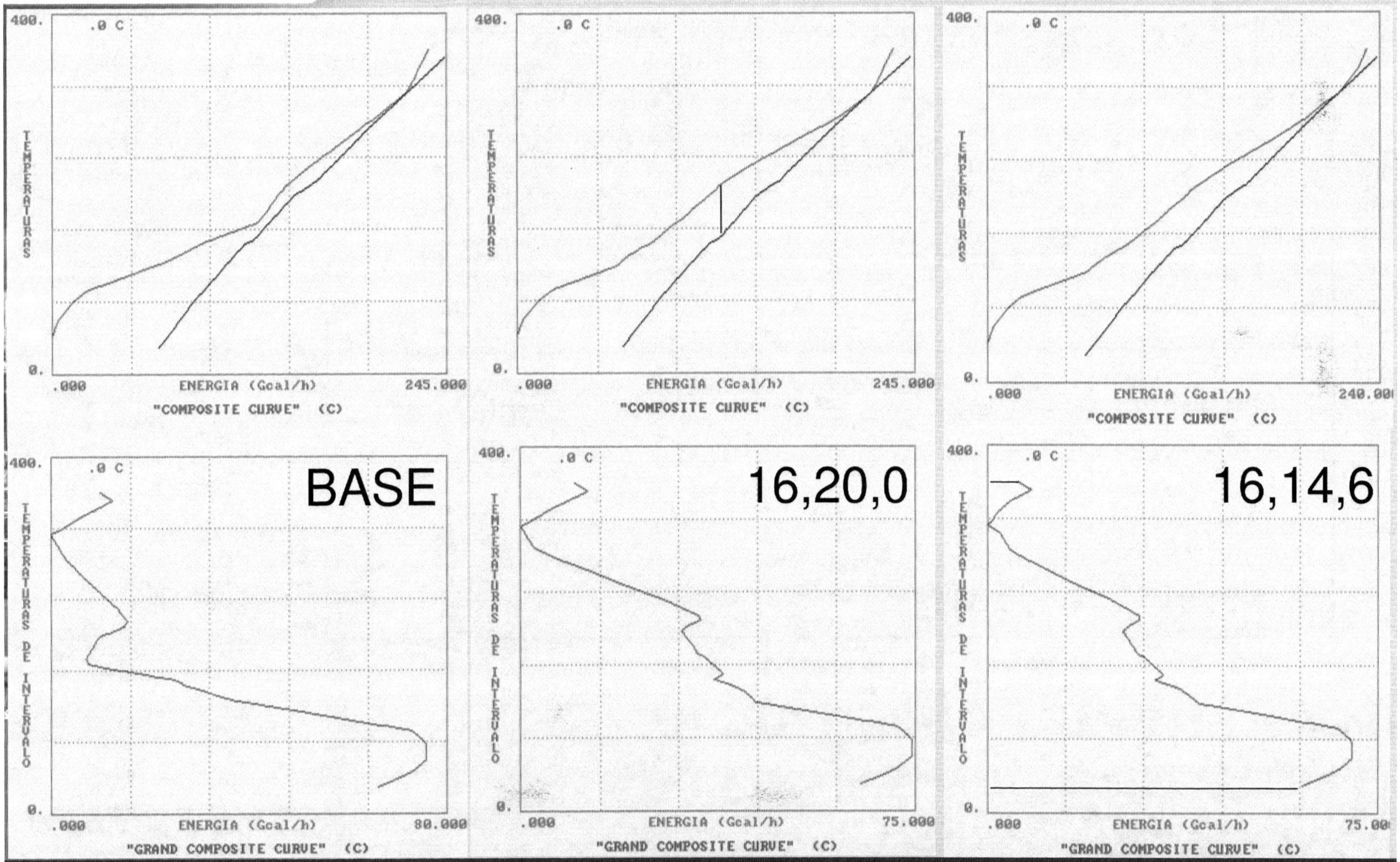

PETROBRAS

# Results – Case Study 1

## Atmospheric Tower

### Duties in $10^6$ kcal/h

| | Base | 16,20,0 | 16,14,6 |
|---|---|---|---|
| **Hot Utility** | 9.79 | 10.01 | 5.74 |
| **Cold Utility** | 66.1 | 64.7 | 60.4 |
| **Δ Hot Utility (Base)** | 0 | + 0.22 | - 4.05 |

- Case 16,14,6:
  - Bottom PA: $6 \times 10^6$ kcal/h; Tout = 338 °C; Treturn = 303 °C
  - Pinch: HVGO; Tpinch = 312 °C
  - Bottom PA: Above the Pinch = 338 – 312 = 26 °C (74,3%)

    6 x 0,743 = $4,45 \times 10^6$ kcal/h ~ $4,27 \times 10^6$ kcal/h (4.05 + 0.22)

**PETROBRAS**

## Results – Case Study 1

# Atmospheric Tower

| | **Base** | **16,20,0** | **16,14,6** |
|---|---|---|---|
| **$\Delta T$ optimum (ºC)** | 26.2 | 14.9 | 19.2 |
| **Utility Cost ($10^6$ US$/yr)** | 8.504 | 7.014 | 6.644 |
| **Capital Cost ($10^6$ US$/yr)** | 6.016 | 6.126 | 5.789 |
| **Overall Cost ($10^6$ US$/yr)** | 14.520 | 13.140 | 12.432 |
| **Savings ($10^6$ US$/yr)** | 0 | 1.380 | 2.088 |

* For Case 16,20,0 at $\Delta T$ =19.2 ºC – Capital Cost = 5.637x$10^6$ US$/yr slightly lower than Case 16,14,6 caused by lower approach near pinch region, but process recovery lead to much lower Utility Cost

# Vacuum Tower – MVGO Draw

- In case study 1: benefit on moving duty from below to above the pinch
- What about moving the pinch by changing process/configuration, keeping specification?

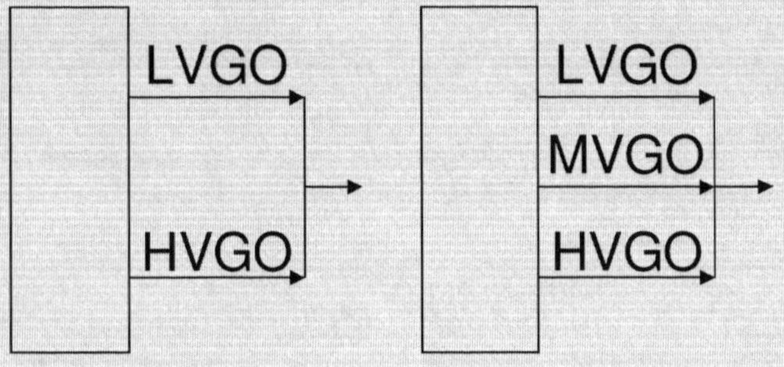


– Add MVGO draw

HVGO : MVGO : LVGO ~ 1 : 4 : 1 (case: MVGO)

High flow rate required to change pinch location.

Atmospheric column configuration constant (best result previously achieved).

process to process recovery above pinch ↑

Hot and cold utility ↓

approach ↓ - Capital cost ↑

(trade-off)

**PETROBRAS**

# Results – Case Study 2

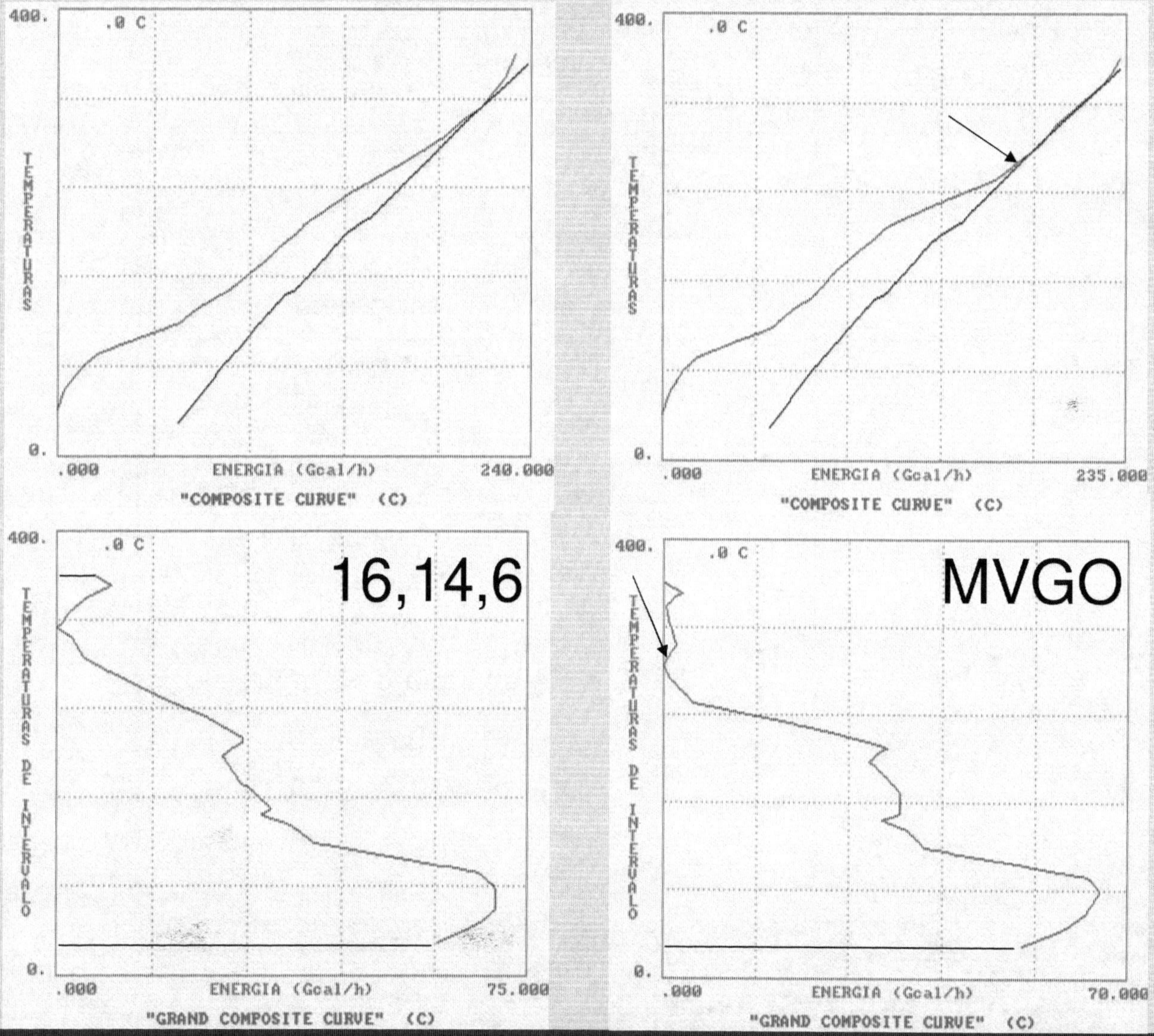

PETROBRAS

# Results – Case Study 2

| | 16,14,6 | MVGO (pinch – Mid PA) |
|---|---|---|
| Hot Utility at pinch ($10^6$ kcal/h) | 5.74 | 0.16 |
| Cold Utility at pinch ($10^6$ kcal/h) | 60.4 | 54.1 |
| Δ Hot Utility (Base) ($10^6$ kcal/h) | - 4.05 | -9.63 |
| ΔT optimum (ºC) | 19.2 | 13.2 |
| Utility Cost ($10^6$ US$/yr) | 6.644 | 4.565 |
| Capital Cost ($10^6$ US$/yr) | 5.789 | 8.555 |
| Overall Cost ($10^6$ US$/yr) | 12.432 | 13.120 |
| Savings ($10^6$ US$/yr) | 2.088 | 1.400 |

As pinch is occurring at MVGO (much higher flow rate than HVGO), there is a large portion of Hot Composite Curve with few variation in flow above the pinch, resulting expressive increment on Capital Cost (penalty too high).

**PETROBRAS**

# Results – Case Study 2

Add MVGO withdraw didn´t present good results, but :

- If products flow rates change?
  - – HVGO : MVGO : LVGO ~ 1 : 2 : 2
  - – T HVGO PA Return = 285 ºC
  - – Named: Case MVGO 285

HVGO kept as pinch stream (same process recovery than Case 16,14,6)

Higher approach (hot x cold composite) – Capital Cost decrease

- **MVGO 285 result only evaluating HEN (capital cost of tower changes not included)**

**PETROBRAS**

# Results – Case Study 2

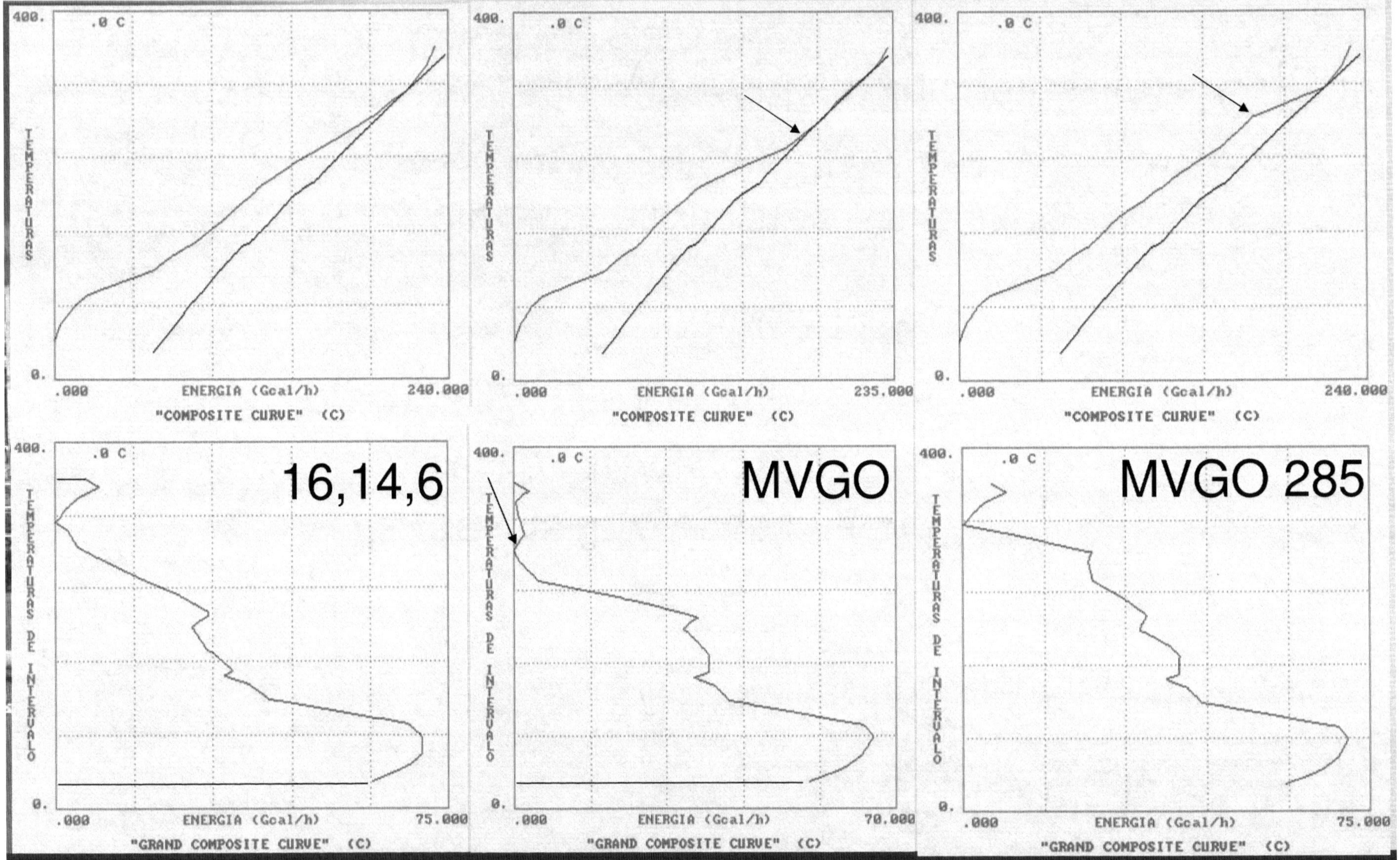

PETROBRAS

# Results – Case Study 2

| | 16,14,6 | MVGO (pinch – Mid PA) | MVGO 285 (pinch – HVGO) |
|---|---|---|---|
| Hot Utility at pinch ($10^6$ kcal/h) | 5.74 | 0.16 | 5.42 |
| Cold Utility at pinch ($10^6$ kcal/h) | 60.4 | 54.1 | 59.4 |
| $\Delta$ Hot Utility (Base) ($10^6$ kcal/h) | - 4.05 | -9.63 | - 4.37 |
| $\Delta$T optimum (ºC) | 19.2 | 13.2 | 13.3 |
| Utility Cost ($10^6$ US$/yr) | 6.644 | 4.565 | 5.784 |
| Capital Cost ($10^6$ US$/yr) | 5.789 | 8.555 | 5.728 |
| Overall Cost ($10^6$ US$/yr) | 12.432 | 13.120 | 11.512 |
| Savings ($10^6$ US$/yr) | 2.088 | 1.400 | 3.008 |

If we keep pinch at HVGO, heat recovery is the same than Case "16,14,6", however the HEN approach is much higher, allowing more heat recovery.

PETROBRAS

## Case Study 3

Pinch Stream Pumparound $\Delta T$

Evaluate modifying pinch stream return temperature (if PA)

HVGO: for low $\Delta T$ – high flow rate (pumping need to be evaluated)

How will thermodynamics respond to flow variation?

**PETROBRAS**

# Results – Case Study 3

## Pinch Stream Pumparound $\Delta T$

| | MVGO 200 | MVGO 230 | MVGO 260 | MVGO 285 |
|---|---|---|---|---|
| T HVGO pan (ºC) | 326 | 324 | 321 | 314 |
| T HVGO PA return (ºC) | 200 | 230 | 260 | 285 |
| Hot Utility ($10^6$ kcal/h) | 3.78 | 4.07 | 4.50 | 5.42 |
| Cold Utility ($10^6$ kcal/h) | 57.8 | 58.1 | 58.5 | 59.4 |

Heat moving across pinch

$$\arctan(\alpha) = \frac{\text{Hot Utility}}{T_{cold}^{end} - T_{pinch}}$$

$$\arctan(\alpha) = 55^{o}$$

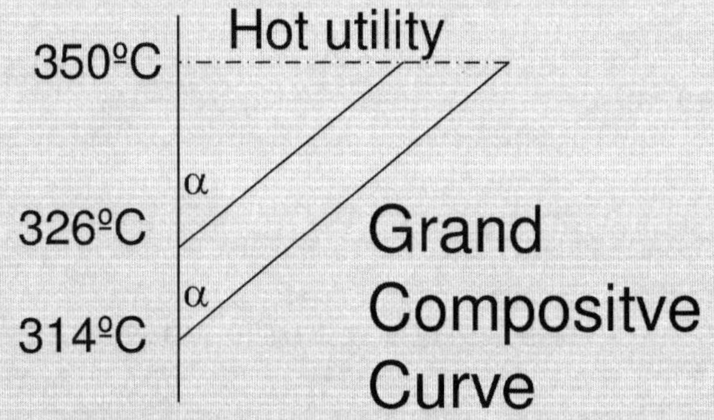

**PETROBRAS**

# Results – Case Study 3

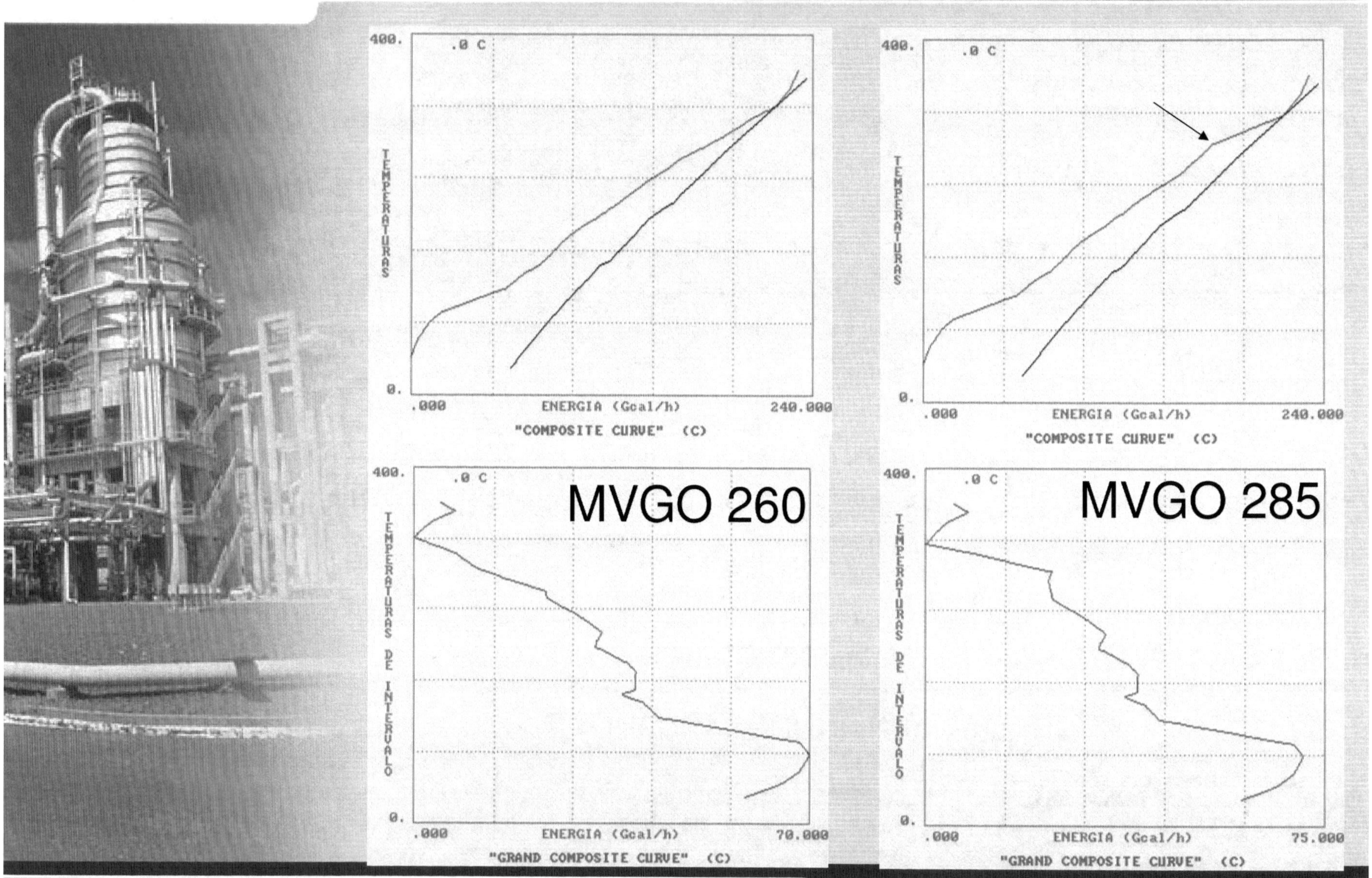

**PETROBRAS**

# Results – Case Study 3

| | MVGO 200 | MVGO 230 | MVGO 260 | MVGO 285 |
|---|---|---|---|---|
| ΔT optimum (ºC) | 19.0 | 17.2 | 14.1 | 13.3 |
| Utility Cost ($10^6$ US$/yr) | 6.181 | 5.994 | 5.685 | 5.784 |
| Capital Cost ($10^6$ US$/yr) | 6.028 | 5.977 | 5.954 | 5.728 |
| Overall Cost ($10^6$ US$/yr) | 12.209 | 11.971 | 11.639 | 11.512 |
| Savings ($10^6$ US$/yr) | 2.311 | 2.549 | 2.881 | 3.008 |

As HVGO flow rate increases, the HEN approach becomes higher, resulting less Capital Cost, allowing more heat integration.

PETROBRAS

# Discussion

- Procedure constraints
  - Pinch analysis assumes direct heat exchange
  - Cost of new sections inside the tower need to be evaluated appart
  - Modification on vacuum and atmospheric collumn simultaneously are not easily evaluated
  - Non optimal design (but close to optimum)

PETROBRAS

# Conclusion

- In Case Study 1, moving duty from below to above the pinch (transfering duty from MPA to BPA) reduced Utility Cost with almost no penalty in Capital Cost.
- In Case Study 2, moving the pinch stream by creating a new drawoff at vacuum tower did not bring benefit initially, as the increase on Capital Cost was too high. However, appropriate flow rate definition for this new stream lead to much higher approaches (lower Capital Cost).
- In Case Study 3, capital cost becomes higher for lower return PA temperature (lower flow rate).

**PETROBRAS**

# Conclusion

- Appropriate variation of process streams observing thermodynamics may result in high process integration (grass root or revamp)
- Optimization taking into account these insights could improve the design.

PETROBRAS

# Thank you very much!

Antonio V. S. de Castro, antonio.castro@petrobras.com.br
Claudio L. M. Kuboski
Carlos Ney da Fonseca
Silvia Waintraub
Washington de O. Geraldelli

# 165c Reducing Energy Consumption by New Distillation Configuration

**Libin Zhang**, University of Illinois at Chicago, 851 S. Morgan St., SEO 218, Chicago, IL 60607, Andreas A. Linninger, Bioengineering, University of Illinois at Chicago, 851 S. Morgan St., Room 218, Chicago, IL 60607, and Rakesh Agrawal, School of Chemical Engineering and Energy Center at Discovery Park, Purdue University, 480 Stadium Mall Drive, West Lafayette, IN 47907.

Separation processes make up 40%-70% of capital and operating costs of chemical manufacturing. Distillation accounts for more than 60% of the total process energy for the manufacture of commodity chemicals [DOE, 2005]. Optimal complex column configuration is estimated to harness energy savings up to 70% [Engelien and Skogestad, 2005]. Therefore, distillation trains are a meaningful target for energy improvements on an industry-wide scale. There is considerable literature on how to separate a feed mixture containing three or more components into pure product streams. Although past work on multicomponent distillation configurations has generated many extremely valuable information, it is still incomplete in providing the low-energy distillation configurations for a given application. There are two major challenges to hamper us providing the solutions. The first reason is that the search space for the distillation networks has been found to lack a few configurations and no systematic method to elucidate all distillation configurations from a network representation. The second reason is that finding the optimum distillation configuration has been the enormous difficulty of the mathematical task and the size of the problem in terms of computational effort.

In this presentation, a systematic procedure to draw distillation column configurations to separate an ideal to near ideal n-component mixture into n product streams each enriched in one of the components will be presented. The method synthesizes all feasible basic configurations using n-1distillation columns with each column having only a condenser at the top and a reboiler at the bottom. It also generates all feasible thermally coupled schemes with classical two-way liquid and vapor communication between the distillation columns. The method is simple and easy to use. After we obtain an array of the optimal configurations, the stochastic genetic method with novel temperature collocation method was incorporated to systematically find an optimum distillation scheme for a given application. Massive problem size reductions due to temperature collocation ensure the realistic composition profiles of each column in the network without sacrificing the computational and thermodynamic rigor. Feasibility of the synthesized sequences is ascertained by means of the MIDI algorithm as a subroutine to the master GA. In contract to existing approaches, this method synthesizes several clusters of designs solutions, each one corresponding to regions of local optimality.

PETROBRAS

# Exploiting the Use of Compact Heat Exchangers on Preheating Trains

(AIChE 2008 Spring Meeting – 165d)


## PETROBRAS R&D Center– CENPES

Antonio V. S. de Castro*, M.Sc.
Alan T. Vilas Boas
Claudio L. M. Kuboski
Diomedes C. da Silva, M.Sc.
Roberto A. Oliveira, M.Sc.
Washington de O. Geraldelli, Ph.D.

# Introduction

- Higher prices of energy and oil
- Crude oil distillation unit is one of the highest energy consumers in a refinery.
- Opportunities for heat integration:
  - In revamps to process advantageous crudes – Special Metallurgy
  - higher heat recovery
- Threats:
  - Special Metallurgy due to naphtenic acids - higher investments at preheating train
  - increase on plot area due to increase on heat recovery - might be a constraint

# Introduction

- Compact Welded Plate Heat Exchanger:
  - Special Metallurgy (Titanium, SMO, etc.) leads to attractive costs compared to a S&T
  - Compact (less plot area)
  - Lower temperature approaches than S&T
  - Innovative solution
  - Refinery designers don't have enough information regarding the performance:
    - Fouling
    - Mechanical resistance
    - Operational cautions, such as start-up and shut-down

**PETROBRAS**

# PETROBRAS Experience

- End of 2001, PETROBRAS Engineers visited a Refinery, in Europe, where a Compabloc CP75 was installed
  - Overhead Condenser with Crude
  - Special Metallurgy required due to Chloride
  - 4 years without maintenance
- August, 2002. Service with two heavy fluids
  - 13 exchangers at refinery since 1994
  - Maintenance during winter shut-down
  - Focus on Process Issues
- Visit to refinery with two parallel units (one stand-by) – Sep,2005
  - HVGO vs. Crude Oil (high temperature)
  - Problems during the start-up (blocked pipe)
  - Operating since November,2004

**PETROBRAS**

# PETROBRAS Experience

- June,2002 – ALFA LAVAL trained PETROBRAS Engineers how to design Compabloc using CAS200 (ALFA LAVAL software)
- September,2006 – Two-phase fluid ALFA LAVAL course, concerning crude oil vaporization
  - – Lack of data for temperature x shear stress x fouling rate
- May,2006 – Visit to ALFA LAVAL factory in France:
  - - Maintenance/Inspection/Cleaning Requirements
  - - Manufacturing Process

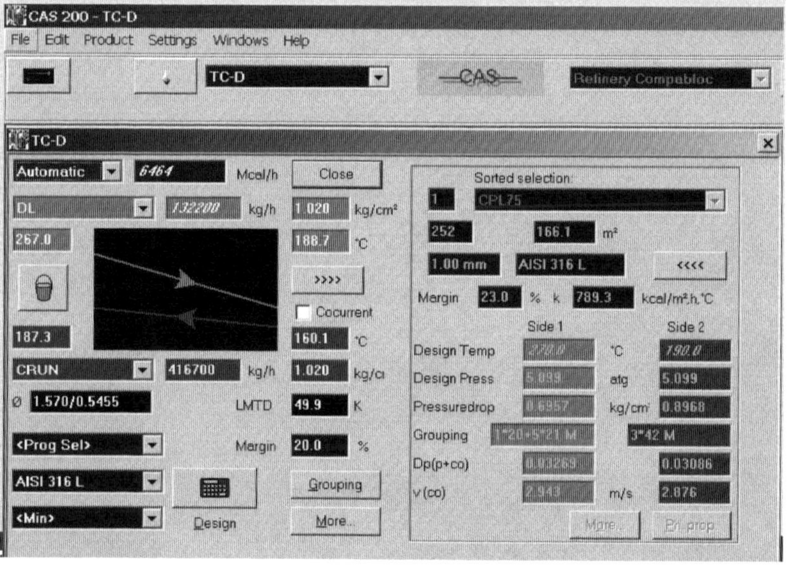

PETROBRAS

# PETROBRAS Cases

## CASE 1

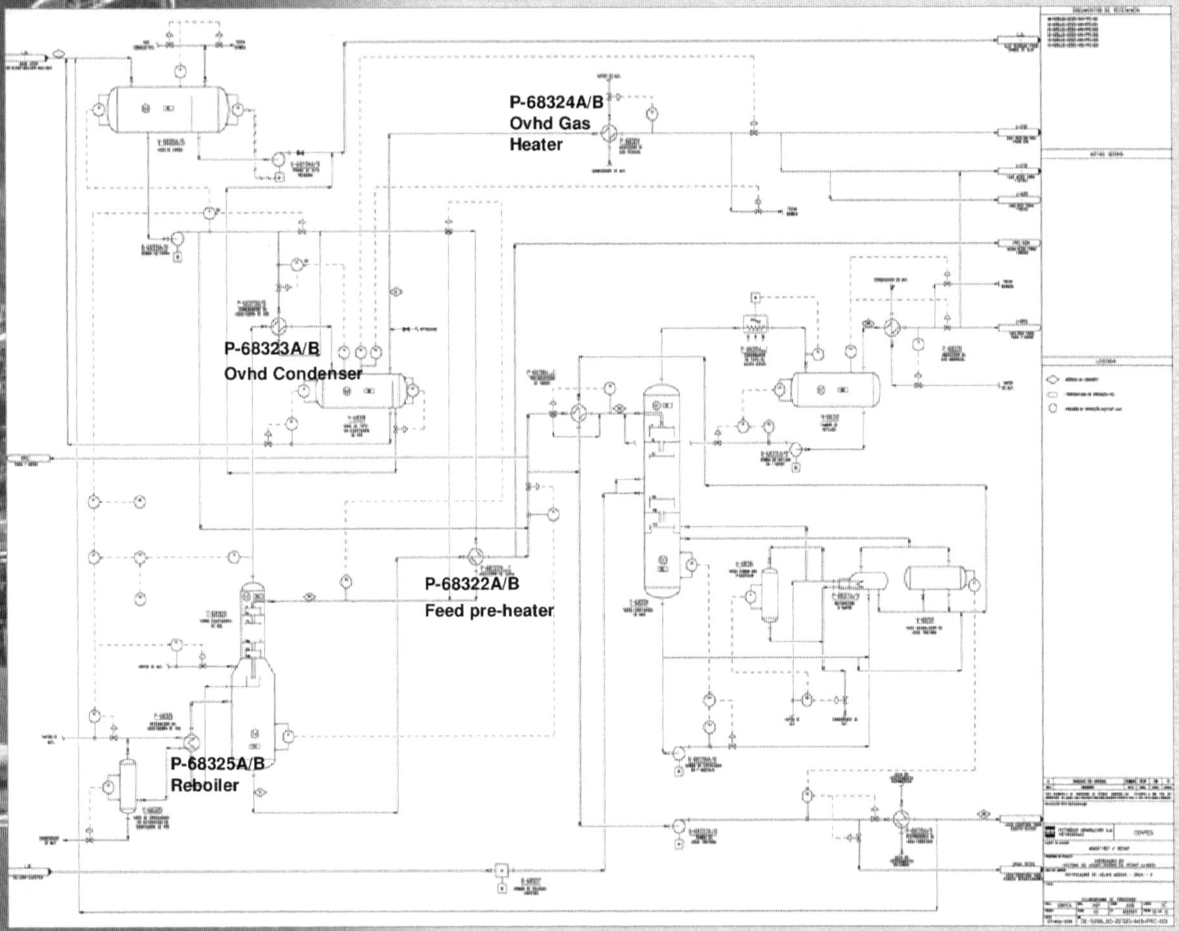


- Sour Water Stripping Unit:
  - Feed preheater
  - Overhead Condenser
  - Ovhd Gas Heater
  - Reboiler
- Plot area constraints
- Fouling at Feed pre-heater
- Leakage at Ovhd Gas Heater

PETROBRAS

# PETROBRAS Cases

**Fouling at Feed preheater**

- Inadequate $H_2S$ Stripping Tower Operation – corrosion
  - Actual Flow Rate ($90m^3/h$) ‹‹ Design ($160m^3/h$)
  - Actual Shear Stress (22Pa) ‹‹ Design (60Pa)
- Mitigation:
  - Increase passes number (↑ Shear Stress)
  - Control Review (by-pass)
  - Self-cleaning Strainers

**Leakage at Ovhd Gas Heater**

- Material replacement:
  - From Titanium to SMO

**CASE 1, cont.**

# PETROBRAS Cases

## CASE 2

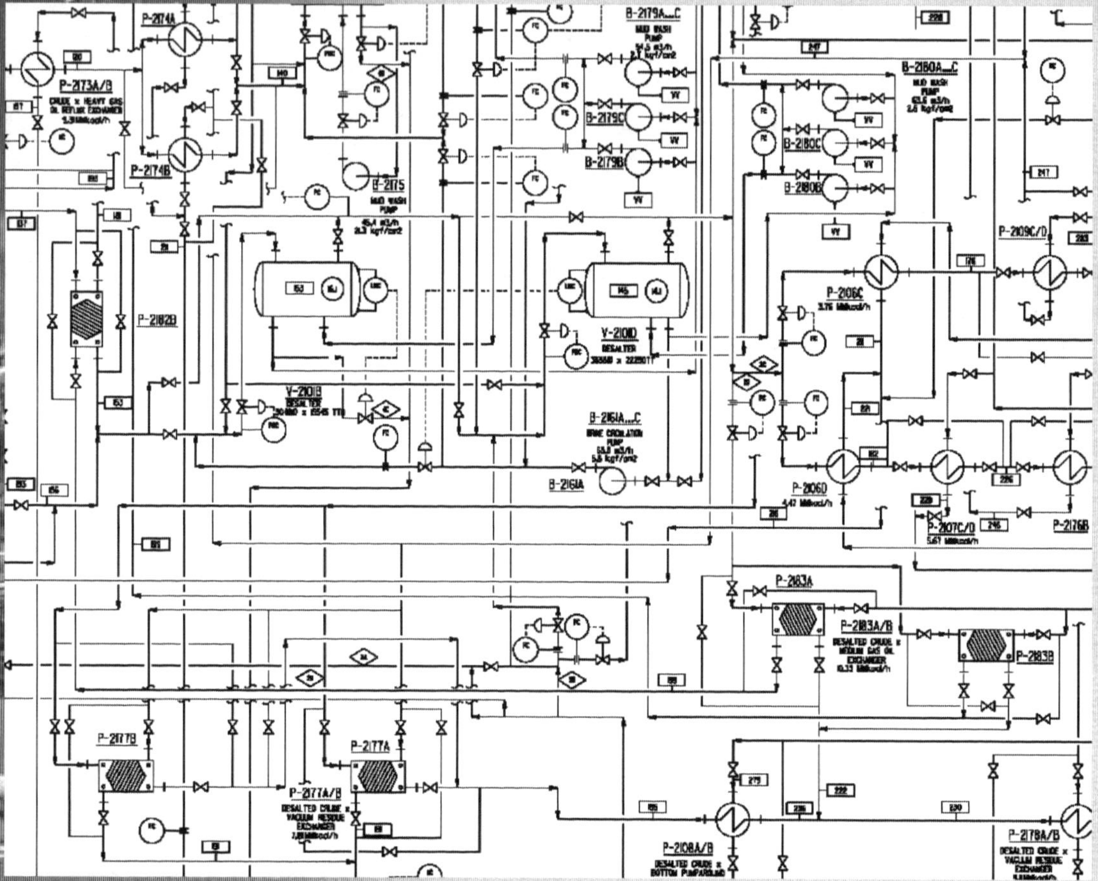


- Basic Engineering Design - Crude Distillation Unit
- Revamp:
  - Increase Feed Rate
  - Heat Integration
  - Plot Area Constraints
- 3 Services: MVGO (2) and VR
- 220,000 bbl/d
- Start-up: 2011

**PETROBRAS**

# PETROBRAS Cases

## CASE 3

### Conceptual Design (Revamp)

### 140,000 bbl/d

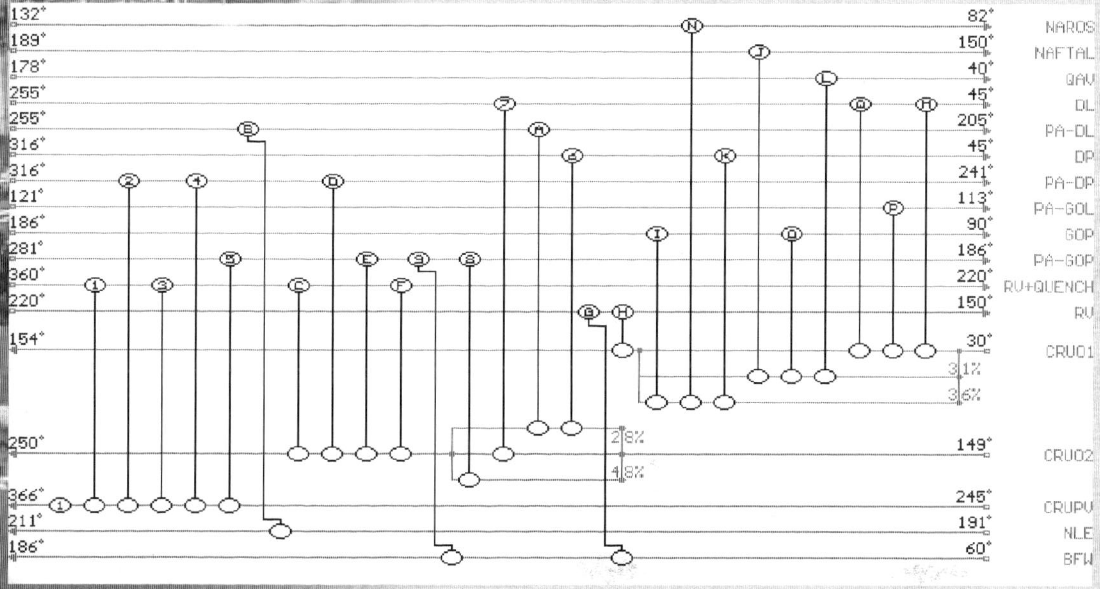


- Metallurgical Adaptation (TAN=2.1)
- Operational Debottlenecking
- Heat Integration Study – Pinch Analysis
- Plot Area and Shutdown time constraints
- Full Compabloc **new** preheat train suggested.

PETROBRAS

# Compabloc Test at Refinery

- During conceptual design, necessity of a performance test.
- A Compabloc CP30 unit has been installed, in order to evaluate the performance of this equipment under unit conditions
- Vacuum Residue preheating Crude before the Desalter
- Low Crude Temperature: ~100 to 115ºC
- Operational and Maintenance experience

**PETROBRAS**

# Compabloc Test at Refinery

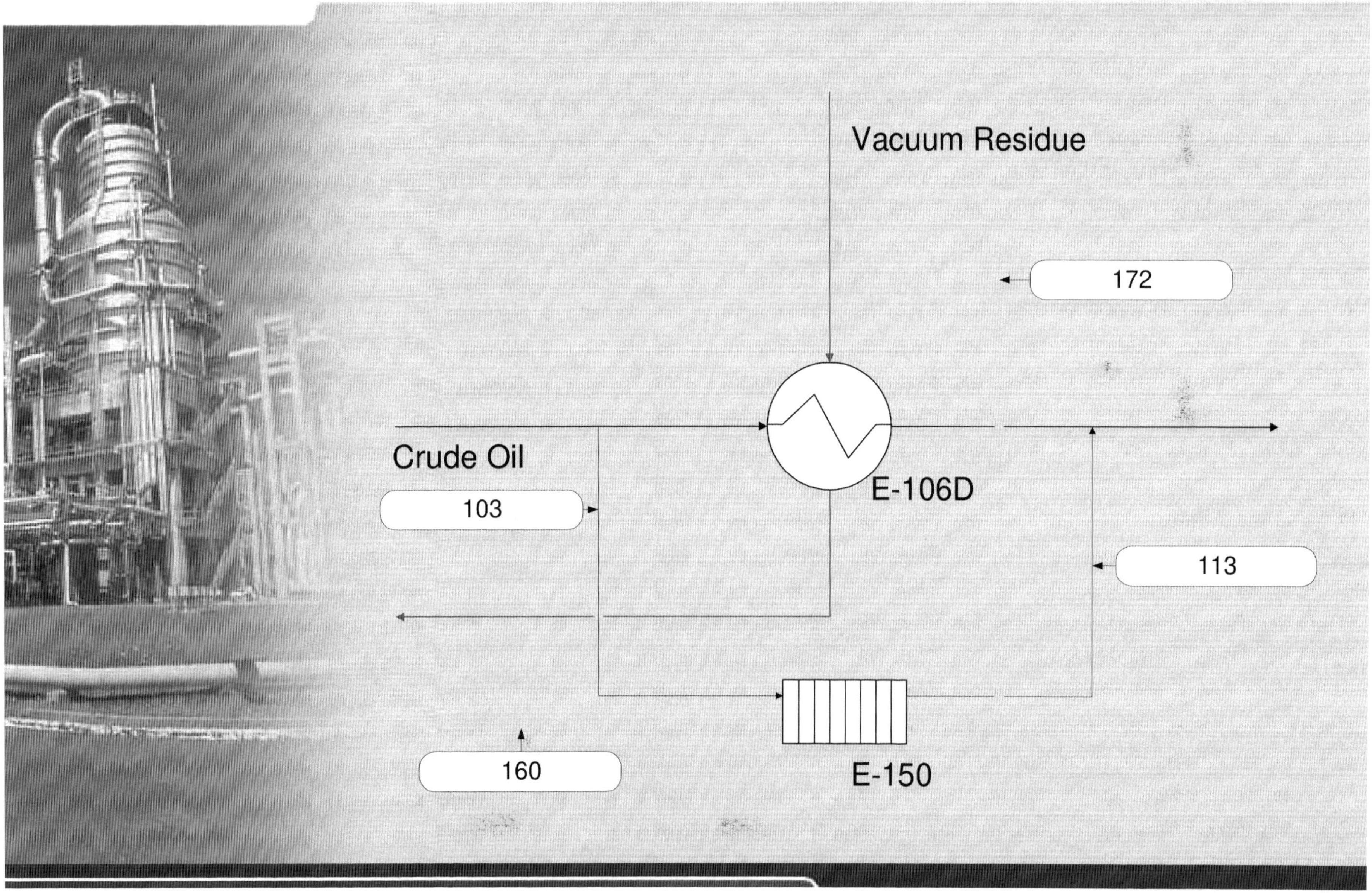

PETROBRAS

# Compabloc Test at Refinery

| Step | Flow rate (design basis) | Period | Objective | Procedure |
|---|---|---|---|---|
| 1 | 100% | 6 months | Thermal performance | a) Check pressure drop;<br>b) Compare design X operational data using CAS200. |
| 2 | 50% | 2 months | Fouling | a) Check pressure drop<br>b) if fouling, go to phase 1;<br>if not, go to phase 4. |
| 3 | 100% | 2 months | Thermal performance and fouling | a) Check pressure drop;<br>b) Auto cleaning capabity. |
| 4 | 25% | 2 months | Fouling | a) Check pressure drop;<br>b) Equipment shut-down for 10 days and restart it. |
| 5 | Cleaning | | | a) Open the equipment to check deposits;<br>b) Cleaning with hidroblasting. |
| 6 | 100% | 3 months | | Similar to phase 1. |

PETROBRAS

## Compabloc Test at Refinery

- 1st Phase Results:
  - Crude side measured pressure drop was from 29 to 47% of calculated values by CAS200;
  - VR side measured pressure drop was from 123 a 167% of calculated values by CAS200;
  - Difference in global heat transfer coefficients (HTC) was much lower. Measured values were from de 88 a 111% of calculated values by CAS200;
  - Problems with Instrumentation could be the cause of such difference in pressure drop

PETROBRAS

# Compabloc Test at Refinery

## Compabloc Opening after 1st step (8 months)

**Deflected Baffle (mechanical resistance)** – it was not designed for that purpose

**Crude Inlet**

**Vacuum Residue Inlet**

PETROBRAS

# Compabloc Test at Refinery

## Pressure drop results - phases 2, 3 and 4

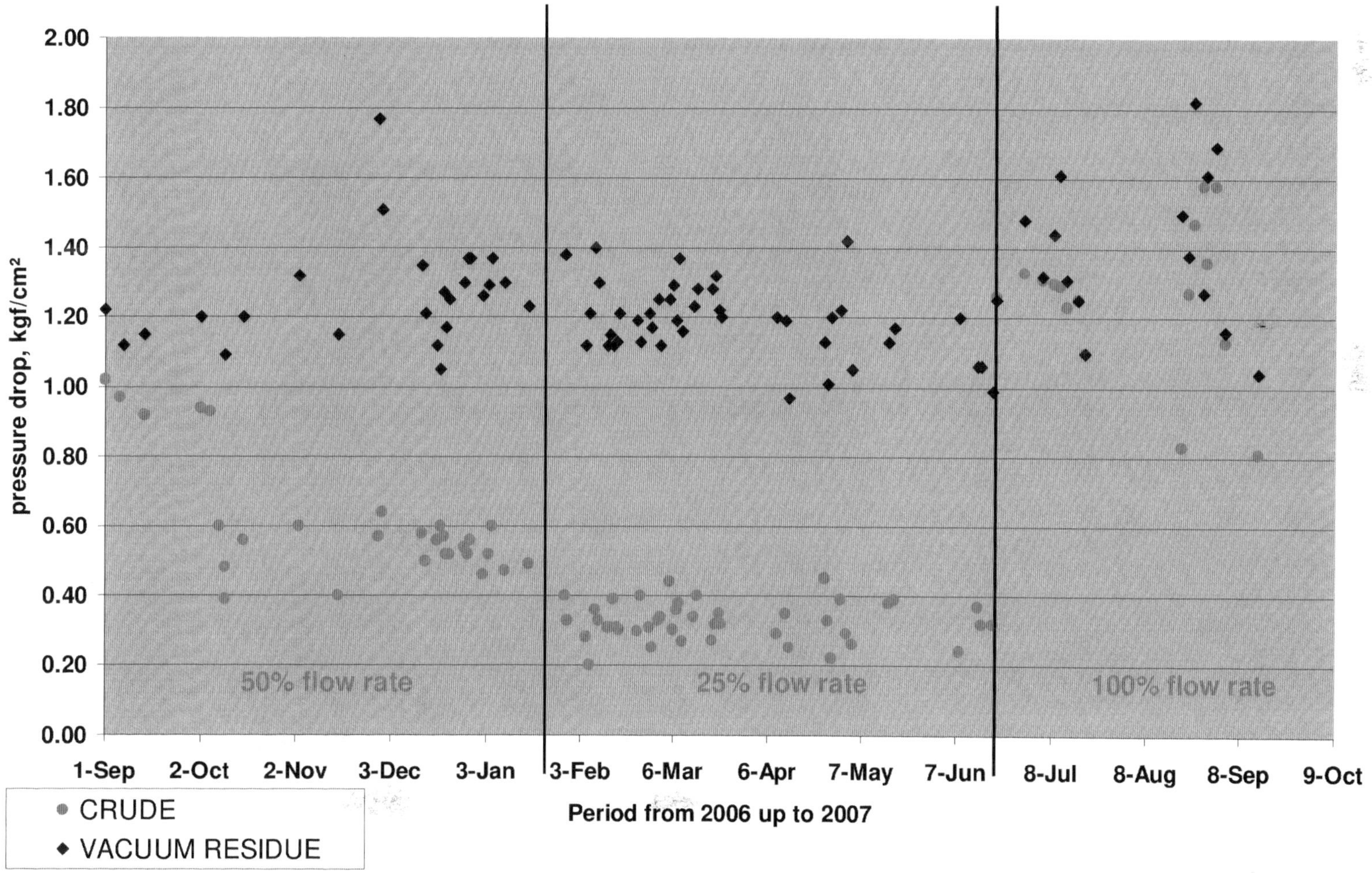

PETROBRAS

# Compabloc Test at Refinery

- Flow Rate Vacuum Residue vs Crude

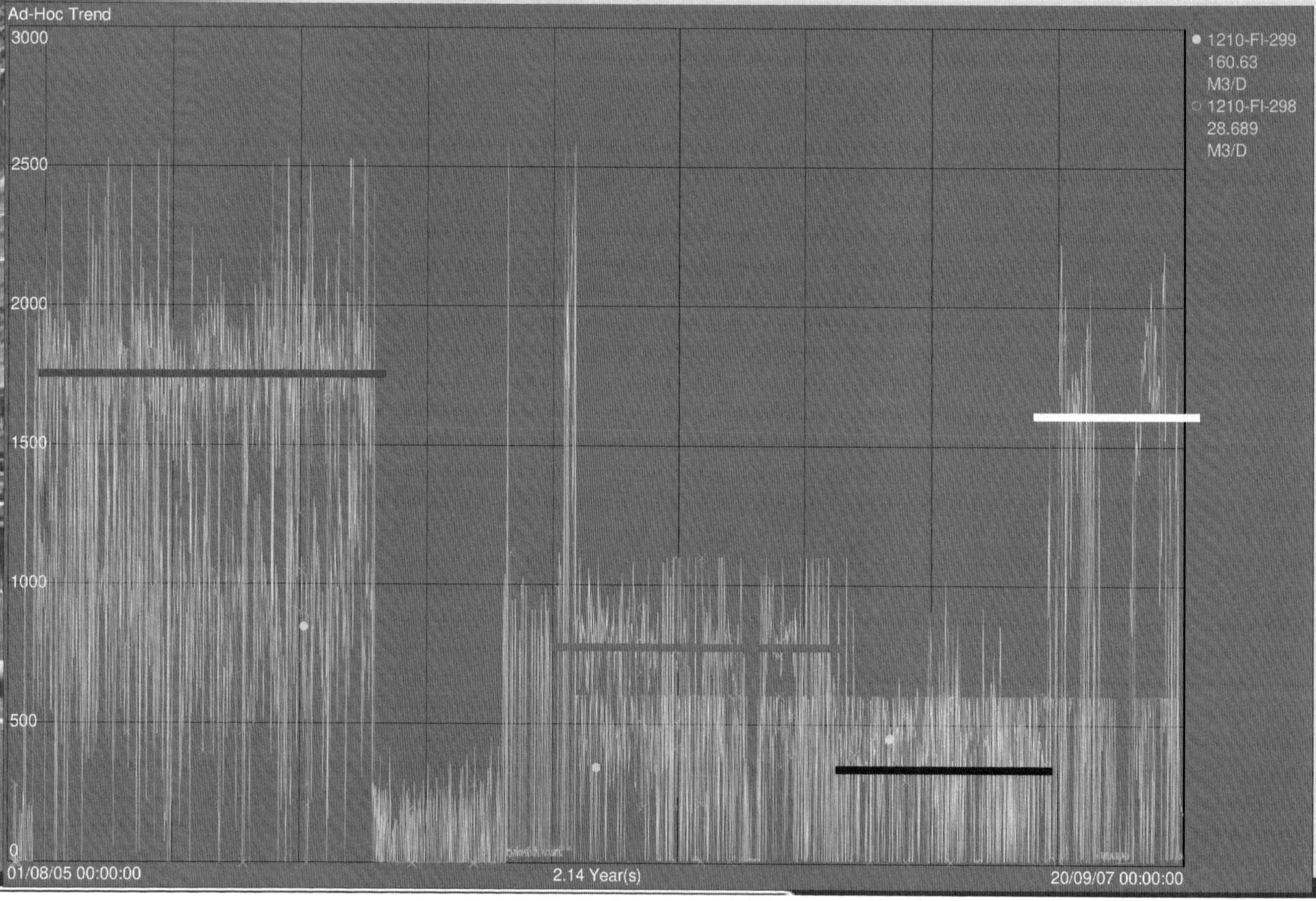

**PETROBRAS**

# Compabloc Test at Refinery

- Positive Aspects:
  - Equipment kept its thermal performance;
  - Design, Operational and Maintenance experience
  - No fouling observed

- Negative Aspects:
  - Mechanical Resistance (Leakage and Baffle);
  - Aspects not related to exchanger (instrumentation, flow fluctuation, …)

PETROBRAS

# Compabloc Test at Refinery

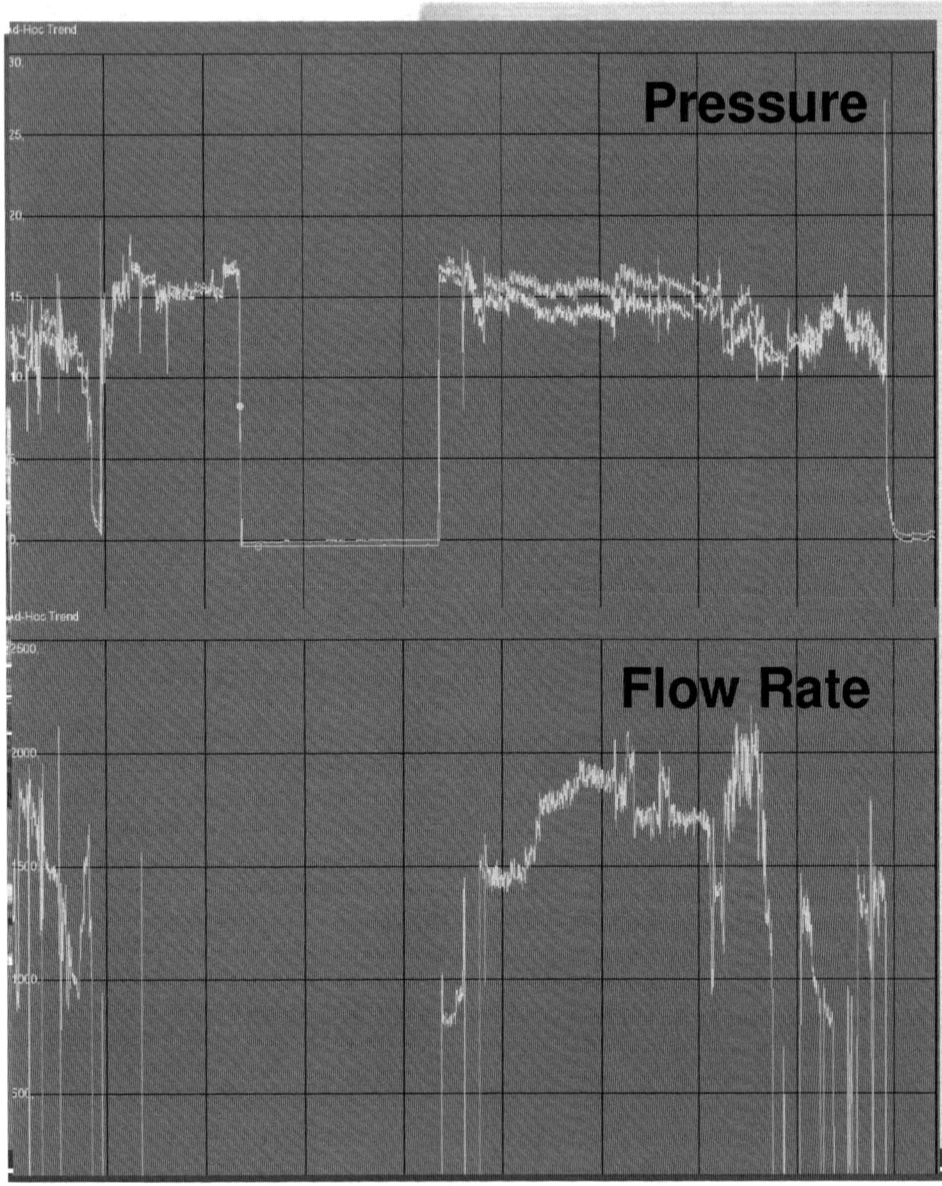


- Crude Side
  - Period: Jul/16/07 up to Sep/20/07
  - Zero Flow Rate for about 3 weeks.
  - No abnormal pressure variation.

**PETROBRAS**

# Compabloc Test at Refinery

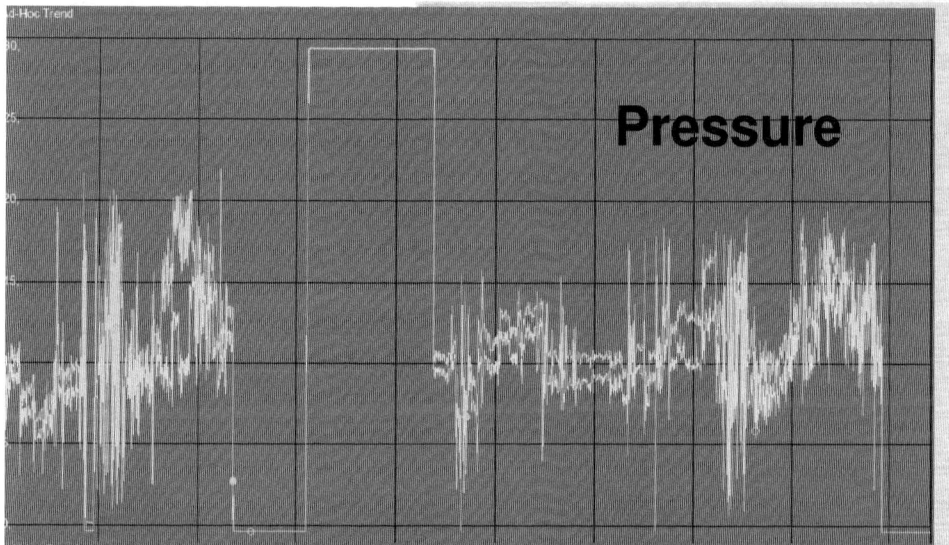


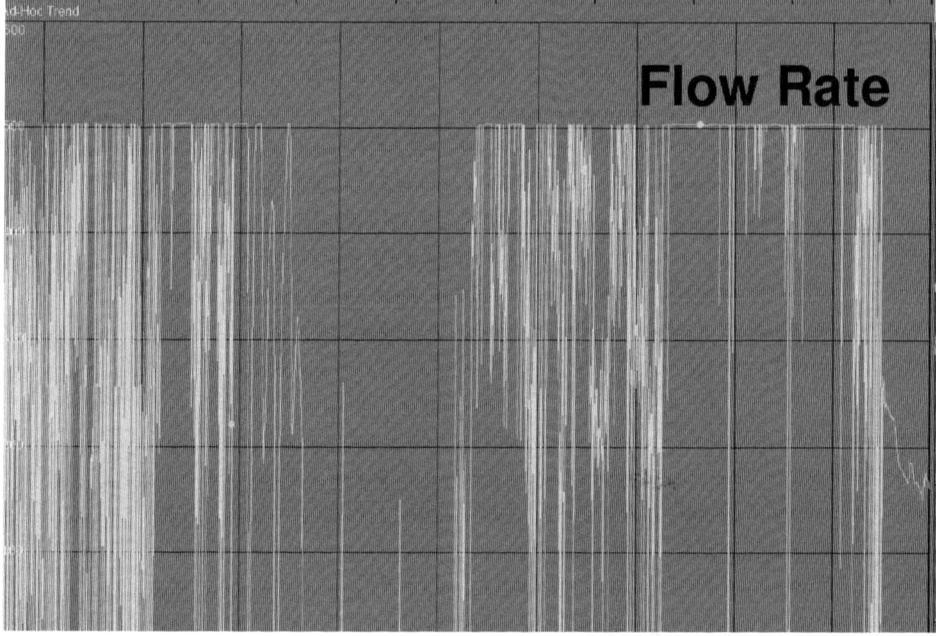


- Vacuum Residue Side
  - Period: 16/Jul/07 up to 20/Sep/07
  - Zero Flow Rate for 10 days.
  - Significative pressure and flow rate variation.

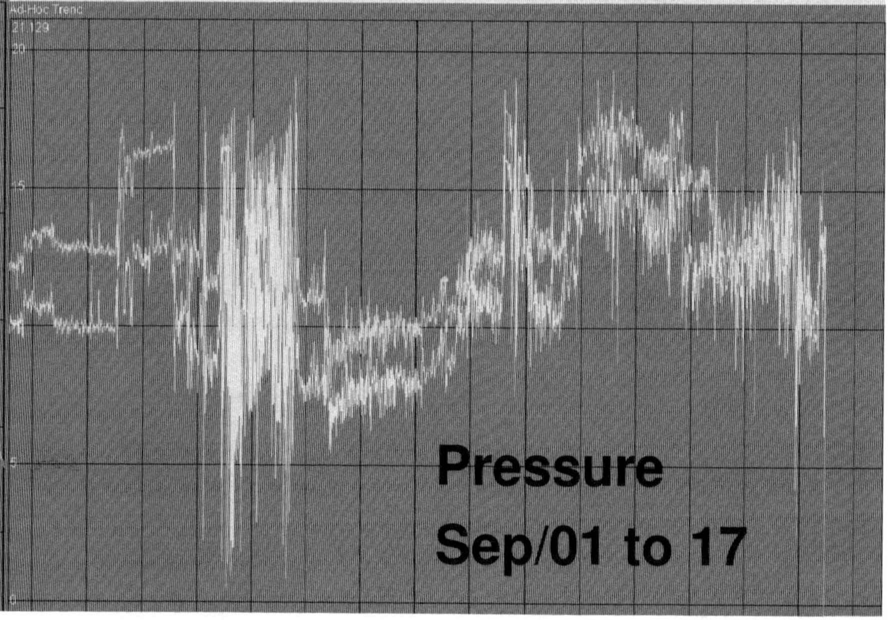

PETROBRAS

# Compabloc Test at Refinery

## Inlet Pressure VR vs Crude (Accordeon Effect)

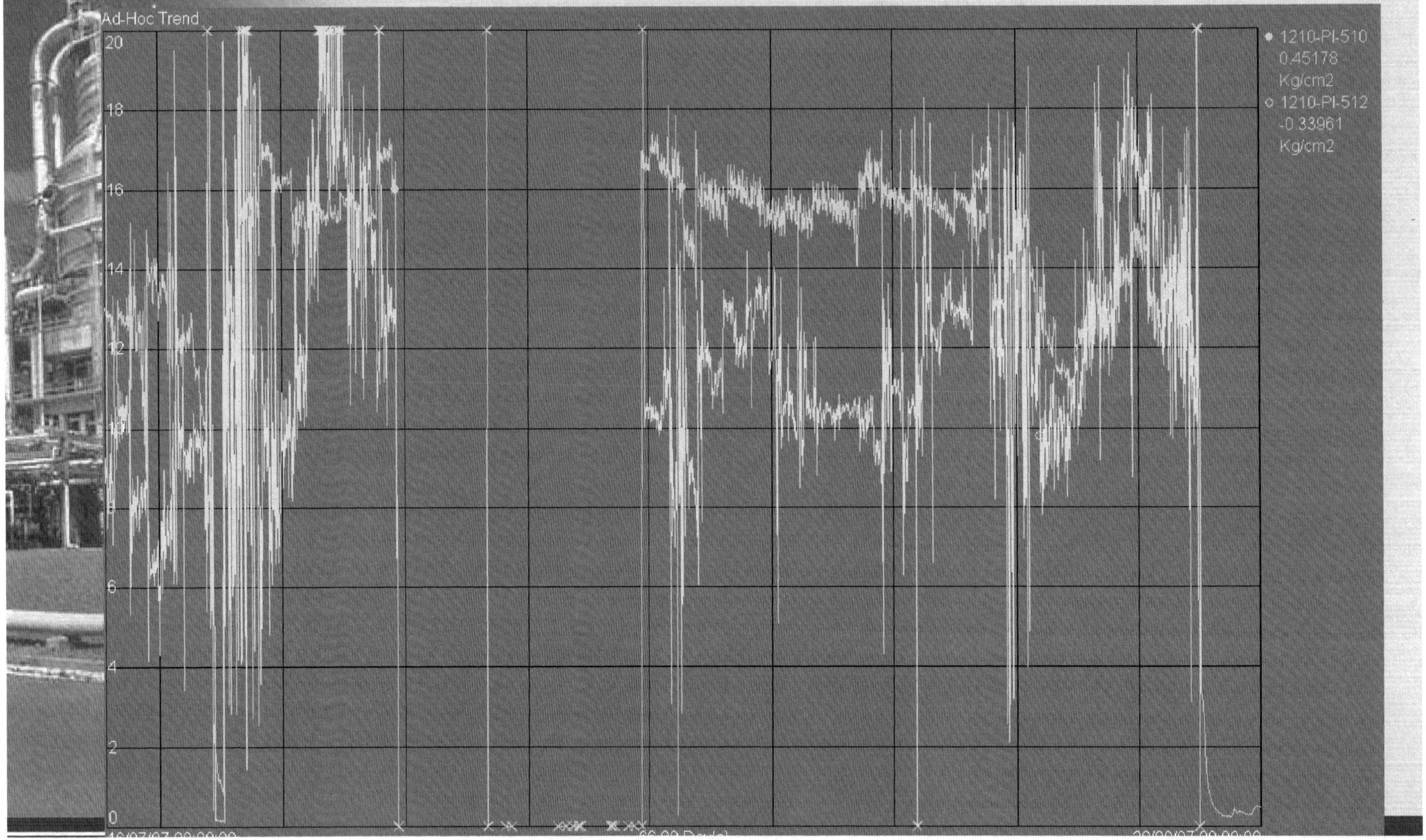

PETROBRAS

# Compabloc Test at Refinery

| Problem | Root Cause | Consequense |
|---|---|---|
| Leakage | Joint Failure | Vacuum Residue Leakage |
| Baffle Deflection | Overpressure or fatigue | Thermal performance loss |

- **ALFA LAVAL**:
  - Development of new gasket;
  - Adoption of reinforced baffles (for high viscous fluids);

- **PETROBRAS:**
  - Avoid surge pressure (hydraulic hammer):
    - Startup and shutdown procedures;
    - Slow valve action;
  - Drain the equipment, in case of several days of shutdown.
  - Panels bolting hot retightening after startup
  - Deposits prevention:
    - Strainers;

PETROBRAS

## PETROBRAS Cases

# CASE 3 Basic Engineering Design (Oct/07)

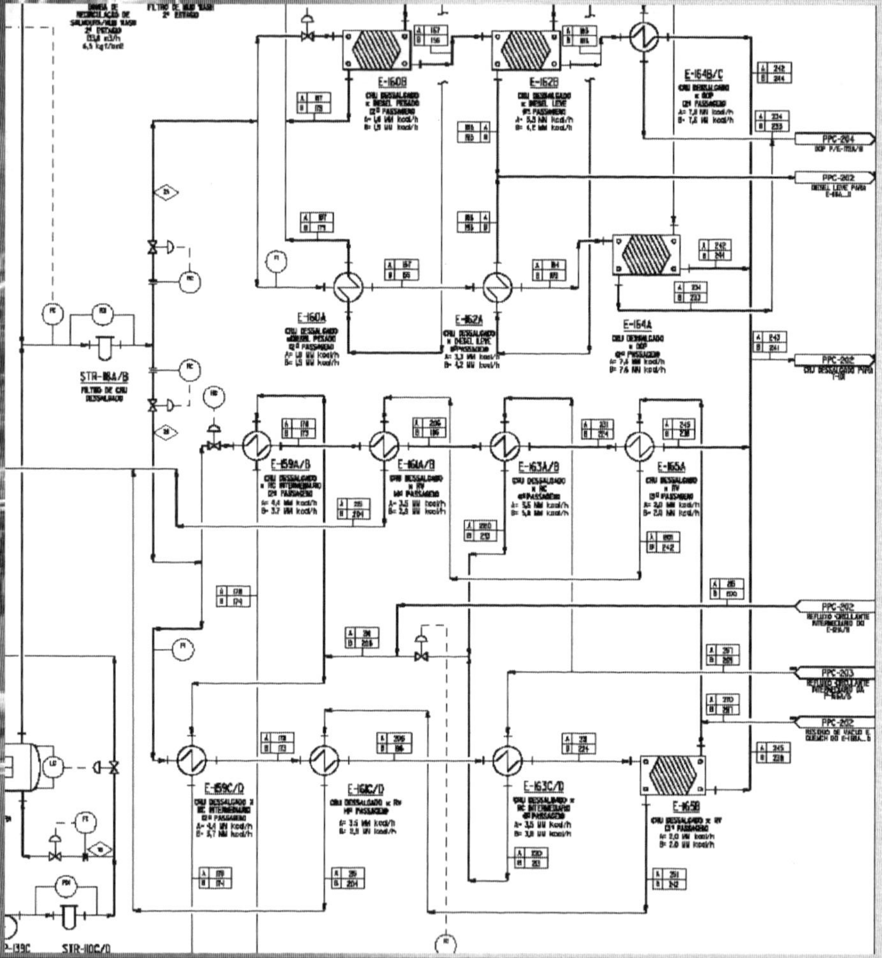


- Preheat Train before Desalter
  - Carbon Steel
  - Equivalent Cost
- Preheat Train between Desalter and PreFlash Tower
  - Crude Vaporization
  - Special Metallurgy
  - Parallel Arrangement

**PETROBRAS**

# PETROBRAS Cases

## CASE 3

## Basic Engineering Design (Oct/07)

### After PreFlash Tower Preheat Train
### (High Temperature – Special Metallurgy)

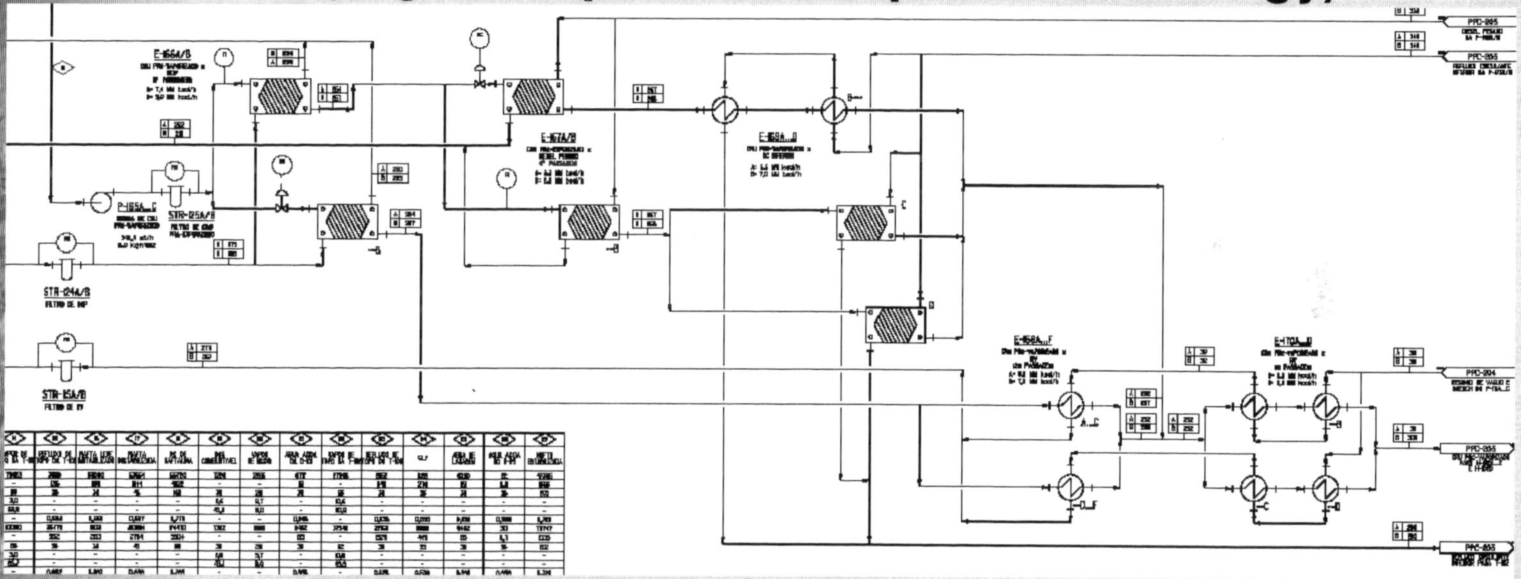


| SERVICE | Shear (Pa) | WALL TEMPERATURE (°C) |
|---|---|---|
| E-166 A/B (x HVGO) | 112 | 264 |
| E-167 A/B (x HD) | 113 | 289 |
| E-169 C/D (x BPA) | 113 | 295 |

# CASE 4

## Basic Engineering Design CDU – Grass root Refinery

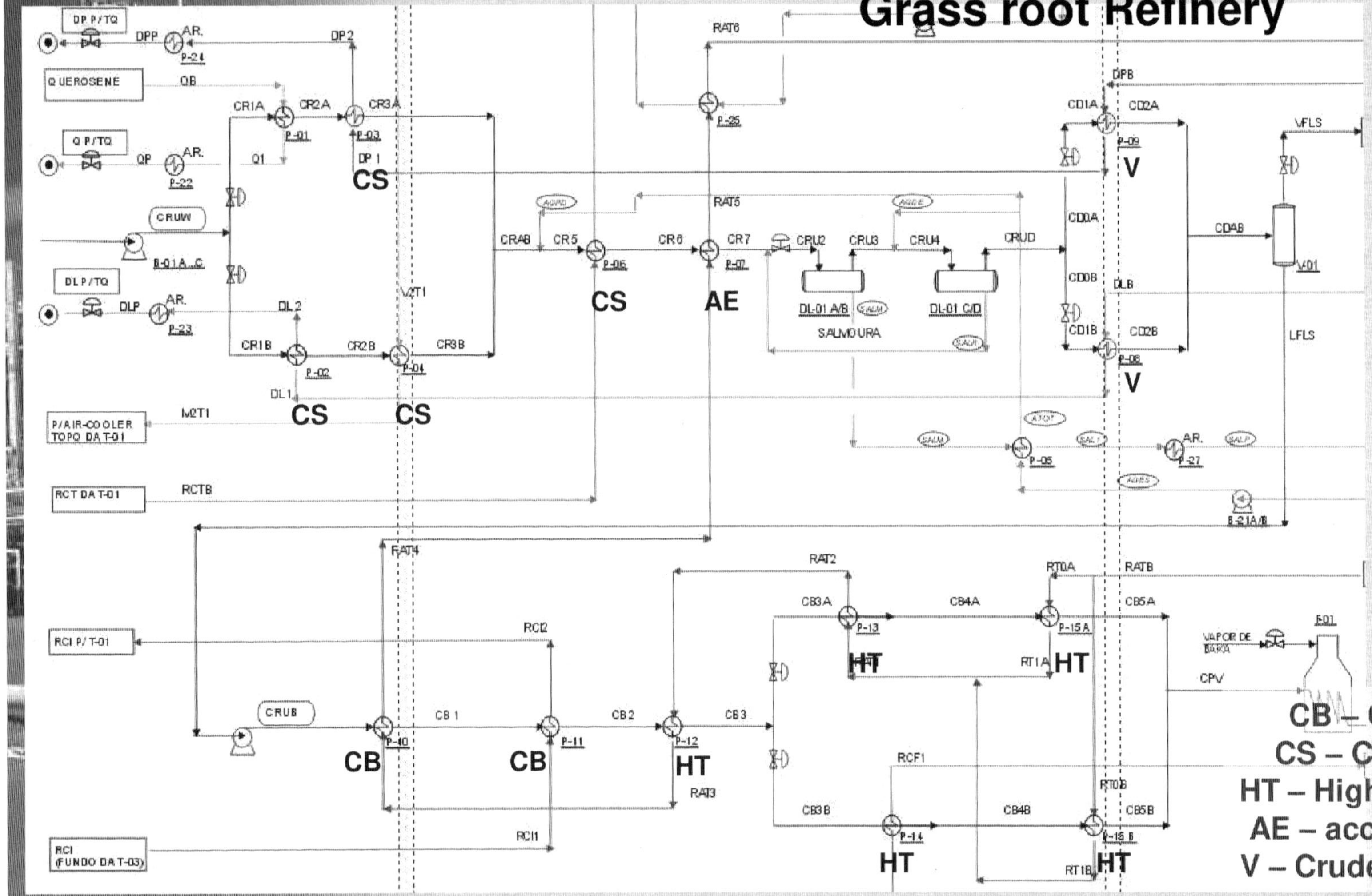


CB – Compabloc
CS – Carbon Steel
HT – High Temperature
AE – accordion Effect
V – Crude Vaporization

PETROBRAS

## PETROBRAS Current Case

# CASE 4

**Basic Engineering Design CDU – Grass root Refinery**

| COMPARISON OF EQUIPMENT COSTS | | |
|---|---|---|
| SERVICE | Compabloc (US$) | Shell and Tube (US$) |
| P-10 | x | ~ 5.6 x |
| P-11 | y | ~ 3.8 y |

- **Installation costs not included;**
- **Equipment designed with the same allowable pressure drop;**
- **Compabloc design and budget by Alfa Laval;**
- **Shell and Tube budget by Aspen Icarus (Brazil Base)**

# Compabloc Design

- Design Parameters:
  - Minimum Shear Stress
  - Higher Pressure drop
  - Countercurrent Pinch Analysis
  - Strainers / Maintenance facilities
  - Parallel Arrangement (flexibility)
  - Accordeon Effect
  - Crude Vaporization
  - High temperature

PETROBRAS

# Next Compabloc Tests

- Next Step at the same refinery:
  - Operational practice and fouling at high temperatures;
  - Service: Not defined yet;
  - Wall temperature above 270ºC (target 300ºC)
- Test at Case 2 refinery :
  - Operational practice and fouling with crude vaporization;
  - Service: Not defined yet (desalted crude vs heavy diesel)
  - Crude Temperature from 150ºC to 220ºC

PETROBRAS

# Thank you very much!

Antonio V. S. de Castro, antonio.castro@petrobras.com.br
Alan T. Vilas Boas
Claudio L. M. Kuboski
Diomedes C. da Silva
Roberto A. Oliveira
Washington de O. Geraldelli

# 165e Modeling Asphaltene Precipitation in Crude Oil Systems

**Walter G. Chapman** and George J. Hirasaki. Rice University, 6100 Main St., Houston, TX 77096

The insolubility of asphaltenes in light paraffin liquids and other incompatible fluids, such as carbon dioxide (CO2), is a source of problems during crude oil production operations. Asphaltene precipitation can cause formation damage and wellbore plugging, requiring expensive treatment and cleanup procedures. In the presence of light gases such methane, ethane, propane and nitrogen (N2), asphaltenes are usually more stable as pressure and temperature increase; however, experimental measurements indicate that in the presence of CO2, asphaltenes become more stable as temperature decreases.

In this work, asphaltene phase behavior in a live reservoir fluid and a dead oil from South America is investigated in a range of pressures and temperatures in presence the of CO2 using the Perturbed Chain-Statistical Associating Fluid Theory (PC-SAFT) equation of state (EOS). This thermodynamic model has been applied to asphaltene precipitation with different crude oil systems, such as those that involve methane, ethane or N2. Simulation results using the PC-SAFT EOS model are in good agreement with experimental measurements. A thermodynamic analysis using a model live oil confirms and explains a crossover behavior observed in the simulation of the precipitation onset with CO2. Simulations show that CO2 can act as an inhibitor or a promoter of asphaltene precipitation depending on the range of temperature, pressure and composition studied. At fixed pressure and live oil composition, CO2 addition increases the asphaltene stability below the crossover temperature, whereas above this point the asphaltene is less stable for increasing CO2 concentration.

## 165f Automated Heat Exchanger Network Design for Process Streams with Temperature-Dependent Thermal Properties

**Lu Chen**, Megan Jobson, and Robin Smith. Centre for Process Integration, School of Chemical Engineering and Analytical Science, The University of Manchester, Manchester, United Kingdom

In this work, a methodology has been developed for the automated design and synthesis of heat exchanger networks (HENs) when the thermal properties of streams vary significantly with temperature.

Rodriguez (2005) developed an optimisation approach to the design of complex HENs. In the approach, Simulated Annealing (SA) was employed as the optimisation algorithm. Both structural options, such as re-piping, re-sequencing of existing exchangers, and continuous options, such as stream- split fractions and exchanger duties, were considered without simplification of cost models. However, the thermal properties (i.e. heat capacity) of streams were assumed constant to maintain a linear problem representation. This assumption will introduce errors when applying the method to streams with a large temperature range, when the thermal properties are not constant, such as in crude oil refining.

The shortcomings are overcome in this work by representing streams using multiple segments, each with constant heat capacity. The optimisation approach of Rodriguez (2005) has been extended by including a non-linear solver that converts the infeasible heat exchanger network designs (generated by the SA algorithm) to feasible ones: the exchanger duties are reallocated within the whole network, where the structure of the HENs is fixed. When evaluating alternative HEN designs, re-piping and re-sequencing of existing exchangers are tracked and costed with reference to the exiting HEN structure. To facilitate the generation of reachable solutions, the number of modifications to an existing HEN can be constrained.

The approach is illustrated its application for the retrofit design and analysis of an existing preheat train for a crude oil distillation column.

Reference

Rodriguez, C. A., Fouling mitigation strategies for heat exchanger networks, PhD Thesis, The University of Manchester, UK (2005)

# Tutorial: Light Naphtha Isomerization

Veronica G. Deak, Richard R. Rosin, and Dana K. Sullivan
UOP LLC
Des Plaines, Illinois

## Abstract

The objectives of light naphtha isomerization have traditionally been to increase the octane of light naphtha streams and to reduce benzene content. These objectives are still relevant today as refiners search for new sources of octane to offset octane losses from MTBE phase-out or gasoline desulfurization and as refiners look for economical ways to comply with the MSAT II limits on gasoline benzene content. Light naphtha isomerization processes have been an effective solution for these objectives for more than forty years. New catalysts and new process technologies are making light naphtha isomerization technology more economical and more effective for meeting the challenges of today's refinery.

This tutorial will review the fundamental thermodynamics and process concepts for light naphtha isomerization. It will then review and compare the common commercial catalysts and process configurations. The tutorial will focus on the importance of light naphtha isomerization as part of a refiner's MSAT II compliance strategy and on the comparative economics of various processes for both new units and unit revamps.

## Introduction

For decades light naphtha isomerization has played an important role in refineries in increasing gasoline octane and managing benzene content. With the current limits on gasoline olefin content, benzene concentration, and endpoint, the paraffinic composition of isomerization product makes it a very useful gasoline blending component. Isomerization catalysts and processes continue to evolve to meet the new demands posed by MTBE phase-out, gasoline desulfurization, compliance with MSAT II regulations, the increase of ethanol in the gasoline pool, and the rapidly increasing cost of platinum.

This paper reviews chemistry, thermodynamics, and processing schemes. It also includes a comparison of the three major types of commercial catalysts, their process configurations, and economics for both new units and conversion of existing equipment. The use of light naphtha isomerization in meeting MSAT II gasoline benzene limits is also discussed.

## Thermodynamics and Reaction Network

The main objective of light naphtha isomerization processes is to increase the octane value of light naphtha by isomerizing the $C_5/C_6$ paraffins. The isomerization of normal pentane to iso-pentane results in a 31 RON increase and the isomerization of normal hexane to methyl-pentanes or dimethyl-butanes results in an average increase of 44 RON or 70 RON, respectively. The octane values of light naphtha paraffins are shown in the table below.

**Table1: Light Naphtha Blending Octane Values**

| *Component* | *RONC* | *Component* | *RONC* |
|---|---|---|---|
| i-pentane | 93.5 | Benzene | 120 |
| n-pentane | 61.7 | Cyclo-hexane | 84.0 |
| 2,2 di-methyl butane | 94.0 | Methyl-cyclo pentane | 96.0 |
| 2,3 di-methyl butane | 105.0 | | |
| 2-methyl pentane | 74.4 | | |
| 3-methyl pentane | 75.5 | | |
| n-hexane | 31.0 | | |

Unlike many reactions in refinery processes, paraffin isomerization reactions are equilibrium limited under most commercial process conditions. A common measure of how close a product or feed stream is to the equilibrium composition is the iso-ratio of a particular paraffin. The iso-ratio is the weight percentage of the particular paraffin divided by the total weight percentage of all isomers of that paraffin. The iso-ratio for $i$-$C_5$ is simply:

**$i$-$C_5$ iso-ratio = wt% $i$-$C_5$ / ( total wt% $C_5$ paraffins ) = wt% $i$-$C_5$ / (wt% $n$-$C_5$ + wt% $i$-$C_5$ )**

For 2,2 di-methylbutane (22DMB) the iso-ratio is:

**22DMB iso-ratio = wt% 22DMB / ( total wt% $C_6$ paraffins )**
**= wt% 22DMB / (wt% $n$-$C_6$ + wt% 2 MP + wt% 3 MP + wt% 23DMB + wt% 22DMB)**

Iso-ratios are stream properties similar to octane, they do not imply a reaction rate or a conversion. Since paraffin isomerization reactions are mildly exothermic, equilibrium conversion to the higher-RON, branched-paraffins is favored by lower reaction temperatures. The figure below shows the iso-ratios for $i$-$C_5$ and 22 DMB for an equilibrium mixture and effect of temperature on the iso-ratios.

**Figure 1: Equilibrium Iso-Ratios versus Temperature**

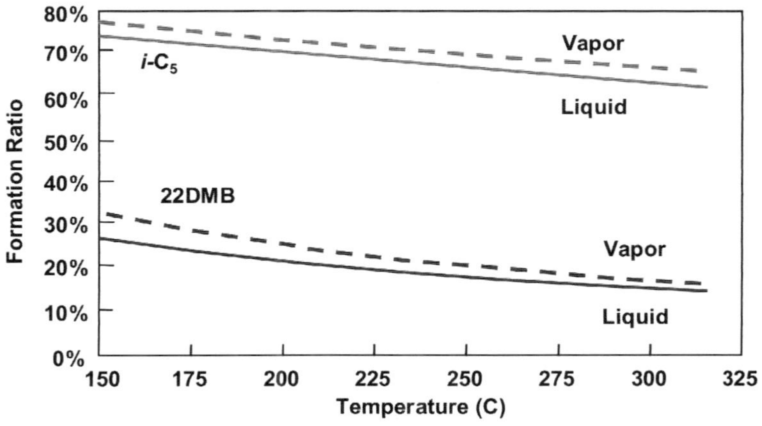


Equilibrium iso-ratios are also slightly influenced by the phase of the mixture, with the vapor phase having a slightly higher equilibrium iso-ratio. Applying the RON blending values to the various iso-ratios, the equilibrium octane for $C_5$ and $C_6$ paraffins is shown below

**Figure 2: Equilibrium RONC for $C_5$ and $C_6$ Paraffins**

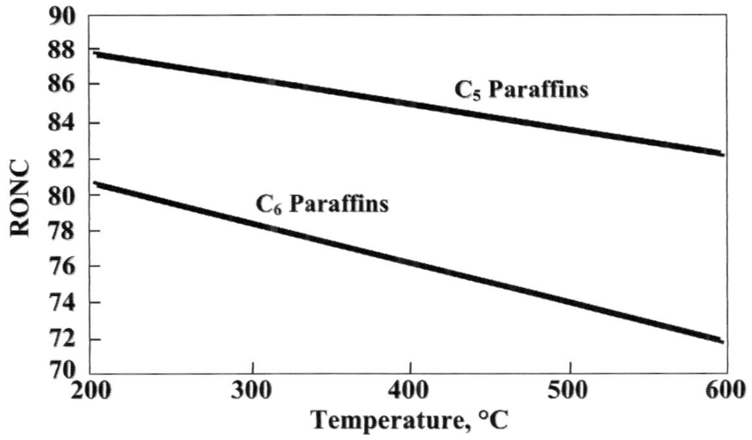


The exact composition of light naphtha can vary widely depending on a refiner's crude slate, processing strategy, and the design and operation of the naphtha splitter. Typically, light naphtha contains more than just $C_5$ and $C_6$ paraffins, it will contain $C_5$ and $C_6$ naphthenes, benzene, and $C_7$ paraffins. The reactions that take place in light naphtha isomerization units fall into three broad categories: paraffin isomerization, cyclic reactions, and cracking. The catalyst formulation of commercial light naphtha isomerization catalysts contains a metal function and a strong acid function. These are the catalytic functions that influence the reactions.

Paraffin isomerization requires both a metal function and acid function. The exact mechanisms are still debated but relative importance of the catalytic functions on reactions is well known. The reactions for $C_5$ and $C_6$ paraffin isomerization are shown below.

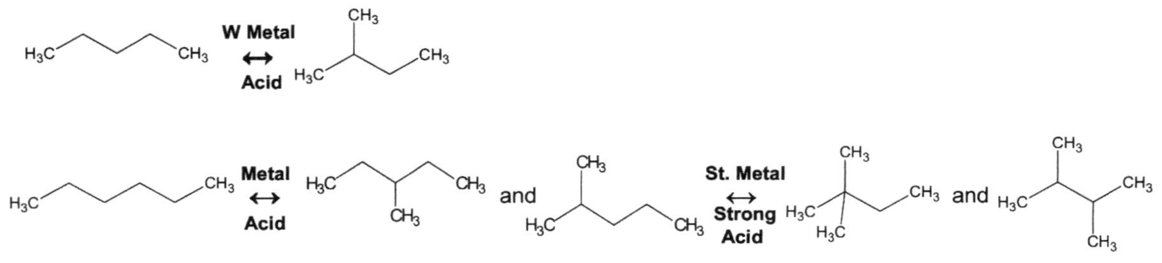


$C_5$ and $C_6$ paraffin isomerization requires both metal and acid catalytic functions. $C_5$ isomerization can take place in the presence of relatively weak metal function. $C_6$ paraffin isomerization requires strong metal function to isomerize *n*-hexane to the high octane di-methyl butanes. Isomerization of *n*-hexane to methyl pentanes does not require as strong of a metal function as isomerization to di-methyl butane.

Thermodynamically, hydrogen does not take part in paraffin isomerization reactions. However, kinetically, some partial pressure of hydrogen is required for all commercial catalysts. The exact mechanism is not clear but the rate determining step for paraffin isomerization over most commercial catalysts appears to be desorption of iso-paraffin carbo-cation. Hydrogen may have a role in the desorption of carbo-cations or in the reduction of coke-like deposits on the catalyst.

Increasingly, refiners are looking to light naphtha isomerization units to manage benzene in the naphtha stream. Light naphtha isomerization catalysts are effective benzene saturation catalysts because of the metal function required for paraffin isomerization. Benzene can significantly effect the performance of a light naphtha isomerization unit thermodynamically because of the very high heat of reaction and kinetically because the basic nature of benzene can interfere with the acid sites of the catalyst.
Cyclohexane formed from benzene saturation is readily isomerized to methyl-cyclopentane in the presence of acid function. In the presence of a strong metal function cyclohexane can ring open to *n*-hexane. In the presence of a weak metal function or low hydrogen partial pressure, cyclohexane and paraffins can alkylate to heavy naphthenes.

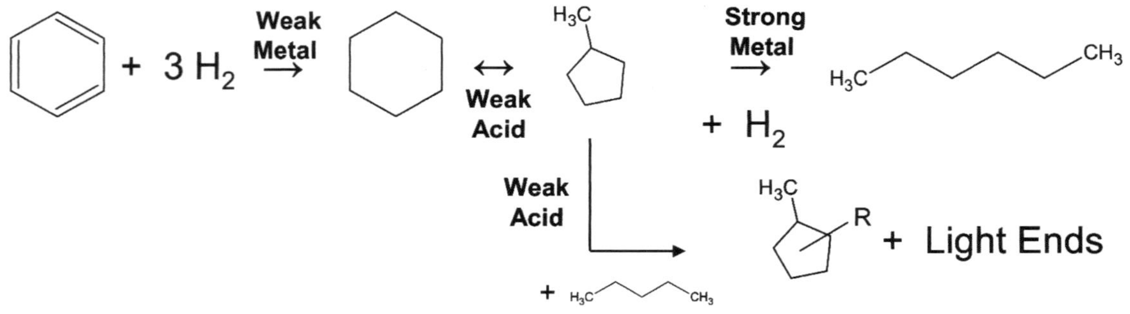

$C_6$ naphthenes do have high octanes, but naphthenes will attenuate the catalyst's ability to isomerize paraffins. For this reason, naphthene ring opening is a desired catalyst feature and heavy naphthene formation is undesirable.

Heavy paraffins, $C_7$'s, will appear in light naphtha to varying degrees depending on the naphtha splitter efficiency. $C_7$ paraffins should ideally be in the heavy naphtha and reformed into toluene and hydrogen. Some portion of $C_7$ paraffins in the light naphtha will crack. The presence of $C_7$ paraffins in light naphtha does not significantly influence the maximum isomerate octane, rather the $C_7$ paraffin cracking will influence the isomerate yield. Any light naphtha isomerization catalyst with sufficient acid strength to isomerize $C_5$ and $C_6$ paraffins will tend to acid crack $C_7$ paraffins to iso-butane and propane.

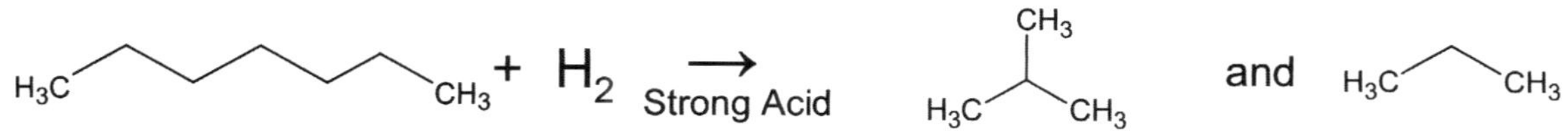


Typically light naphtha operation at maximum octane and maximum yield results in half of the $C_7$ paraffins cracked to light ends. Operations can be tuned to lower severity resulting in higher $C_7$+ yields but lower octanes will result.

Light naphtha isomerization units catalyze many reactions in addition to light paraffin isomerization reactions. It is the content of compounds other than $C_5/C_6$ paraffins in light naphtha that will affect the performance of the light naphtha isomerization unit. It is common to express the severity of a light naphtha feed in terms of the concentration of these compounds. A common measure is the X-factor of the feed.

**X-Factor = wt% Feed Benzene + wt% Feed $C_6$ Naphthenes + wt% Feed $C_7$+ compounds**

## Light Naphtha Isomerization Catalysts

There are three different commercial catalyst types for light naphtha isomerization: zeolitic catalyst, amorphous chlorided alumina catalyst, and sulfated zirconia catalyst. These catalyst types differ by the chemistry of the acid function. UOP is the only vendor who develops and manufactures all three types of catalysts to address the widest range of processing needs. Paraffin isomerization reactions require a combination of Bronsted and Lewis acid sites which promote varying levels of protonation activity needed for isomerization reactions to proceed. Light naphtha isomerization units carry out many different reactions, many of which require a metal function, as described above. All of the commercial isomerization catalysts include platinum in the formulation to create metal function. Platinum is preferred for its sulfur tolerance. With platinum, sulfur is not a permanent poison since it can be stripped.

As discussed above, isomerization of paraffins in $C_5/C_6$ streams is an equilibrium limited reaction where branched paraffin isomers are generally favored by low temperatures. The most active catalysts, capable of operating at the lowest temperatures, will produce the highest octane products. Only chlorided alumina type catalysts have enough acid activity for a commercially viable $C_4$ isomerization process.

To assist in aromatics saturation, there are a variety of elements that can be used. UOP offers platinum based benzene saturation catalysts not only because of the superior sulfur resistance of platinum but also because platinum based benzene saturation catalysts can operate at significantly higher temperatures than, for example, nickel catalysts. The use of platinum results in a more robust benzene saturation process with higher yields and smaller recycle streams.

***Zeolitic Catalysts***

Zeolitic catalysts have the lowest activity of the isomerization catalysts, so must operate at the highest temperatures. High temperature operation results in lower product octane due to less favorable equilibrium for branched paraffins. In addition to operating at high temperatures, zeolitic catalysts are the least selective isomerization catalysts, giving lower product yields. UOP HS-10™ catalyst is a platinum impregnated zeolite. The main benefit of zeolitic catalysts is that they are not permanently deactivated by water or other oxygenates and are fully regenerable. Long catalysts lives have been commercially realized for this type of catalyst, typically exceeding 10 to 15 years. Feed hydrotreating is not an absolute requirement, but it is recommended for optimum catalyst performance. Zeolitic catalysts do not require a halide promoter, so there is no chloride injection or caustic scrubber.

***Chlorided Alumina Catalysts***

Amorphous chlorided alumina catalysts, such as UOP's I-82™ catalyst, I-84 catalyst, and I-122 catalyst are the most active light naphtha isomerization catalysts currently available. Their high activity means that they operate at the lowest isomerization temperatures and thus achieve the highest octane products and highest yields. However, these catalysts are permanently deactivated by oxygenate compounds, like water and CO, so feed and makeup gas driers are needed. Well maintained driers can facilitate a long catalyst life, in excess of 10 years. Process units designed for chlorided alumina catalysts also require continuous chloride injection to maintain high activity. A caustic scrubber is needed to neutralize resulting HCl in the off-gas. These catalysts are not regenerable.

Platinum helps to saturate aromatics and ring-open naphthenes, so high platinum containing catalysts are needed only for feeds that have more aromatics and naphthenes, higher X-factor feeds, or for protection from sulfur upsets. In 2006, UOP introduced I-84 and I-122 catalysts as lower platinum alternatives for low severity or low X-factor isomerization feeds for light naphtha isomerization.

Chlorided alumina catalysts are irreversibly deactivated when acid sites are converted by oxygenates. Catalysts with the highest acid site densities will have the longest catalyst life. UOP favors chlorided alumina catalysts with higher support densities to achieve higher acid site density for highest initial activity and longer life.

***Sulfated Zirconia Catalysts***

Sulfated zirconia catalysts can vary, but typically have activities between zeolitic and chlorided alumina. UOP's first generation of sulfated zirconia catalyst, LPI-100™ catalyst, was commercialized in 1996. In 2003 UOP introduced its second generation, PI-242™ catalyst, which approaches the activity of chlorided alumina catalysts. Sulfated zirconia catalysts are tolerant of moderate levels of oxygenates in the feed and in the makeup gas, like water and CO, eliminating the need for driers. These catalysts do not require a halide promoter, so there is no chloride injection or caustic scrubber. Sulfated zirconia catalysts are fully regenerable using a simple oxidation procedure comparable to the procedure used for zeolitic based isomerization catalysts. These properties make PI-242 catalyst an ideal candidate for the conversion of existing fixed-bed reforming units or for increasing the capacity or performance of older zeolite based units. This catalyst has also been chosen for new unit designs where the full performance benefit of the chlorided alumina based catalysts is not required or where there is a concern about feed contaminant levels.

***Benzene Saturation Catalysts***

For saturation of benzene in light naphthas, the most commonly used catalysts are either nickel or platinum based catalysts. UOP recommends platinum based catalysts. Platinum-based catalysts are not permanently poisoned by sulfur upsets and will saturate benzene with very little cracking of the naphtha to light ends, consuming minimal hydrogen. Over 25 years ago UOP commercialized H-8™ catalyst, an alumina supported platinum catalyst. This catalyst is used in UOP's BenSat™ process. Because of the increasing price of platinum and the unique demands of motor fuel benzene saturation, UOP introduced H-18 catalyst in 2007, an alumina supported catalyst with significantly less platinum than H-8, which also allows smaller catalyst volume.

## Process Flow schemes

There are a variety of flow schemes used commercially for light naphtha isomerization. The choice of which to use depends on the feed composition, desired product octane, and available capital. Hydrocarbon-once-through flow schemes produce isomerate with research octane numbers ranging from 78 to 85 depending on the catalyst system and feed composition. Hydrocarbon-recycle flow schemes deliver increased octane but lower yield by recycling lower octane components back to the reactor section inlet using fractionation or molecular sieve adsorption. Net product research octane numbers range from 87 to 93 depending on flow scheme, catalyst system, and feed composition.

A process is offered by UOP for each of the three catalyst types, as shown in Table 2. For once-through hydrocarbon flow schemes the chlorided alumina Penex™ process has the highest octane product and highest product yields. The Penex unit is the most widely used process in commercial operation today. The Par-Isom™ process utilizes PI-242 catalyst. This sulfated zirconia catalyst is more resistant to contaminants than chlorided alumina but achieves a somewhat lower product octane and yield. Once-through hydrocarbon units using a zeolitic material such as UOP's HS-10 catalyst have the lowest product octane and yield of the three types.

Fractionation is the most popular method to recycle low octane components, especially deisohexanizer (DIH) and deisopentanizer (DIP) columns. Adsorptive separation units such as the liquid phase UOP Molex™ process or vapor phase UOP IsoSiv™ process are also used to recycle hydrocarbons to the reactor. It should be noted that flow schemes which recycle normal pentane will yield a higher vapor pressure net product because of the increased isopentane content of the isomerate product.

More detailed discussions of the various flow schemes and their advantages can be found in previous publications.[1,2]

**Table 2: Unit Comparison**

| *Catalyst Type* | *Chlorided Alumina* | *Sulfated Zirconia* | *Zeolitic* |
|---|---|---|---|
| UOP Process | Penex | Par-Isom | |
| UOP Catalyst | I-82, I-84, I-122 | PI-242 | HS-10 |
| Product RON | 82 - 85 | 80 - 83 | 78 - 80 |
| Recycle Compressor | No | Yes | Yes |
| Charge Heater Type | Steam Exchanger | Steam Exchanger | Fired Heater |
| Feed and Makeup Gas Driers | Yes | No | No |
| Neutralization Facilities | Yes | No | No |

## Economics

The performance and economics of typical Penex, Par-Isom, Penex with a DIH column (Penex-DIH), Par-Isom with a DIH column (Par-Isom-DIH), and Once-Through Zeolitic units are compared in Table 3. The values are for a 10,000 BPD unit with a 10X feed, which is considered light. Studies with high X, heavier feeds have shown the same relative results.

Chlorided alumina units have the best performance but also the highest Estimated Erected Cost (EEC). In spite of the higher cost Penex units typically have the highest return on investment and profitability of the three types of process units. A Par-Isom unit provides somewhat lower performance but also has a lower EEC which is only about 75% that of a Penex unit. The Par-Isom-DIH unit has nearly the same octane product as a Penex-DIH because the lower activity of the sulfated zirconia catalyst can be compensated for with a larger recycle of low octane hydrocarbons from the column. The lower product yield with the Par-Isom-DIH flow scheme compared to the Penex-DIH is due to the cracking activity of the catalyst and is magnified by the larger hydrocarbon recycle stream.

Zeolitic catalyst units are rarely chosen for new isomerization projects because of their relatively poor product octane and yield and high operating and new unit construction costs.

A common way to meet processing needs and reduce EEC is to convert another type of unit to a Light Naphtha Isomerization unit. The most popular conversion is of a fixed bed reforming (FBR) unit, and EEC values for this type of conversion are included in Table 3. The reactors, recycle compressor, stabilizer, separator, and most heat exchangers can be employed in the new process flow scheme. Conversion of a FBR to a Par-Isom unit requires only minor piping and control system changes. Conversion to a Penex unit would require the addition of feed and makeup gas driers as well as a net gas neutralization section.

**Table 3: Isomerization Unit Economics for 10,000 BPD unit with a 10X (Light) Feed**

| Process Unit | Catalyst | C5+ Product RON | Product Yield RON*BBLS | Estimated Erected Cost $MM | |
|---|---|---|---|---|---|
| | | | | New Unit | Converted FBR |
| Penex | I-82 | 84.6 | 828,000 | 13.5 | 6.3 |
| Par-Isom | PI-242 | 82.9 | 806,000 | 10.0 | 0.32 |
| Penex-DIH | I-82 | 87.1 | 865,000 | 19.1 | 12.4 |
| Par-Isom-DIH | PI-242 | 86.7 | 834,000 | 14.7 | 5.4 |
| Zeolitic | HS-10 | 79.6 | 783,000 | | 0.32 |

## Light Naphtha Isomerization and Benzene Management

The U.S. Environmental Protection Agency MSAT II (Mobile Sources Air Toxics Phase 2) regulations will require decreasing the annual average U.S. gasoline pool benzene content to 0.62 vol% as of January 1, 2011 for most refiners. This is considerably lower than the current limit of 1.0 vol% for Reformulated Gasoline (RFG). Light naphtha isomerization is expected to play a key role in gasoline benzene management to meet the new specifications.

Many refiners are already routing benzene and benzene precursors to a light naphtha isomerization unit via the overhead of the their naphtha splitter columns to avoid sending them to the reforming unit. While this approach can reduce gasoline benzene to less than 1.0 vol%, in many cases it will be insufficient to meet the new 0.62 vol% benzene specification. To further reduce benzene many refiners will need to consider adding a reformate splitter column and sending the benzene containing light reformate for further processing. An existing light naphtha isomerization unit may well be able to handle the light reformate as an additional feed stream. This approach will not only reduce the benzene but will also help recover the octane lost from saturating the benzene. If the new isomerization unit feed contains more than about 5 - 7% benzene, the exotherm across the isomerization reactor section may be too high and a benzene saturation reactor may be required upstream. For the Penex process this combination is known as the PenexPlus™ process, which can handle as much as 30% benzene in the feed. The PenexPlus process can help refiners optimize their platinum inventory. The presence of aromatics will inhibit isomerization reactions, so feedstocks that have had the benzene saturated upstream of the isomerization reactors will need less platinum on the isomerization catalyst. A benzene saturation reactor can also be installed upstream of a Par-Isom unit. Sending a high benzene feed to a zeolitic catalyst unit, even with a preceding benzene saturation reactor, is very problematic due to this catalyst's extremely poor performance with very high X-factor feeds.

For those refiners long in octane and with no existing light naphtha isomerization unit, a stand-alone benzene saturation unit such as the UOP BenSat process can be employed. High levels of benzene in the feed (up to 30%) can be handled with virtually complete benzene saturation if needed.

## Conclusions

Light naphtha isomerization has played an important part in gasoline production for many years, and it will continue to be a key process in helping refiners meet new gasoline formulation challenges, particularly reducing benzene content while maintaining octane. New developments in catalysts and processes will help to maintain its role in producing a high octane, non-aromatic, low sulfur gasoline blending stream.

The paraffin isomerization reactions are unusual among refinery processes in being equilibrium limited, with the equilibrium favored by lower temperatures. In addition the catalysts are very effective at saturating aromatics including benzene, making the process an important option in meeting the upcoming MSAT II regulations.

The three types of catalysts and many different flow schemes give great flexibility in providing a process which can meet a refiner's need for producing octane-barrels and reducing benzene while fitting within capital and operating cost targets.

## References


1. Hunter, M.J. (2003), "Light Naphtha Isomerisation to Meet 21$^{st}$ Century Gasoline Specifications", OIL GAS European Magazine, February

2. Kuchar, P.J. and others (1993), "Paraffin Isomerization Innovations", Fuels Processing Technology, 35, 183-200

3. J. Meister, T. Crowe, D. Banks, W. Keesom and M. Stine, "Gasoline Production in the Post Energy Bill, Benzene Restricted World," 2005 NPRA Annual Meeting, AM-06-39.

4. G. C. Anderson, R. R. Rosin, M. A. Stine, and M. J. Hunter, "New Solutions for Light Paraffin Isomerization", 2004 NPRA Annual Meeting, AM-04-46.

5. V. M. Godfrey, R. R. Rosin, J. G. Vassilakis, "Advances in Light Naphtha Catalysts, Optimizing Precious Metal Investment", Hydrocarbon Engineering, March 2007.

6. Richard R. Rosin, Chris J. Anderle, Gary C. Sturtevant, "Optimizing Your Naphtha Complex in a Low Benzene World", 2007 NPRA Annual Meeting, AM-07-46.

# OPTIONS FOR GASOLINE POOL BENZENE MANAGEMENT

David J. Shecterle

UOP LLC, A Honeywell Company
Des Plaines, Illinois, U.S.A.

## INTRODUCTION

Once again refiners must address benzene in gasoline. As a result of new regulations enacted by the EPA, all U.S. gasoline will be subject to restrictions on benzene at levels far lower than those now applicable only to reformulated gasoline (RFG). Benzene reduction solutions used for RFG production may not be sufficient to meet the new regulations. Refiners will need to examine solutions in the context of the entire refinery. Meeting these demands with maximum profitability will require optimization across multiple units of a refiner's naphtha complex.

New regulations enacted by the EPA under Mobile Sources Air Toxics Phase 2 (MSAT2) will require reduction of benzene in the U.S. gasoline pool to 0.62 vol-% by 2011. Through an Averaging, Banking and Trading or ABT program refiners can use credits to comply with the 0.62 vol% benzene standard. However, there is a "maximum average standard for refineries of 1.3 vol% beginning on July 1, 2012, which acts as an upper limit on benzene content when credits are used to meet the 0.62 vol% standard."[1] These new regulations will require most refiners to address benzene issues because their current methods of controlling gasoline benzene content may not be adequate to meet the new lower benzene limit.

The main source of benzene in gasoline is reformate and most benzene reduction solutions in use today focus on preventing benzene formation in the reforming unit by removing benzene precursors from the reforming unit feed. The most common solutions for benzene management of reformate are either separation for petrochemical production or chemical conversion followed by return to the gasoline pool.

The new lower benzene limits applicable to all gasoline require re-examining benzene reduction solutions. Will benzene precursor removal be sufficient to meet the targets or will refiners have to remove benzene from reformate? For most refiners benzene reduction will be a requirement. The challenge will be to implement a solution that provides the best value, which will often be the low cost solution.

Benzene is currently being managed to meet RFG limits and to help manage toxic emissions in the gasoline pool. Reviewing where benzene comes from within the refinery is an important step in determining the best solution for further benzene reduction. Typically, reformate is the largest process stream contributing benzene to the gasoline pool. As shown in Table 1, reformate can make up about 75 to 80 vol-% of the gasoline pool benzene.

Table 1 shows contributions of benzene to a typical FCC based refiner's gasoline pool from various processing streams. The main source of benzene in gasoline is reformate.

**Table 1 - Typical Sources of Benzene Contributions to Gasoline**

| ***Process Stream*** | ***Range of Benzene Concentrations, vol-%*** | ***Influencers*** | ***Benzene Contribution to the Pool, %*** |
|---|---|---|---|
| Reformate | 0.2-8 | Feed PONA, Feed IBP, pressure | **75 – 80** |
| FCC naphtha | 0.5 to 1.3 | Mode of operation and crude source | 10 – 15 |
| Light Straight Run Naphtha $C_5$-220ºF | 0.8-8 | Crude source | 2 – 5 |
| VGO Hydrocracked to $C_5$+-380ºF Naphtha | 1 to 5 | Conversion of VGO to naphtha, feedstock | 1 – 5 |

In order to meet the MSAT2 limits, refiners must concentrate on the benzene content of the reformate. Refiners with an FCC unit will need to achieve very low benzene contents in their reformate to compensate for the relatively high benzene content of the FCC naphtha. Reducing reformate benzene content can be accomplished by either reducing benzene precursors in the reformer feed via naphtha prefractionation or by post-fractionation to remove benzene from the reformate. Both of these approaches and permutations there of, are applied in the industry today. Future benzene solutions that maximize complex profitability will require solutions that consider not only fractionation but further processing of the reformate for benzene removal and octane upgrade. This paper will concentrate on light naphtha and light reformate processing schemes that can be used to reduce gasoline pool benzene content and recover octane lost due to benzene removal.

## BENZENE MANAGEMENT OPTIONS

Prefractionation is a powerful method for significantly lowering reformate benzene levels. Depending on the design of existing naphtha splitters, refinery gasoline benzene levels may be reduced by as much as 50% or more by changing naphtha splitter specifications. Many existing naphtha splitters were designed as dehexanizer columns or deisohexanizer columns to remove $C_6$ paraffins or isoparaffins from reforming unit feedstocks. The light naphtha overhead product from the naphtha splitter was often sent to light naphtha isomerization units. $C_6$ naphthenes, cyclohexane and methylcyclopentane, were concentrated in the naphtha splitter bottoms and sent to the reforming unit, where they were converted to benzene to maximize reformate octane and hydrogen production. Naphtha splitters built in the 1990s were often designed to send a portion of the $C_6$ naphthenes overhead to reduce reformate benzene content allowing refiners to meet RFG specifications. The new MSAT2 benzene regulations will require tighter product specs for the naphtha splitter columns to minimize the reforming unit feed $C_6$ naphthene content. It is also desirable to maximize $C_7$ content of the of the naphtha splitter column bottoms stream to maximize toluene and hydrogen production in the downstream reforming unit. More trays and more heat input will be needed for naphtha splitter columns to meet the new requirements.

Removing $C_6$ naphthenes and benzene from reforming unit feed stocks will reduce the overall benzene content of the reformate but not eliminate benzene in reformate as benzene is formed in reforming units by dealkylation of heavier aromatics. Dealkylation reactions increase with increasing reforming unit operating pressure. More heavy aromatics will be dealkylated to benzene in higher pressure reforming units which will offset the benefits achieved from reducing the $C_6$ naphthene content of the reforming unit feed.

Prefractionation results in a higher concentration of benzene in the light naphtha. Depending on which other streams are blended into the gasoline pool the light naphtha may be either blended directly into the gasoline pool or processed for octane enhancement or benzene removal.

Figure 1 shows the possible naphtha processing options with prefractionation.

- Blending: Fractionating $C_6$ naphthenes into the light naphtha fraction and blending the light naphtha directly into the gasoline pool is the least expensive way to reduce overall pool benzene levels.
- Saturation: If the benzene contribution from the light naphtha is too high, the benzene in the light naphtha can be saturated before blending into the gasoline pool. There will be a reduction of pool octane due to the benzene saturation.
- Isomerization: Light naphtha isomerization can be used to saturate benzene and convert low octane normal paraffins to higher octane isoparaffins.

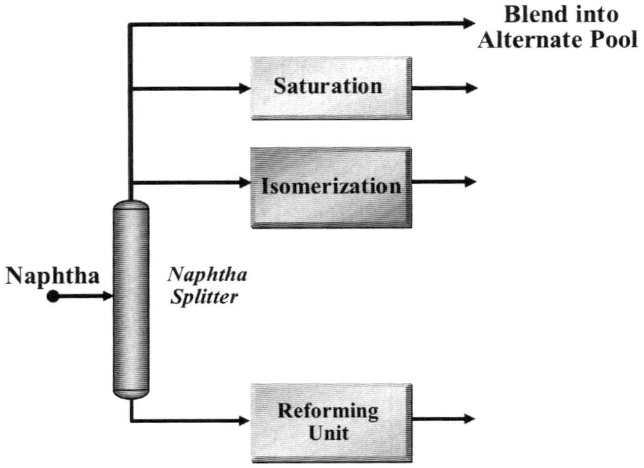


**Figure 1 Prefractionation for Benzene Reduction**

Fractionating benzene out of reformate, referred to in this paper as post-fractionation, gives a refiner a higher level of control over gasoline benzene content. In post fractionation schemes, a reformate splitter is used to separate a benzene rich $C_6$ fraction from the heavy reformate. Size and energy requirements of reformate splitter columns vary depending on the downstream processing of the light reformate stream. The heavy reformate stream from the Reformate Splitter bottom is typically blended directly into the gasoline pool. If the light reformate stream is being sent to an isomerization unit, it is desirable to limit its $C_7$ paraffin content to reduce yield losses due to cracking. Reformate splitter columns typically contain 40 to 60 trays depending on the overhead benzene recovery and overhead $C_7$ paraffin content. Figure 2 illustrates processing options to manage the benzene content of light reformate.

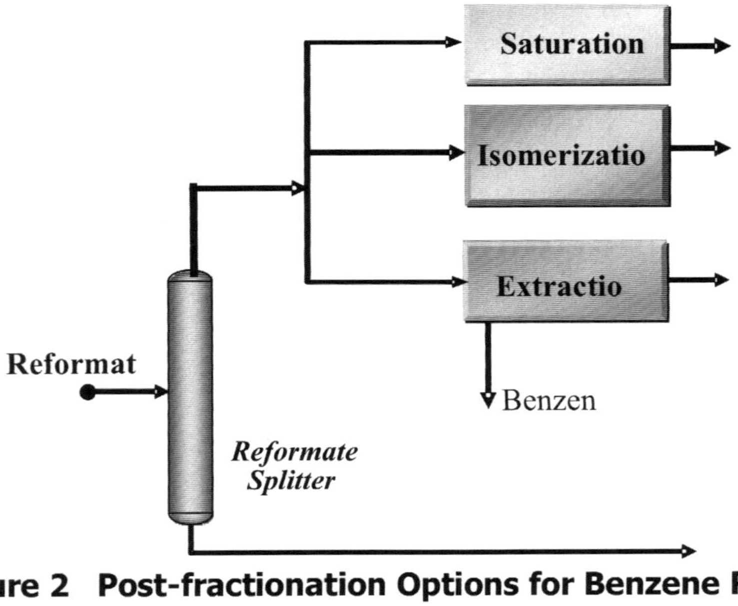


**Figure 2 Post-fractionation Options for Benzene Reduction**

The simplest method for managing benzene is saturation of the benzene. The UOP BenSat™ process with UOP H-8™ catalyst has been the industry standard for benzene saturation for over 25 years. Due to expected high demand for application of the BenSat process in MSAT2 benzene reduction solutions, UOP has updated the process to improve economics. A new catalyst has been developed, UOP H-18™ catalyst, which permits lower reactor catalyst volumes and lower catalyst precious metal content. The new H-18 catalyst still uses platinum for maximum reliability, highest yield, and minimal hydrogen consumption. Platinum based saturation catalysts are not permanently poisoned by sulfur or feed heavies upsets and do not cause cracking to light ends like nickel based catalysts. Over ten BenSat process units are in operation. No BenSat unit has ever required a catalyst reload due to catalyst deactivation. Many units have operated with the same catalyst load for over 11 years. A simplified flow scheme of the BenSat process is shown in figure 3.

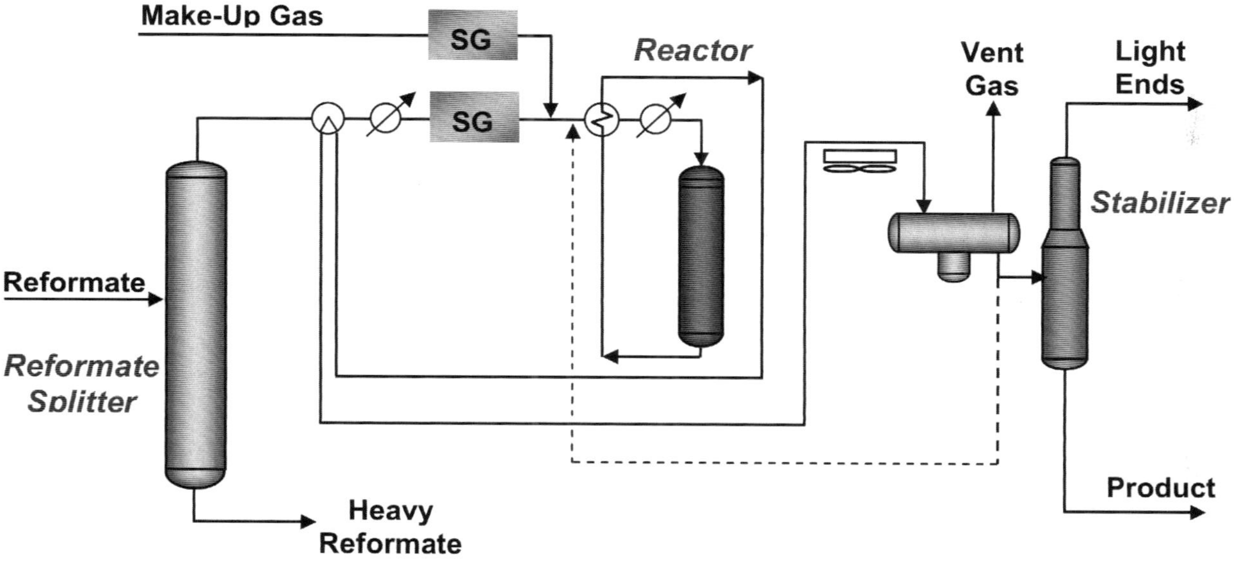


**Figure 3 Flow Scheme of BenSat Process with Reformate Splitter**

The BenSat process flow scheme is a simple hydrogen once through scheme requiring no recycle gas compressor. Light reformate may be combined with light naphtha before being preheated and sent to a sulfur guard bed. Treated feed from the sulfur guard bed may be diluted with separator liquid, if necessary, to reduce reactor differential temperatures. Recycle liquid is typically not required for feeds containing less than approximately 12 lv% benzene. Makeup gas is treated in a sulfur guard bed, if necessary, before combining with the feed. Combined feed is exchanged against reactor effluent and then sent to the down flow reactor. Small steam heaters are provided for the feed sulfur guard bed and reactor. These steam heated exchangers are only used for startup. During normal operation the benzene saturation exotherm provides all heat necessary for the reactor section. Reactor effluent is exchanged against combined feed and fresh feed before being condensed and sent to the separator. Excess hydrogen is vented off of the separator at 400 psig. This hydrogen vent stream is typically cascaded to refinery hydrotreating units as part of refinery hydrogen management. BenSat unit separator liquid is sent to a stabilizer to remove dissolved hydrogen and light ends that came in with the makeup gas. Since the BenSat unit catalyst does not crack heavy material in the feed, light ends production is minimized and the stabilizer is operated as a stripper with vapor only net overhead product.

BenSat units are often designed to produce product benzene levels below 0.1 lv%. This gives refiners flexibility in the operation of other gasoline producing units in the refinery and permits them to target benzene levels well below the 0.62 vol% limit of MSAT2. Because of the volumetric expansion of benzene to cyclohexane, the BenSat volumetric yield increases. Depending on the feed benzene levels, 100+ vol% $C_5$+ yields can be expected.

Benzene separation and saturation will result in an octane loss that can be offset by ethanol blending or light naphtha isomerization.

A common solution for the prefractionation approach is to separate the SRN benzene and benzene precursors into the light naphtha stream. A light naphtha isomerization unit is then employed for benzene saturation and to offset the octane loss by production of high octane isomerate. This solution works equally well for the post-fractionation approach. UOP offers a wide range of light naphtha isomerization technologies. The most common process is the UOP Penex™ process with over 120 units in operation. The Penex process can process up to 5 vol% benzene in the feed. A typical Penex process flow scheme is shown in figure 4.

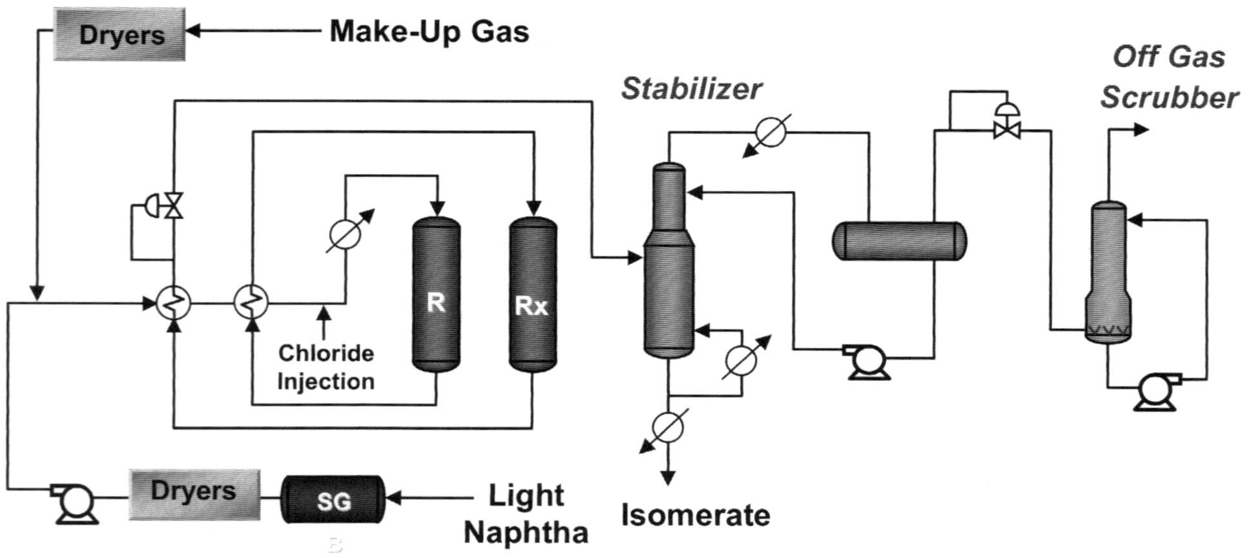


**Figure 4 Penex Process Flow Scheme**

Figure 4 illustrates a hydrocarbon once through Penex process unit. The Penex process uses the highest activity chlorided alumina UOP catalysts and, therefore, operates at lower temperatures than other Isomerization processes. UOP offers three chlorided alumina catalysts, (UOP I-82™, UOP I-84™, and UOP I-122™ catalysts), with differing platinum levels, for Penex service allowing refiners to lower their platinum investment when processing lower severity feeds. The combination of low operating temperature and high activity catalyst enables the process to achieve the highest normal paraffin conversion and selectivity to high octane isopentane and dimethylbutanes of any current commercially demonstrated Isomerization technology. A hydrocarbon once through Penex unit is capable of producing isomerate with a RONC in the range 82 – 85 with $C_5$+ yields close to 100 vol%. Higher isomerate octane can be achieved by adding a deisohexanizer column to recycle low octane $C_6$ isomers from the product back to the reactors where they are converted to high octane dimethylbutanes. The standard design for the Penex process is for Hydrogen Once-Through (HOT) operation. Due to the low operating temperature of the Penex unit's reactors, cracking reactions are minimized and there is no significant coke laydown on the catalyst. This enables the reactors to be run at a low Hydrogen/Hydrocarbon ratio which removes the need for recycle gas operation. Thus the product coolers, separator and recycle gas compressor are removed from the flowscheme reducing significantly the capital and operating cost of the unit.

The Penex process uses molecular sieve driers to dry both the hydrocarbon feed and the make-up gas. The driers are crucial to the Penex process since water is a permanent poison for the catalyst. The dry environment inside the plant also allows low cost carbon steel construction despite the addition of organic chloride and the formation of HCl in the reactors. The use of carbon steel minimizes the capital cost of the unit.

The Penex process utilizes two reactors, usually of equal size, piped and valved to allow switching of their relative processing positions. The catalyst deactivates in a plug flow manner due to water deactivation, not coke laydown. Catalyst life depends on removing contaminants from the reactor feed stream. Refiners with excellent contaminant control have experienced catalyst lives of over 10 years. When the catalyst in the lead reactor is deactivated then that reactor is taken off-line and the catalyst is replaced. During this time the unit remains on-stream at the design throughput operating with only the lag reactor. The two reactor system also allows the lag reactor, in normal operation, to operate at a lower temperature than the lead reactor in order to take full advantage of the more favorable equilibrium ratios at lower temperatures.

For specific solutions that result in greater than 5 vol% benzene in the Penex unit feed, a benzene saturation unit can be incorporated into the Penex process. This configuration is known as the UOP Penex-Plus™ process. There are two different flow scheme configurations that can be used for the Penex-Plus process. Either of these two schemes can also be used to convert an existing Penex process unit to a Penex-Plus unit which can process feeds with benzene contents ranging from 5 to over 30 lv%. The Penex-Plus product stream contains no benzene. The conventional Penex-Plus flow scheme is shown in Figure 5.

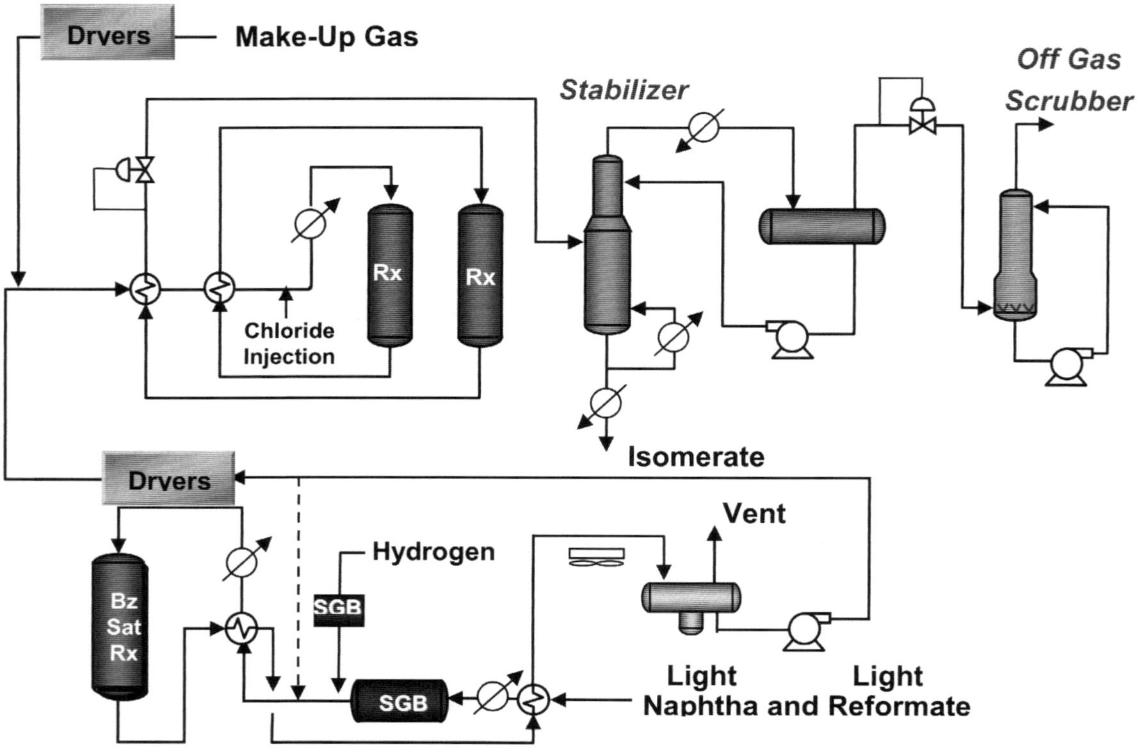


**Figure 5 Conventional Penex-Plus Process Flow Scheme**

The Penex-Plus process is specifically designed to process high benzene content feeds. $C_5$/ $C_6$ Isomerization with the Penex-Plus process has the ability to reduce the benzene content and upgrade the light naphtha or light reformate streams.

Conventional isomerization technology is not suitable for very high benzene feedstocks. Benzene saturation is highly exothermic resulting in high reactor temperatures that are detrimental to Isomerization product iso-ratios which are greater at lower temperatures. The Penex-Plus process is designed to optimize conditions for both benzene saturation and isomerization reactions within the same integrated unit. The Penex-Plus process can be applied to light straight run or light reformate feedstocks and is usually used for feeds containing both light naphtha and light reformate. Within the Penex-Plus unit, UOP has separated the benzene saturation reactor from the Isomerization reactors to optimize the overall performance of both. Integrating the benzene saturation reactor into a Penex unit to form the Penex-Plus process results in a reduction of equipment over stand-alone benzene saturation and isomerization units.

A comparison of Figures 4 and 5 shows the added equipment necessary when converting an existing Penex process unit to a conventional Penex-Plus. The conventional Penex-Plus has a separate benzene reactor section upstream of the Penex unit. BenSat receiver net product is sent directly to the Penex section feed driers. This flow scheme is used for feedstocks with benzene contents above approximately 10 lv%. It allows for dilution of the benzene content of the saturation reactor feed stream without increasing the size of the Penex section. An alternate Penex-Plus flow scheme is shown in Figure 6.

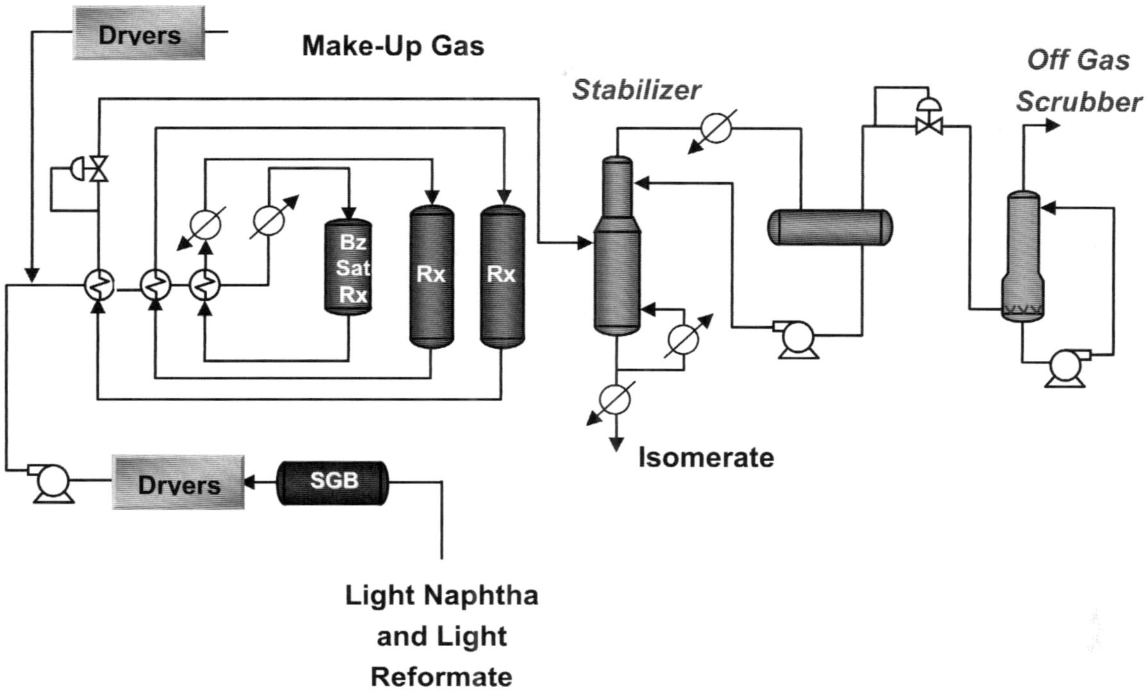


**Figure 6 Alternate Penex-Plus Process Flow Scheme**

The alternate Penex-Plus process flow scheme shown in Figure 6 is used for feeds containing less than 10-12 LV% benzene. In this flow scheme the benzene saturation reactor is placed just upstream of the isomerization reactors allowing a further reduction in equipment count. This flow scheme is especially well suited to the conversion of existing Penex units to Penex-Plus service. The only new equipment required is the benzene saturation reactor, combined feed exchanger and reactor cooler. In the conventional Penex-Plus flow scheme shown in Figure 5, the feed driers are operated at high pressure to keep the liquid from the benzene saturation section separator in the liquid phase. For existing units, this would usually require replacing the feed driers. In the alternate Penex-Plus flow scheme, the feed driers are operated at the relatively low pressures needed for the Penex unit and the existing driers can usually be reused. The benzene separation section separator and makeup gas sulfur guard bed shown in the conventional Penex-Plus flow scheme are also not required for the alternate Penex-Plus flow scheme.

For refiners that do not currently have a light naphtha isomerization unit, but wish to consider isomerization as part of their benzene management strategy, the reuse of idle equipment offers a cost effective option. The conversion of a fixed bed reforming unit to a light naphtha isomerization unit is particularly attractive due to the significant re-use of equipment. The UOP Par-Isom™ process with UOP PI-242 ™ catalyst is particularly well suited for unit conversions. Additional feed dryers are not needed and since chloride promoters are not required for PI-242 catalyst, caustic scrubbing is also not needed. Figure 7 shows a typical fixed bed reforming unit and Figure 8 shows the equipment from the fixed bed reforming unit that can be reused for the Par-Isom conversion. Figure 8 shows reuse of one of the reforming unit fired heaters. Due to the low operating temperatures of the Par-Isom unit, a fired heater is not required for normal operation. A steam charge heater is usually used for the Par-Isom unit. One of the existing fired heaters is often retained for catalyst regeneration operations.

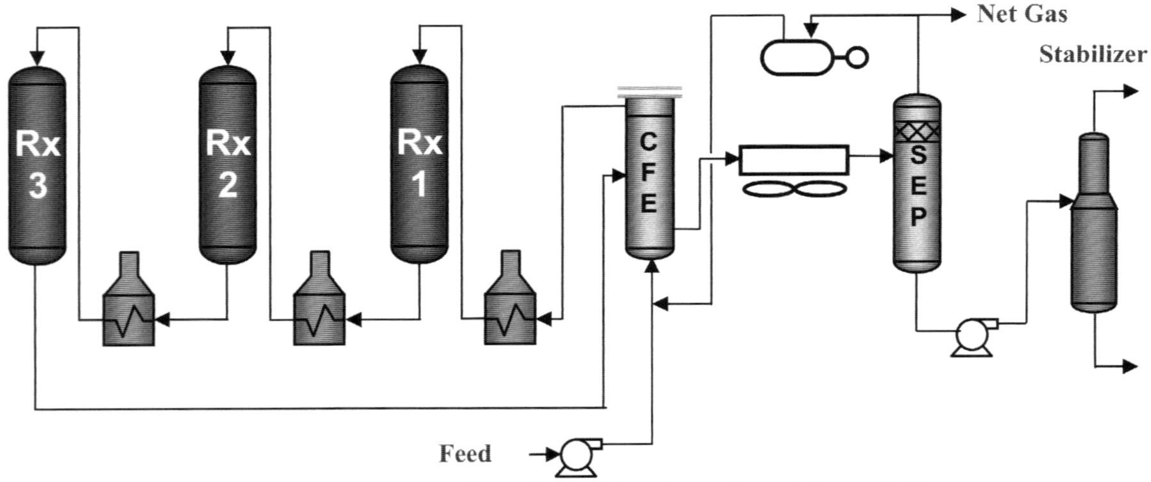


**Figure 7 Fixed Bed Reforming Flow Scheme**

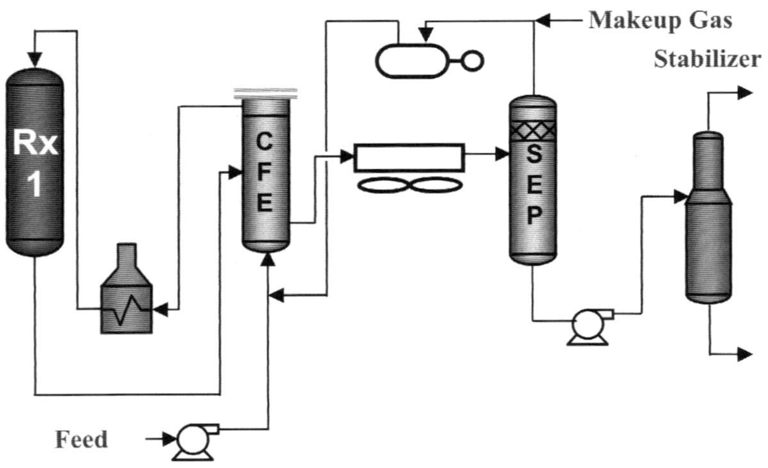


**Figure 8 Par-Isom Process Flow Scheme**

Table 2 compares the estimated erected cost estimates for converting a fixed bed reforming unit to a Par-Isom unit processing 10,000 BPSD of high severity feed. Table 2 shows that the equipment cost is minimal and the payback is less than a year. The addition of product separation and recycle in the form of a deisohexanizer significantly improves the isomerate octane while still maintaining less than a two year payback.

| *Process* | *Product Quality RON* | *Erected Equipment Cost* | *Payback* |
|---|---|---|---|
| **Par-Isom Process** | **81.4 RON** | **$0.25 MM** | **0.5 yr** |
| **Par-Isom/DIH Process** | **87.0 RON** | **$5.1 MM** | **1.7 yr** |

**Table 2 Economics Comparison of Conversion of Fixed Reformer to Par-Isom Process or Par-Isom/DIH Process**

Destruction of benzene via saturation or isomerization processes are the most common and usually the most economical means of removing benzene from light reformate streams. Other post-fractionation solutions include benzene extraction.

Extraction of benzene from light reformate for chemical production requires significant investment as well as ready access to a petrochemical market and thus will be considered by a limited number of refiners.

## Refinery Example

For illustration purposes, consider the base case refinery in Figure 9. This refinery is processing 150,000 BPD of a mixture of crudes: Arabian medium, Maya, and WTI. Gasoline production is around 87,000 BPD with 11% premium and 89% regular in both RFG and conventional gasoline. In the base case, the feed to the reforming unit is deisohexanized. Benzene and $C_6$ naphthenes remain in the heavy naphtha. The light straight run naphtha is de-pentanized to the extent that RVP limitations will allow. Any pentanes and the light straight run naphtha $C_6$+ from the depentanizer are blended to gasoline with no additional processing. The minimum reformate RON is 98. This base case configuration results in a gasoline pool containing around 2 vol-% benzene, with either a low pressure or high pressure reforming unit. The majority of the benzene is from the reformate and the LSR $C_6^+$streams.

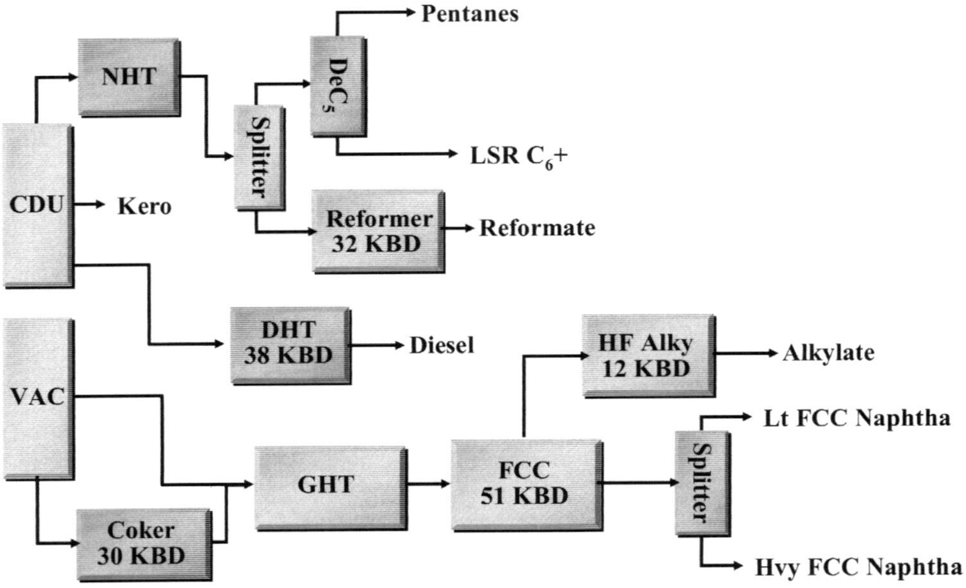


**Figure 9 Example Refinery Flowscheme**

Prefractionation of the LSR naphtha stream to remove $C_6$ cyclic components from the reforming unit feed will significantly reduce gasoline pool benzene content but not enough to meet the 0.62 vol% MSAT2 regulation. Figure 10 illustrates that adjusting the split on the naphtha splitter to send methylcyclopentane, cyclohexane and benzene overhead reduces the gasoline pool benzene concentration to 0.8 to 0.90 vol% when low pressure reforming is used. Gasoline pool benzene levels are approximately 1.0 vol% when the reforming unit operates at high pressure. The benzene present in the light naphtha must be saturated in order to achieve the benzene target.

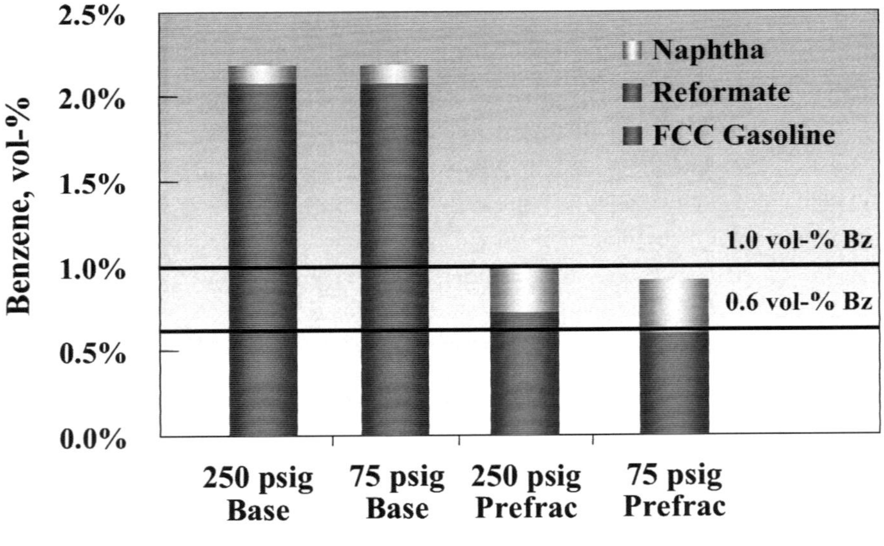


**Figure 10 Stream Contribution to Gasoline Pool Benzene**

Figure 10 shows that pre-fractionation with benzene saturation of the naphtha stream and low pressure reforming meets the MSAT2 benzene level with little or no compliance margin. A change in feed or product slate could force this refinery out of compliance. Most refiners, even those with low pressure reformers, will need to consider the post-fractionation option.

For this example, a high pressure reforming unit produces enough benzene through dealkylation reactions that any pre-fractionation solution will not meet the MSAT2 targets. Separation of benzene from the reformate using a reformate splitter will be needed for high pressure reforming units. Refiners with low pressure reforming units will benefit from the additional compliance margin that post-fractionation will provide.

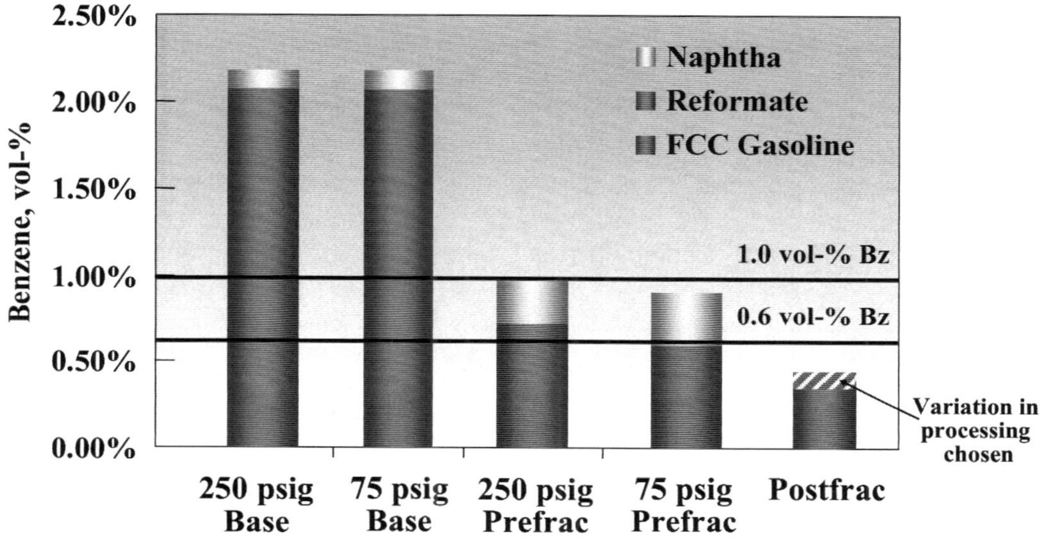


**Figure 11 Effectiveness of Benzene Removal Strategies**

Figure 11 illustrates the levels of benzene that can be achieved using prefractionation and post-fractionation solutions. Post-fractionation with benzene saturation is required if a high pressure reforming unit is used in order to meet MSAT2 benzene limits. The benzene range depicted for the post-fractionation option brackets the benzene levels obtained using the various benzene reduction methods: saturation, extraction, or isomerization.

## CONCLUSIONS

The lower gasoline benzene specifications mandated by MSAT2 will require processing unit modifications for most refineries. The 0.62 vol% target benzene levels can be obtained by focusing on the benzene produced in the reforming unit. Solutions will be refinery-specific and will be determined by the current refinery configuration, the type of reforming unit, the amount of benzene contributed from other blendstocks, the amount of ethanol blended, and the use of credits. Many refiners will end up using both prefractionation and post-fractionation solutions in combination with benzene reduction processes such as benzene saturation or isomerization of light naphtha and light reformate in order to comply with the MSAT2 regulations by January 2011. Benzene saturation either in a stand-alone benzene saturation unit such as the BenSat process or in a light naphtha isomerization unit such as the Penex process or the Par-Isom process are well established, robust, economical technologies for new benzene management solutions.

## REFERENCES

[1]Environmental Protection Agency, "40 CFR Parts 59, 80, 85, and 86 Control of Hazardous Air Pollutants From Mobile Sources; Final Rule" February 26, 2007

R. R. Rosin, C. J. Anderle, G. C. Sturtevant, "Optimizing Your Naphtha Complex in a Low Benzene World", 2007 NPRA Ann Meeting, AM-07-46

J. Meister, T. Crowe, D. Banks, W. Keesom and M. Stine, "Gasoline Production in the Post Energy Bill, Benzene Restricted World," 2006 NPRA Annual Meeting, AM-06-39.

G. C. Anderson, R. R. Rosin, M. A. Stine, and M. J. Hunter, "New Solutions for Light Paraffin Isomerization", 2004 NPRA Annual Meeting, AM-04-46.

## 169c HF Alkylation: A New Age

**Larry Shoemaker**, UOP LLC, 25 E. Algonquin Road, Des Plaines, IL 60017 and James P. Glavin, UOP, 25 E. Algonquin Road, Des Plaines, IL 60017.

The demand for alkylate gasoline in North America is expected to continue to grow. Alkylate is the optimum blending material in the gasoline pool due to its excellent blending characterizes: no olefins, low vapor pressure, high octane, and absence of aromatic compounds. At the same time there is increased pressure from local communities to improve the safety of the existing and new HF Alkyaltion units.

This paper will summarize the existing options available for meeting the demands of increased alkylate production and improved safety features of HF Alkyaltion units.

# 169d Benzene Regulations and the Refinery H2 Balance

**Alan Zagoria**, CEOC, Des Plaines, IL 60017

Maximum benzene limits in gasoline are posing a number of challenges to refiners. One of the less obvious challenges is negative impact on refinery H2 balance. Whether addressing the issue by pre-fractionating naphtha or post-fractionating reformate, the refinery will have less H2 than it has today. The impact of the various processing routes to comply with new regulations on the refinery H2 balance will be discussed. Also discussed is a general approach to address and improve the refinery's H2 balance through systematic, refinery-wide H2 management.

# 202a Flaring Minimization to Reduce Fuel Consumption - 2 Case Studies in Tabriz Petroleum Refinery and Khangiran Gas Refinery

**Omid Zadakbar**[1], Ali Vatani[1], Kianoosh Karimpour[1], and Armin Zadakbar[2]. (1) Faculty of Chemical Engineering, University of Tehran, Enghelab St., Tehran, Iran, (2) Faculty of Industrial Engineering, University of Science and Culture, Park St. Ashrafi Blvd., Tehran, Iran

ABSTRACT

Flaring is a common method of disposal of flammable waste gases in the upstream facilities, downstream refining, and petrochemical processing industries. During recent years in Iran, all of projects have included collection of associated gases, and flare gas recovery in oil and gas refineries has been neglected. Therefore we try to investigate the operational conditions of 11 important refineries and petrochemical plants at University of Tehran. After the comprehensive evaluations, we advised the practical methods to reduce, recover and reuse flare gases for each petroleum refinery, natural gas refinery and petrochemical plants. Adequate process evaluation of plants especially units that produce flare gases, comprehensive monitoring of flow and composition of flare gases, investigation of existing flare systems and finding alternative choices to reuse flare gases have been done in 11 petroleum refineries, natural gas refineries and petrochemical plants.

This paper presents two case studies of reducing, recovery and reusing flare gases in Tabriz petroleum refinery and Khangiran gas refinery. Expected results, system operation and control, design guidelines and economics of the process have been discussed. Flare gases are compressed and returned to fuel gas header for immediate use as fuel gas. Flare gas recovery system can reduce flaring noise, operating and maintenance cost, air pollution and emission, fuel gas and steam consumption while increase process stability and flare tip life without any impact on existing safety relief system.

## 202b Deliver the Challenges - Design and Operate Refining Plants More Efficiently to Minimize CO2 Footprint

**X. X. Frank Zhu**, UOP LLC, 25 East Algonquin Road, Des Plaines, IL 60071

Global warming is one of the most serious challenges facing us today. To protect the health and economic well-being of current and future generations, we must minimize green gas emissions by using the practical solutions at our disposal and developing innovative solutions.

Refining and petrochemical industry produces a large amount of green gas emissions from operation which contributes to global warming. CO2 regulation and tax which expect to come out in the future in US could have impact on refining operation and business. This presentation will assess the potential impacts and provide ways for reducing CO2 emissions from operation. In summary, there are three major ways for minimizing CO2 emissions from the industry. The first one is to revamp existing facility by better heat recovery and using appropriate technologies. The second one is to optimize operation by using process know-how supported with effective tools while the third one is to look for opportunities for CO2 capturing and trading.

When doing the above, one needs to understand the similarity and difference between improving energy efficiency and reducing green gas emissions. The difference lies in the fact that energy efficiency is to do with the amount of MMBtu burnt while gas emissions is to do with both MMBtu and carbon content in emitted gases. This difference calls for different solutions to be applied for abating green gas emissions. This presentation will demonstrate UOP's effort in this area and successful stories in achieving the goal for minimizing green gas emissions for the industry.

# Optimising heat exchanger design to reduce size, cost and energy consumption for plant expansion and new projects

Peter Ellerby, Cal Gavin Limited, Alcester, UK
Martin Gough, Cal Gavin Limited, Alcester, UK
Peter Drögemüller, Cal Gavin Limited, Alcester, UK

## Introduction

The focus for process equipment design is becoming increasingly centred on energy efficiency and reduced carbon emissions. At the same time the constraints of reduced size and cost continue to be important in a competitive market. However the roles of process engineer and equipment designer are often assigned such that neither is in a position to optimise on all these areas. For example an exchanger network may be optimised for lowest energy consumption but the resulting exchanger requirements may be impractically large and expensive. In another case a designer may be provided a specification including limits on size, pressure drop or fouling factors which are arbitrary or based on typical values which were set many years ago with different considerations..

Ideally the process engineer and designer would have freedom to work together and consider all the options and weight the relative importance of size, cost and energy - in reality these are often separate disciplines looking at different aspects. With the increasing emphasis on energy reduction and limiting carbon emissions it is important that engineers are fully aware of all the tools available to create practical designs which reduce size, cost and energy usage and carbon emissions.

Technologies offering better performance or reduced size have often been adopted slowly or in limited areas to assess their effectiveness and allow identification of any potential problems. This is particularly true for more difficult applications involving two phase heat transfer. Revamps or debottlenecking projects can be ideal opportunities to demonstrate how technologies can be used to optimise the design.

## Debottleneck to Design

The aims for utilising heat transfer equipment will be similar for both a revamp or debottlenecking project and a new design. However the problems faced can be more severe as constraints can be physical and real, rather than preferred, and options may be limited or have starkly different costs where existing equipment is in place. Such scenarios lead to greater thought being given to enhancement technologies as a means of solving difficult problems.

hiTRAN® Matrix Elements are now well established as a tool for enhancing heat transfer in laminar and transitional flow regimes. Application to a large range of applications over a period of 25 years backed by extensive testing and research means they can be specified with confidence for new exchangers in applicable services. Such research has allowed hiTRAN to benefit applications in turbulent flow also. Even in some services where fouling has been an issue, hiTRAN® has been adopted to provide a benefit for both heat transfer and fouling. The mechanism of heat transfer enhancement and fouling mitigation promoted by hiTRAN® for these applications is well established and has been discussed extensively.

The use of hiTRAN® in two phase applications has been more limited and with a few exceptions has been largely based on revamping existing units. Two phase applications can be more complex with the flow regimes and heat transfer mechanisms changing through the exchanger. They, of course, require a greater range of data and careful application of this data to optimise the heat transfer benefit within, usually, a low allowable pressure drop. Two phase systems are often sensitive to pressure drop and a change can affect the operation of associated equipment such as a distillation column. Large changes may also affect the heat release curve such that the temperature difference is reduced. In this scenario the benefit from enhancement could be reduced or even negated. However for the right application significant benefits are achievable.

In this paper we will demonstrate how tubeside enhancement can offer significant benefits through consideration of actual cases. Further to this we will also discuss the potential benefits adoption of this technology would bring in new designs.

## Case Studies

### 1. Texas Tower Feed/ Effluent Exchanger– German Refinery

'Texas Tower' is the name given to tall vertical combined feed/ effluent service exchangers which are common in refineries throughout the world. As in this particular case, they are usually critical items in the plant operation, in continuous operation and throughput limiting.

The objective for this plant was an increase in throughput which meant that the reactor feed loop would need more heat input in order to maintain the required inlet temperature. The main heat input was provided by the reactor effluent through the 'Texas Tower' with a fired heater being used to meet the desired temperature at the reactor. In order to meet the new condition the heat duty of this exchanger would need to increase by about 15% relative to its current performance. Some increase would be achieved through increased heat transfer (due to higher velocities) and existing margin in the original designs but this was insufficient to provide all the additional heat transfer required.

The primary options available were to replace one of these items. Replacing a fired heater would involve significant capital cost and result in increased energy usage and carbon emissions- therefore the preference was to replace the heat exchanger. However as the original heater was vertical with nearly 2000 tubes at 12000mm length the capital outlay of the replacement combined with potential civil engineering costs to accommodate the additional weight meant that this would be a significant part of the project cost.

On this basis the operators were very amenable to looking at alternative solutions, though the key criteria was that the performance must be met as failure would jeopardise the financial success of the whole revamp.

The design and operation of this unit is typical of most feed/ effluent exchangers – in order to maximise the heat recovery from the effluent there is a large temperature cross (Fig 1.) which is handled by employing a single pass design. Long tubes are then preferred in order to provide the required area whilst maximising the coefficients. This type of service offers limited scope for increased duty as increasing the flowrates cause the temperatures to close even further, reducing the Mean Temperature Difference and therefore requiring a significant increase in overall coefficient. Figures 1 and 2 indicate how the temperatures pinch together and the effect this has on the MTD.

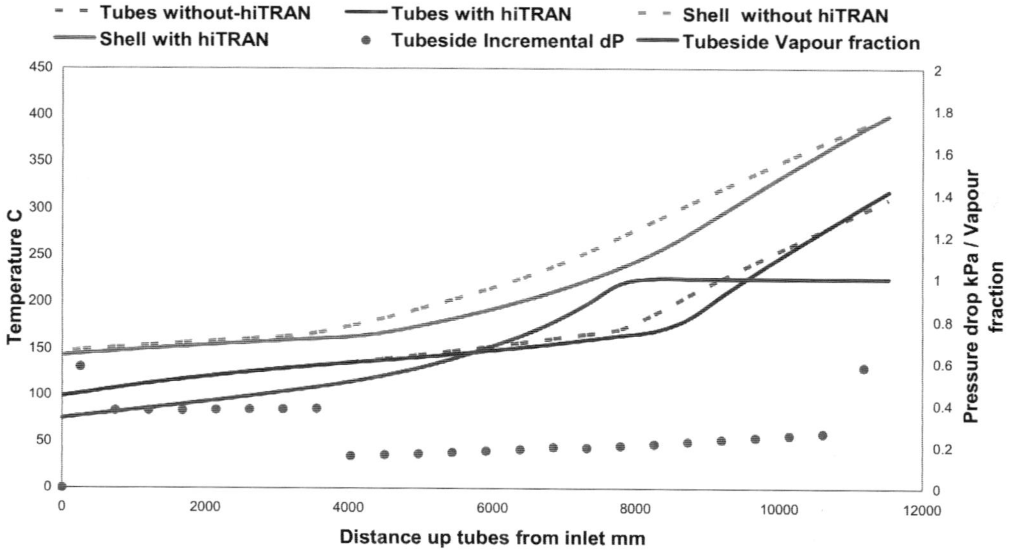


**Figure 1- Temperatures and pressure drop along tubelength**

Also, typical of many 'Texas Towers', both streams are two phase with the tubeside vapourising as it flows up and the shellside condensing. This means the flow regimes and coefficients change significantly along the length of the tubes with part of the area being limited by the tubeside resistance and some by the shellside - Figure 2 shows it results in a fluctuating overall coefficient. In this case some additional pressure drop was available but the most effective use of this pressure drop necessitated that the enhancement be tailored to provide most enhancement where the tubeside resistance was the controlling one - Figure 1 shows that the pressure drop is higher towards the inlet end where the fluid is two phase. hiTRAN® Matrix Elements afford this possibility as they can be fitted as a partial Matrix length or the geometry, in terms of Matrix density etc., can be varied to provide different levels of pressure drop/ enhancement.

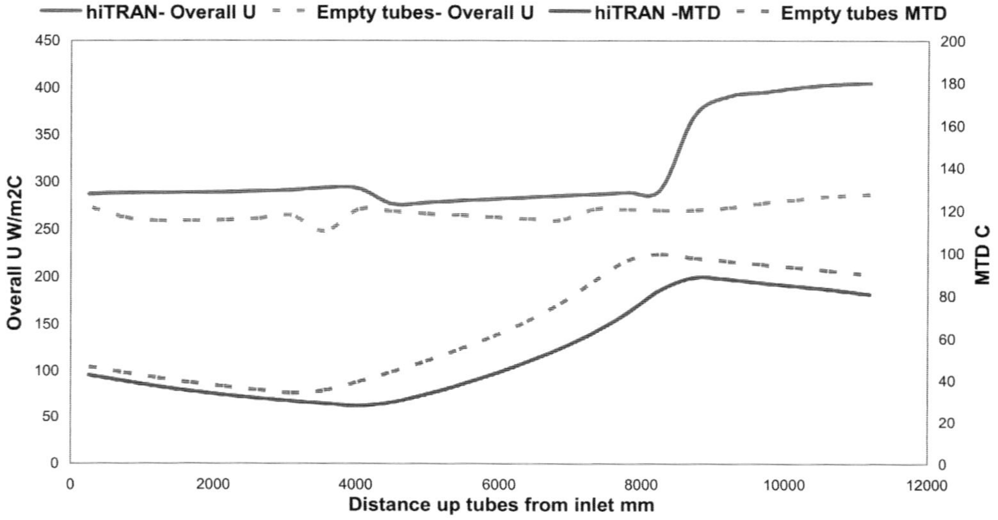


**Figure 2- Coefficient and MTD along tubelength**

For this particular case the fluid entered partially vapourised and continued to boil until about 60% up the tubes with the remainder of the tube superheating the vapour. Whilst the sensible vapour heating in the top of the tubes was the controlling heat transfer resistance in that area, the potential for both film boiling and mist flow (indicated by the simulation) to occur in the lower section meant that enhancement could be beneficial there also. Film boiling occurs when the wall to bulk temperature difference exceeds a threshold value which leads to a vapour film forming at the tube wall. This film has low conductivity and hence leads to a low heat transfer rate in that area. Whilst film boiling is avoided where possible, in some processes, such a cryogenics, or in this case it may be unavoidable. Mist flow is also prevalent in cryogenic applications but can also occur in other tubular vapourising cases. Poor mixing allows separation of the vapour and liquid phases with the liquid being carried in fine droplet form. Whilst carry over of the mist out of the exchanger would not occur here, as the amount of superheat is large, the poor mixing and mist flow creates a section of low heat transfer. hiTRAN® has been established as successful in interrupting the tubewall film and increasing turbulence and mixing thereby improving heat transfer in both these areas.

However the main enhancement is of the convective boiling and vapour superheating regions. Here improvements were achieved even with the tube side condition being in fully turbulent flow (Re up to 90000). This is illustrated in Figure 2 where we see the increase in overall coefficient up the tube with the largest increase being towards the top.

After careful analysis of the options the solution was optimised as a lower enhancement Element in the 2 phase section than for the superheating section. Particular attention was given to the calculation of pressure drop to ensure it was within the allowable.

Installing hiTRAN® Matrix Elements into 12m tubes in-situ in the vertical exchanger was, of course, a challenge but was completed without problem. The overall success came several weeks later when the restarted plant reached full flow and confirmed the performance improvements were as predicted. At the time of writing the Elements have been in operation for over six months without problem.

## Benefits and the Implications for Design

Whilst the target here was to achieve a specific flowrate whilst maintaining the required outlet temperature for the reaction the performance improvements obtained suggest a clear benefit for use of hiTRAN® in similar applications. Enhanced heat transfer can be used to reduce the required area and thereby size and weight of the exchanger. Here the additional heat recovery of 0.8MW from the effluent stream, if applied to a new design, would be a direct reduction in energy input and carbon emissions. Figure 3 indicates the saving in surface area achieved. In order to meet the duty without hiTRAN® a new exchanger with almost 20% additional area would have been required.

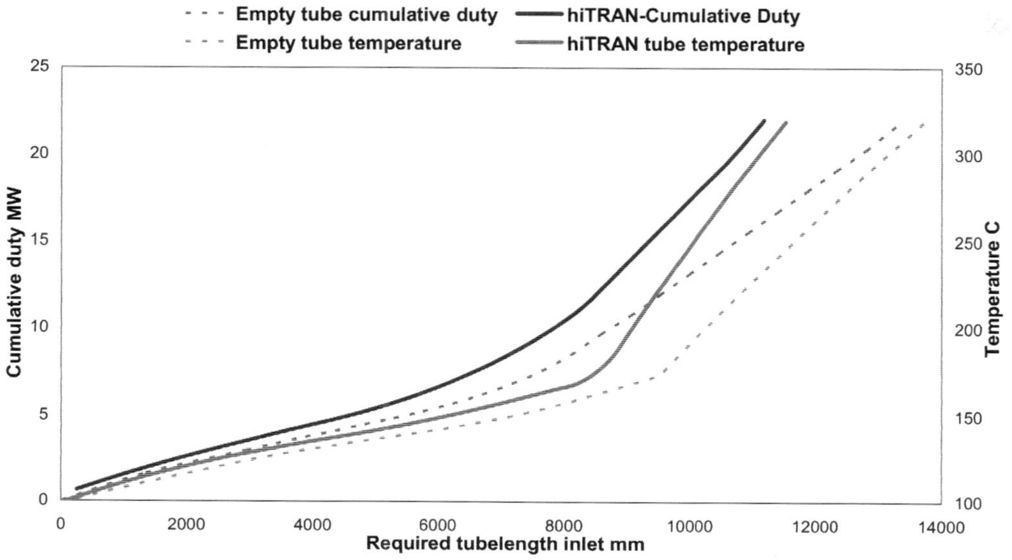


**Figure 3- Comparison of required tubelength**

## 2. Horizontal Tubeside Condenser- Germany

Condensation often occurs with good heat transfer rates, meaning it can often be assigned to the shellside for low pressure drop. However, where condensing is multicomponent or takes place in the presence of inerts the coefficients can become significantly lower as condensation proceeds. In such cases assigning the fluid to the tubeside may allow higher velocities and better coefficients in this region. They also allow the application of tubeside enhancement which can offer a substantial improvement in performance.

Pressure drop is again a key issue as the operating pressures are often low and operation of the equipment is then sensitive to relatively small changes. Therefore successful implementation of enhancement again involves analysis of the flow regimes and coefficients through the exchanger to apply it in a way which offers greatest enhancement/ pressure drop ratio.

To illustrate the above and the options for use of hiTRAN® we will look at a small horizontal process condenser. The tubeside fluid is a wide condensing range mixture where the tubeside coefficient reduces as the condensation proceeds. To maximise the benefit within the allowable pressure drop hiTRAN® is applied to last 60% of the tubelength. Considering Figure 4 we can see the reason for this - the tubeside coefficient falls quickly along the shell as the vapour fraction and consequently the vapour velocity reduce. Likewise the pressure drop impact is less in this region where the velocities are lower.

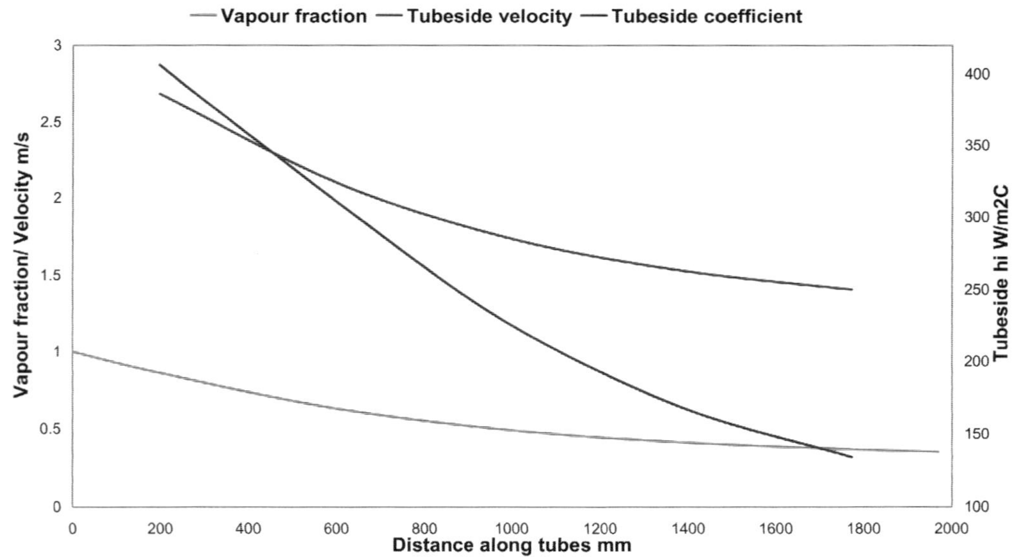


**Figure 4- Coefficient and important factors along tubelength**

One of the factors in the heat transfer resistance in tubeside condensation is what happens to the condensate film. This is dependant on the flow regime and vapour velocity but as the condensation proceeds through the tube this will typically be reduced – for a vertical arrangement this will lead to a annular film increasing in thickness down the tube and for horizontal tubes there may be stratification with a pool of liquid flowing along the bottom of the tube. The action of hiTRAN® in breaking up films at the tube wall and increasing mixing will enhance the coefficient in this regime. The increased turbulence and mixing also ensures that the liquid and vapour phases are fully in equilibrium- this maximises the amount of condensate produced as the bulk temperature reduces. The rate of reduction of the bulk vapour temperature can be a limiting factor in the condensation process and enhancing it with hiTRAN® can significantly improve the performance. Figure 5 charts the increase in tubeside coefficient achieved in the section where hiTRAN® was installed.

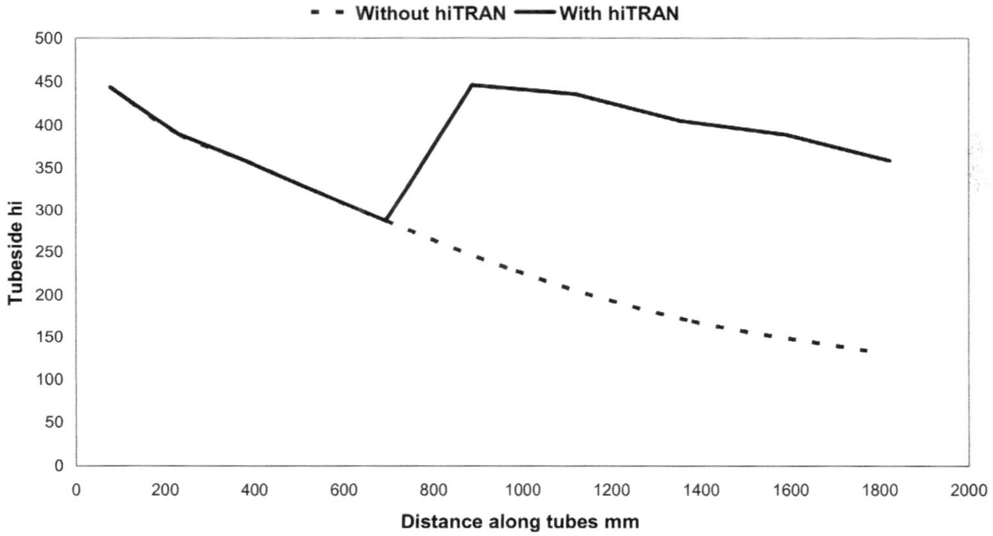


**Figure 5 - Effect of hiTRAN on tubeside coefficient**

## Benefits and the Implications for Design

The area reduction available for this unit is significant. Figure 6 shows that in order to achieve the same performance without hiTRAN® the exchanger would have been more than double the length. Such a reduction in length also offsets any pressure drop increase incurred in the section with hiTRAN®.

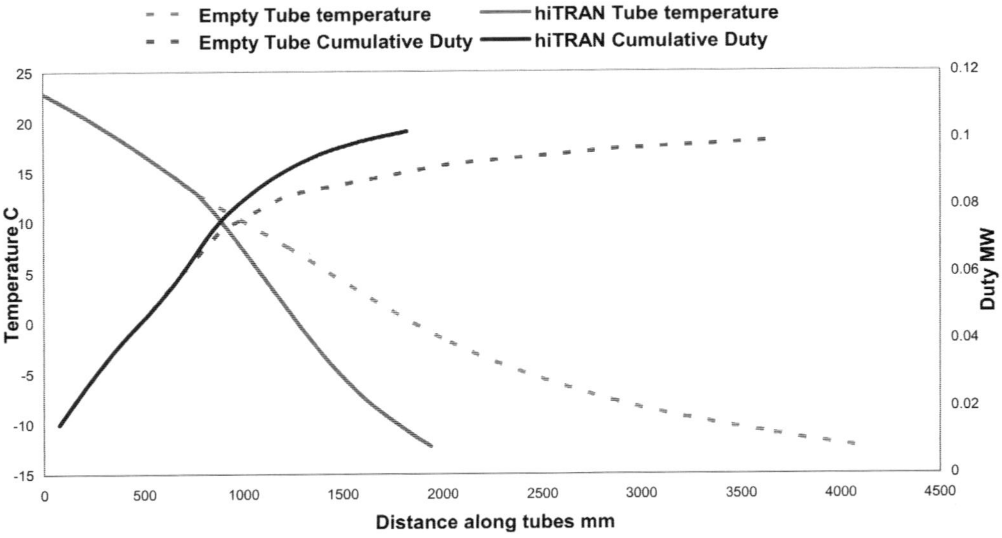


**Figure 6 - Comparison of required tubelength**

hiTRAN® is shown to be applicable for both horizontal and vertical tubeside condensers for wide condensing range fluids. These include process and column condensers as well as vent condensers. The introduction of hiTRAN® for such applications has allowed smaller exchangers or more effective units within the same space. The benefit of increased product and energy recovery from the condensing stream is clear, reducing the cost of dealing with the vapour downstream and improving energy efficiency. Likewise a more efficient vent condenser is a simple way to recover both material and energy and reduce the need for additional equipment to limit plant emissions.

## Conclusion

At a time when meeting the competing requirements of lowest size, cost and reduced energy and carbon emissions using all the means available to achieve these aims is vital. The enhancement of tubeside heat transfer through hiTRAN® Matrix Elements is shown to be one of these means for applicable services. Having become established in providing such enhancement in laminar/ transitional liquid flow regimes that list of applicable services can now be extended to include certain two phase applications.

For enhancement of either vapourising or condensing services to be effective the tubeside must be controlling the heat transfer rate somewhat. For vapourising cases they tend to be wide boiling range mixtures and the prevailing boiling mechanism will probably be convective rather than nucleation. Problems including film boiling and mist flow may also be applicable. Condensing services will typically be wide condensing range or condensation in the presence of a significant level of inerts.

The foregoing examples indicate that the data and methods are available for application of hiTRAN® in these areas and the potential benefits are significant. These options can now be taken into account at the design stage to meet the competing goals of low cost and low energy consumption.

**References**


**1.** Martin J Gough, Peter Droegemueller, Peter Ellerby and Alex Smeethe, Cal Gavin Limited, 'The theory, historical application and current practice of using wire matrix elements to enhance heat transfer and reduce tube side fouling in hydrocarbon streams' AIChE 2004.

**2.** Peter Droegemueller, Peter Ellerby, Martin J. Gough, Cal Gavin Ltd, Mann Lee, Bashir I. Master, Krishnan S. Chunangad, ABB Lummus Heat Transfer, 'Dual Enhanced Heat Exchangers using Cal Gavin hiTRAN® Wire Matrix Tube Inserts along with the ABB-LGHT HELIXCHANGER® Technology' HEE Asia, 2007

**3.** Smeethe, A., P. Drogemueller, J. Wood, W. Bujalski, 2004, 'Fluid dynamics in a tube equipped with wire matrix inserts'. Proc. 4th European Thermal Sciences Conference, Birmingham.

**4.** Gough, M.J., I.J. Gibbard, G.T. Polley and A.S. McMullan, 1995, 'Case studies of refinery fouling reduction', Proc. Engineering Foundation Conference on Fouling Mitigation of Industrial Heat Exchangers, California.

**5.** J.M. Ritchie, P. Droegemueller, M.J.H. Simmons, ' hiTRAN® wire matrix inserts in fouling applications' , Fouling Conference, Tomar, 2007

# How to Save Energy Through Enhanced Automation

AIChE Spring Meeting
2008
Douglas White
Emerson Process
Management

# Presenter

Doug White

Principal Consultant and Vice President, APC Services

Process Systems and Solutions

Emerson Process Management

Houston, Texas

EMERSON™ Process Management

# Natural Gas Prices

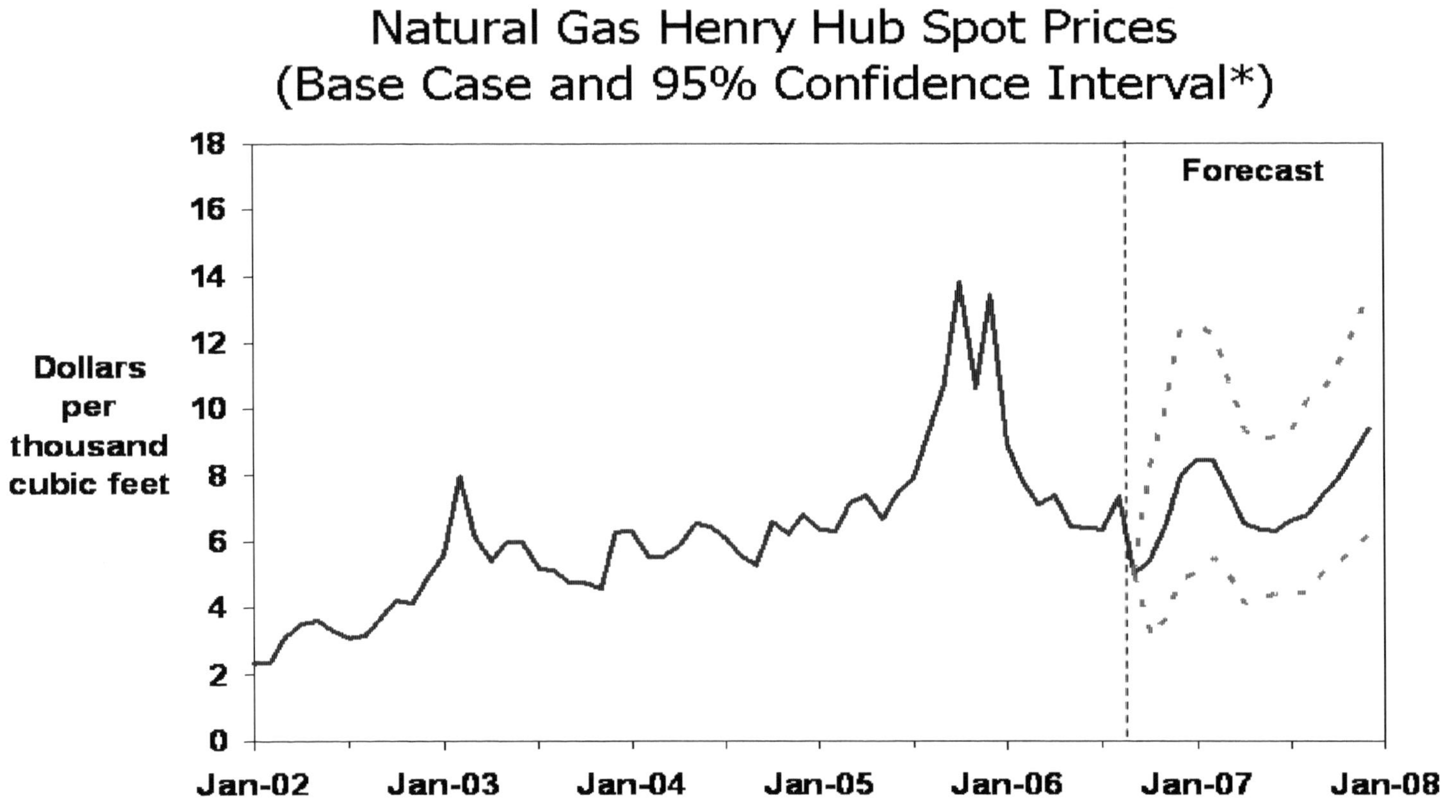


*The confidence intervals show +/- 2 standard errors based on the properties of the model.

Short-Term Energy Outlook, October 2006

www.eia.doe.gov

# Process Energy Usage

| | Process Energy; MM BTU/ Ton | Value; 10% Energy Reduction; $/ Ton ($7/ MMBTU) |
|---|---|---|
| Petroleum Refining | 4.4 | 3.1 |
| Integrated Pulp/Paper Mill | 29.0 | 20.3 |
| Cement Production | 7.9 | 5.5 |
| Chemicals | | |
| Ethylene | 7.1 | 5.0 |
| Polyethylene | 6.7 | 4.7 |
| EDC | 9.4 | 6.6 |
| PVC | 4.0 | 2.8 |
| EO | 6.2 | 4.3 |
| EG | 7.1 | 5.0 |
| Ethylbenzene | 2.9 | 2.1 |
| Styrene | 38.8 | 27.2 |

Largest Controllable Cost in Most Plants!

# Session Objective

Present some case studies of the many ways that automation, advanced automation and asset management can save energy in process plants

# Outline

- How is energy used in process plants?
- How can automation help save energy?
- How do we implement an energy reduction program?

# General Process Site Energy Flow

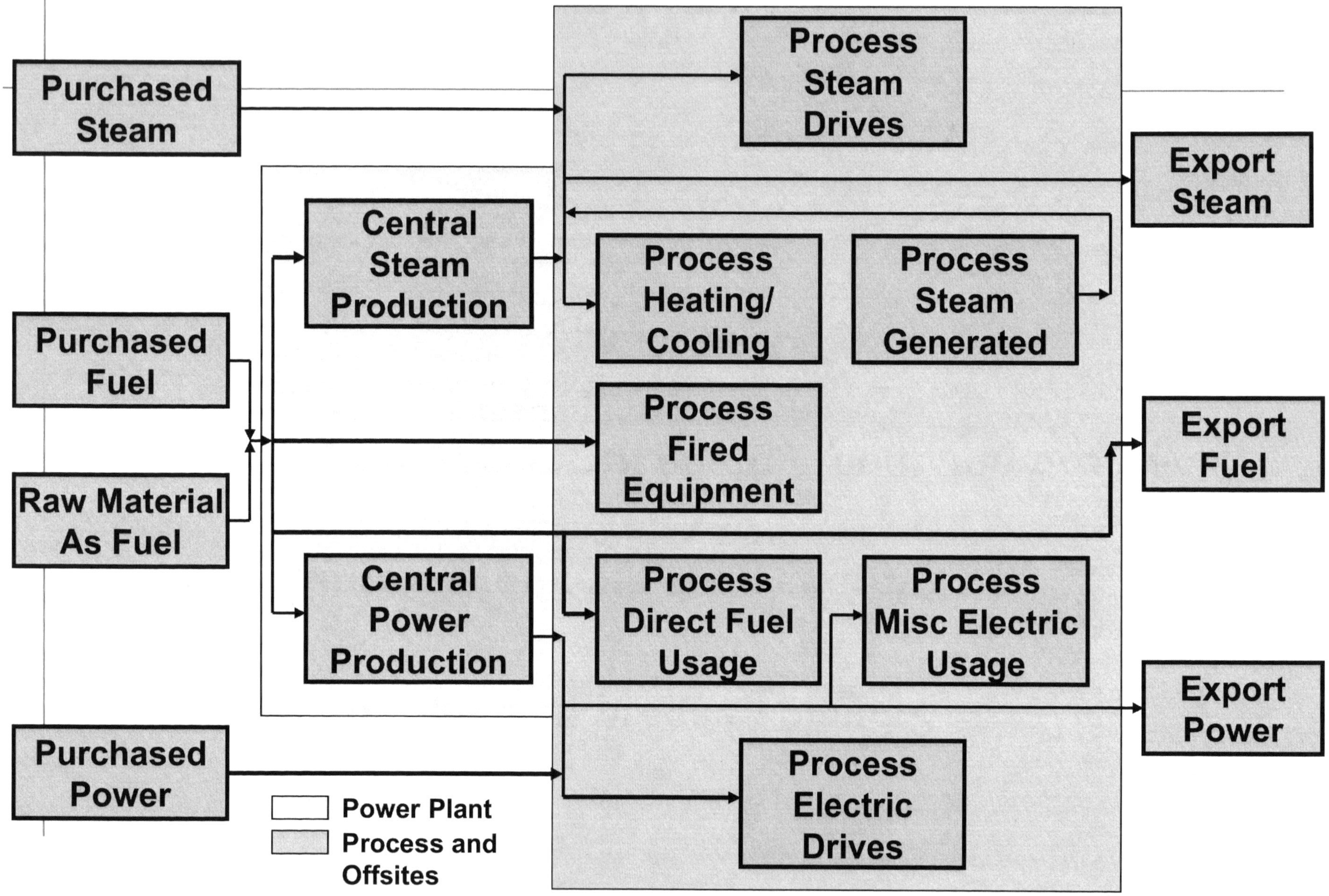

# "Average" Chemical Site Energy Flow

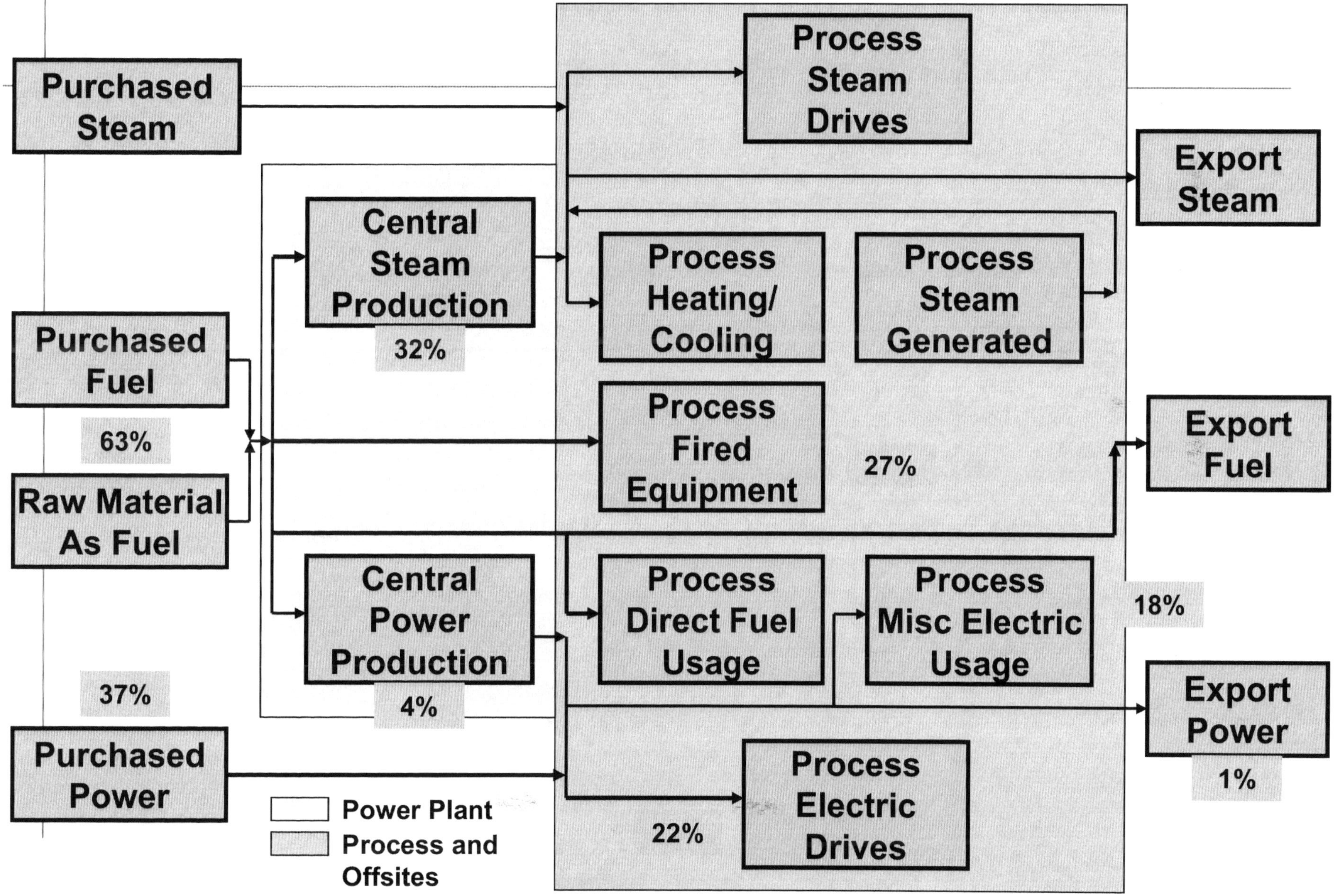


%-Equivalent BTU basis (Including losses) on total input

# Integrated Pulping Paper Mill Energy Flow

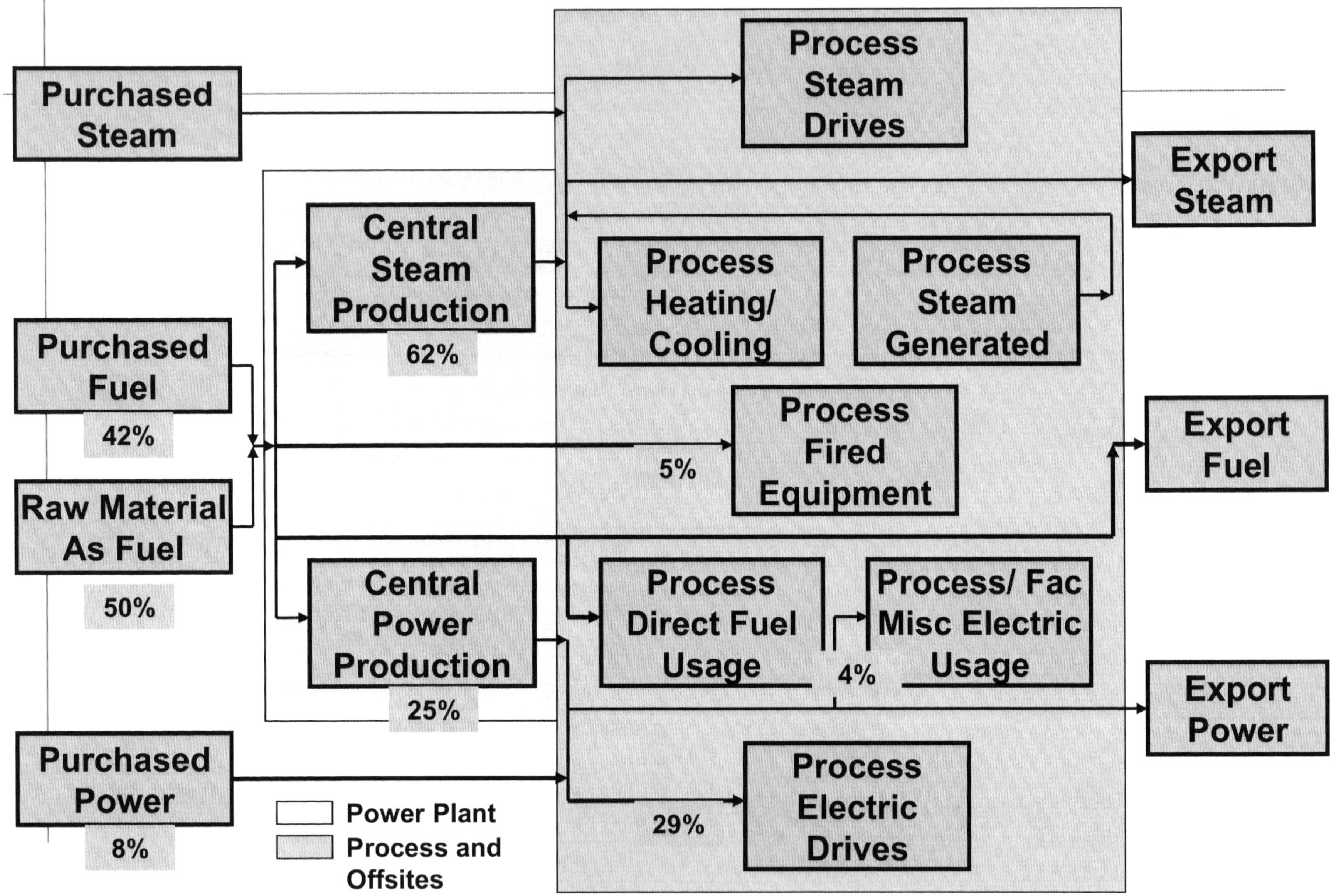


%-Equivalent BTU basis (Including losses) on total input

# Oil Refinery Energy Flow

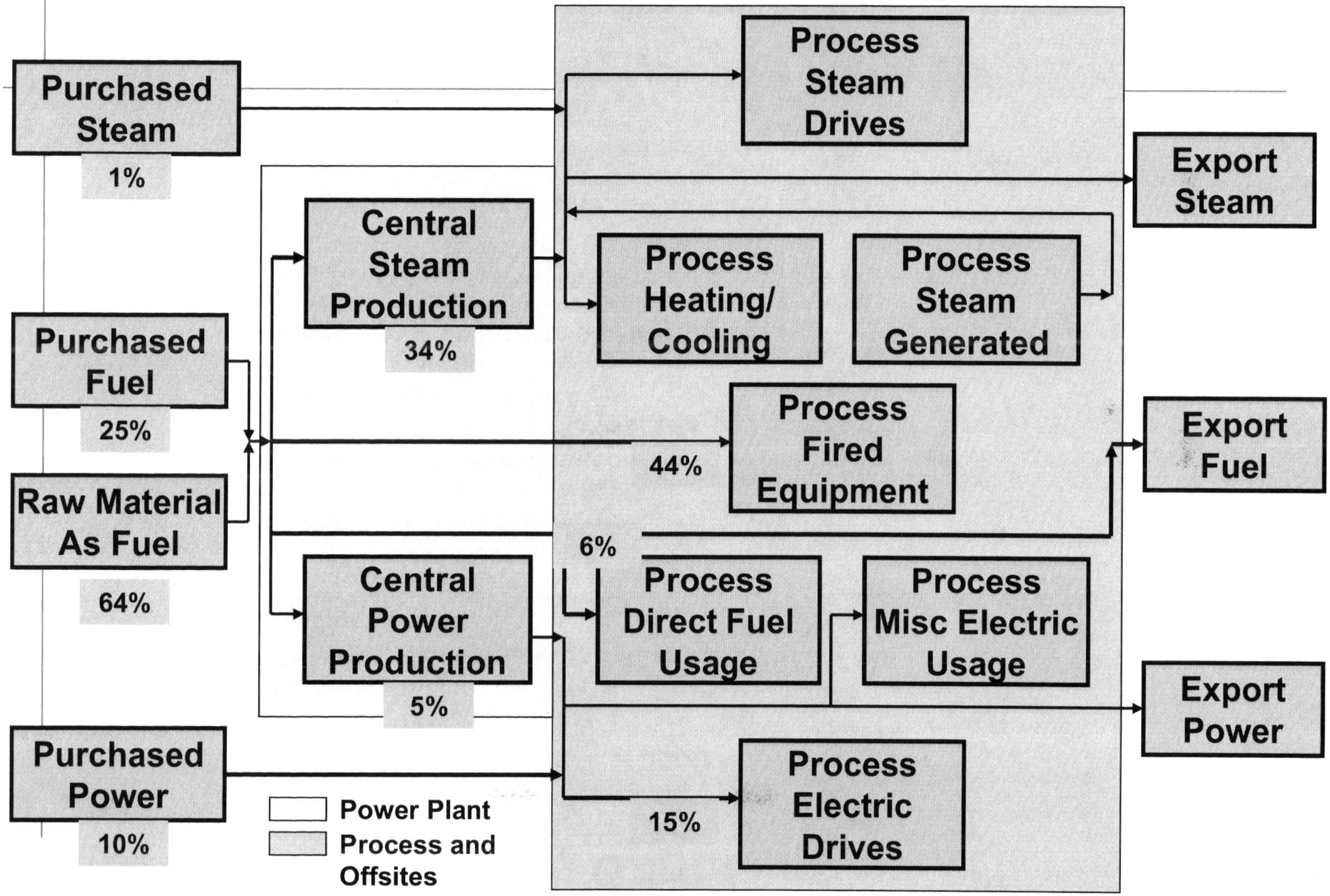


%-Equivalent BTU basis (Including losses) on total input

# Reducing Plant Energy Costs

- Reduce Usage
  - Individual Equipment
    - Improve Efficiencies – Boilers, Heaters, kilns
    - Maximize Useful Recovery - Preheat
    - Minimize Losses
      - Cooling water
      - Minimize Motor Losses
  - Unit Savings
    - Optimize Process Unit Operations
      - Distillation/ Fractionation
    - Maximize Waste Heat Recovery
    - Minimize waste/ off spec
  - Site/ Multi – Unit Savings
    - Minimize Steam Losses and Downgrading
    - Switch of steam drives for electric or vice versa
    - Seasonal effects

- Reduce Cost of Production and Purchase
  - Fuel Substitution
  - Generation Maximization
  - Boiler and Turbine Allocation
  - Electric Purchase Optimization

***Automation and Advanced Automation are the keys to effective operation and minimum ongoing energy usage***

# Process Industry Energy Saving Primary Targets

- Fired Heaters
- Distillation/ Fractionation
- Central Power and Steam Production

# How can Automation Reduce Energy Usage?

# Process Industry Energy Saving Primary Targets

- Fired Heaters
- Distillation/ Fractionation
- Central Power and Steam Production

# Reducing Plant Energy Costs

- Reduce Usage
  - Individual Equipment
    - Improve Efficiencies – Boilers, Heaters, kilns
    - Maximize Useful Recovery - Preheat
    - Minimize Losses
      - Cooling water
      - Minimize Motor Losses
  - Unit Savings
    - Optimize Process Unit Operations
      - Distillation/ Fractionation
    - Maximize Waste Heat Recovery
    - Minimize waste/ off spec
  - Site/ Multi – Unit Savings
    - Minimize Steam Losses and Downgrading
    - Switch of steam drives for electric or vice versa
    - Seasonal effects

- Reduce Cost of Production and Purchase
  - Fuel Substitution
  - Generation Maximization
  - Boiler and Turbine Allocation
  - Electric Purchase Optimization

***Automation and Advanced Automation are the keys to effective operation and minimum ongoing energy usage***

# Oil Refinery Energy Flow

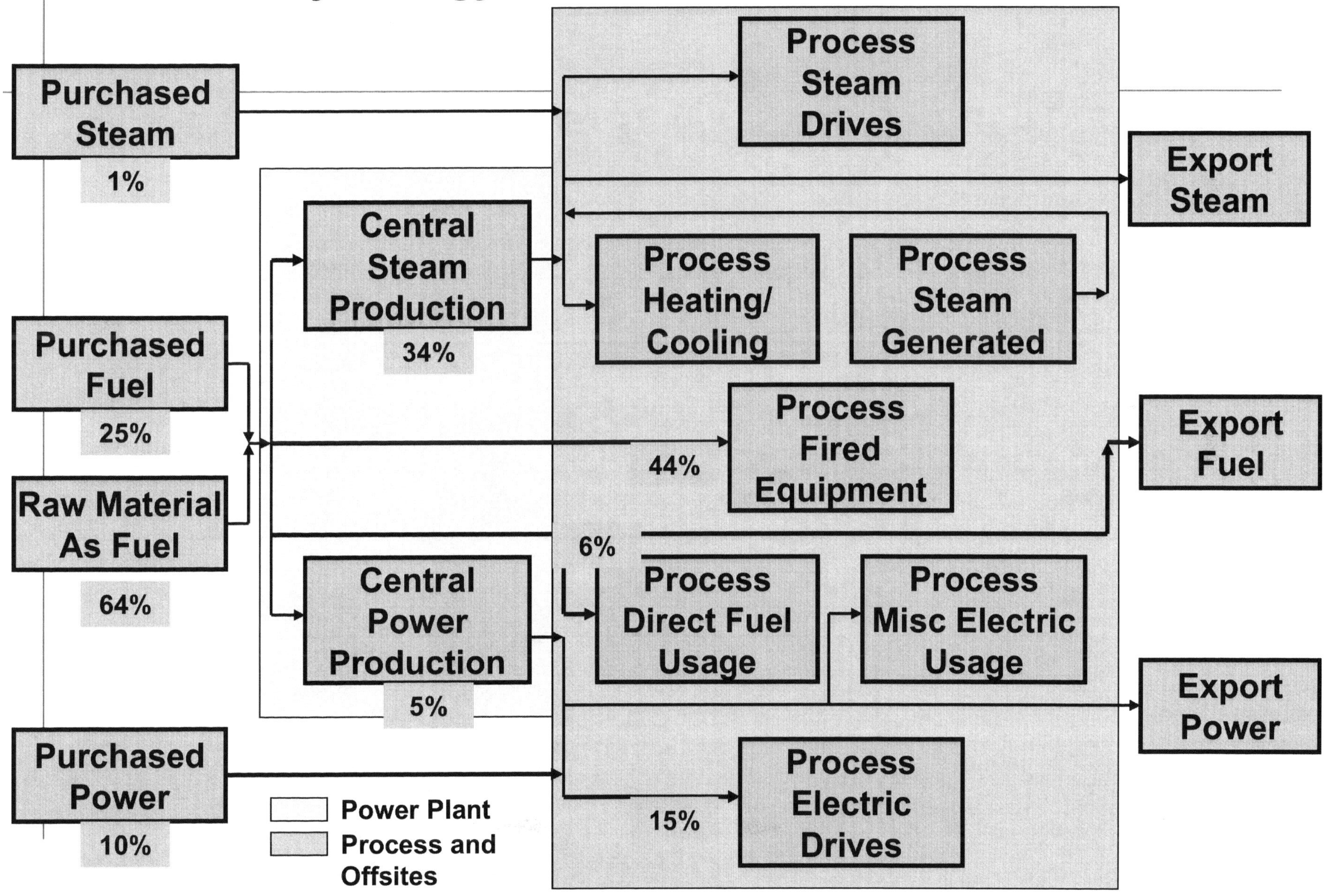


%-Equivalent BTU basis (Including losses) on total input

# Component Heating Values

| Fuel Gas Component Heating Value | |
|---|---|
| Component | Heat of Combustion kcal/ NM3 (gross) |
| Hydrogen | 3020 |
| Methane | 9520 |
| Ethane | 16820 |
| Propane | 24320 |
| Butane | 32010 |

***Control Fuel Flows By Mass Instead of Volume***

# Energy Savings From Improved Measurements – Hydrogen Plant

*Figure 1 Hydrogen Production in SMR Unit*

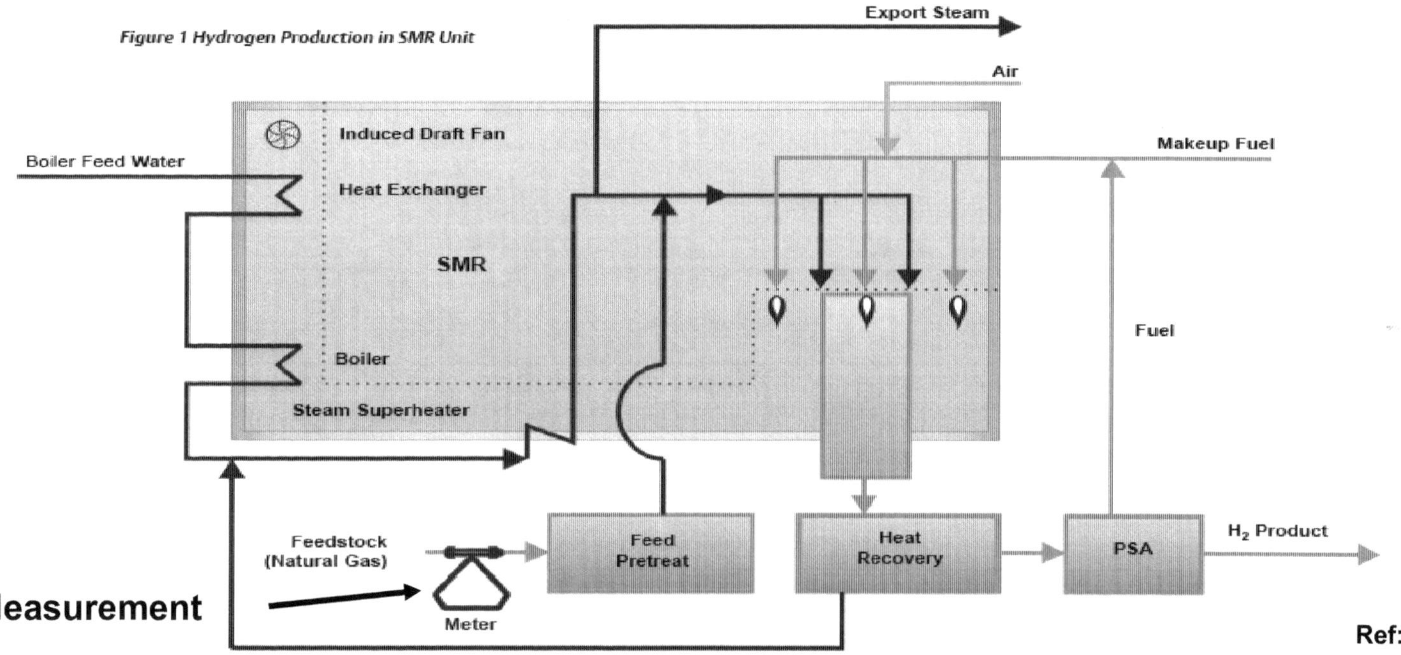


**Mass Measurement**

**Ref: MMI WP_00724**

**Objective: Control S/C ratio as close to 3.2 as possible but avoid going below**

**Disturbance: Fuel gas C1 77 – 85%; C2 6.8 – 15; N2, CO also fluctuate**

**Test: Normal orifice plus GC – max error 0.2; MMI – max error – 0.02**

**Benefits: Moving 0.2 ratio closer to limit worth 8 BTU/SCF of H2; 80 MMSCFD plant; $7 MM BTU gas –**

***$1.6 MM per year***

# Steam System Control Issues

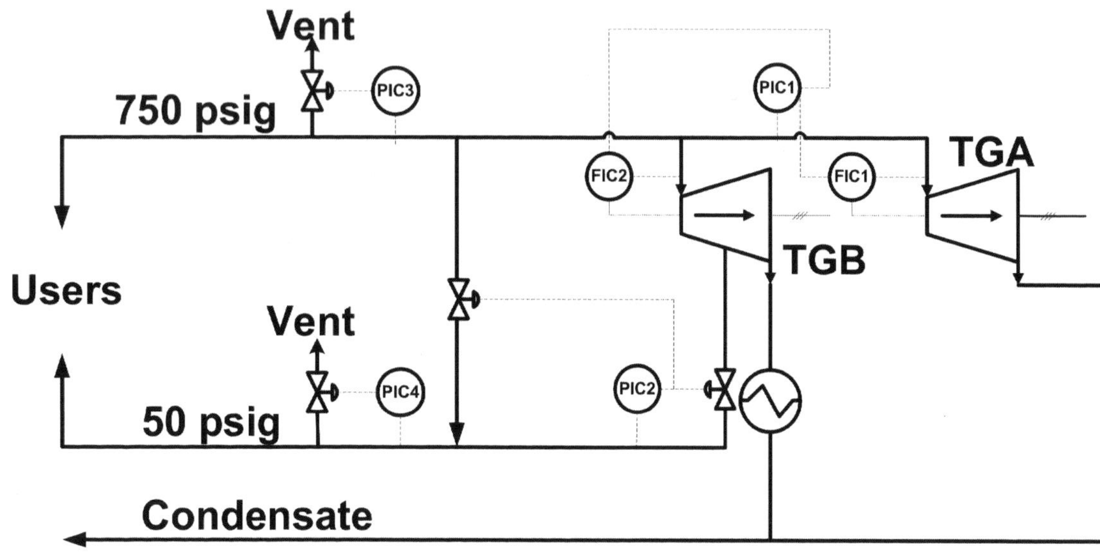


**Objective: Maximum Flow to TGA**

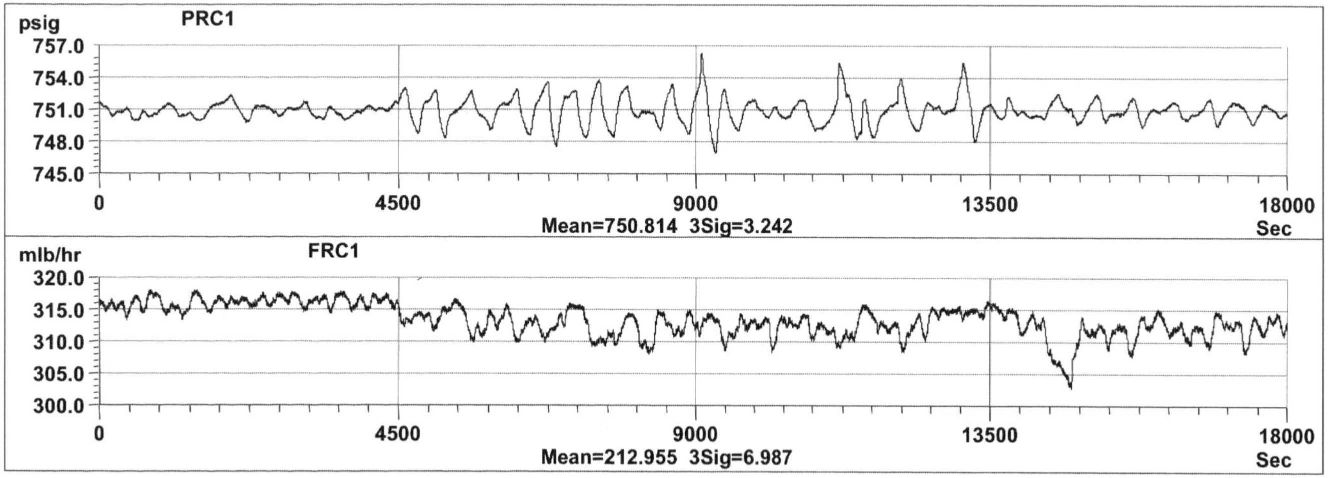


**Problem: Pressure Instability in Header Limited Flow to TGA**

# Steam System Diagnosis – Valves and Tuning

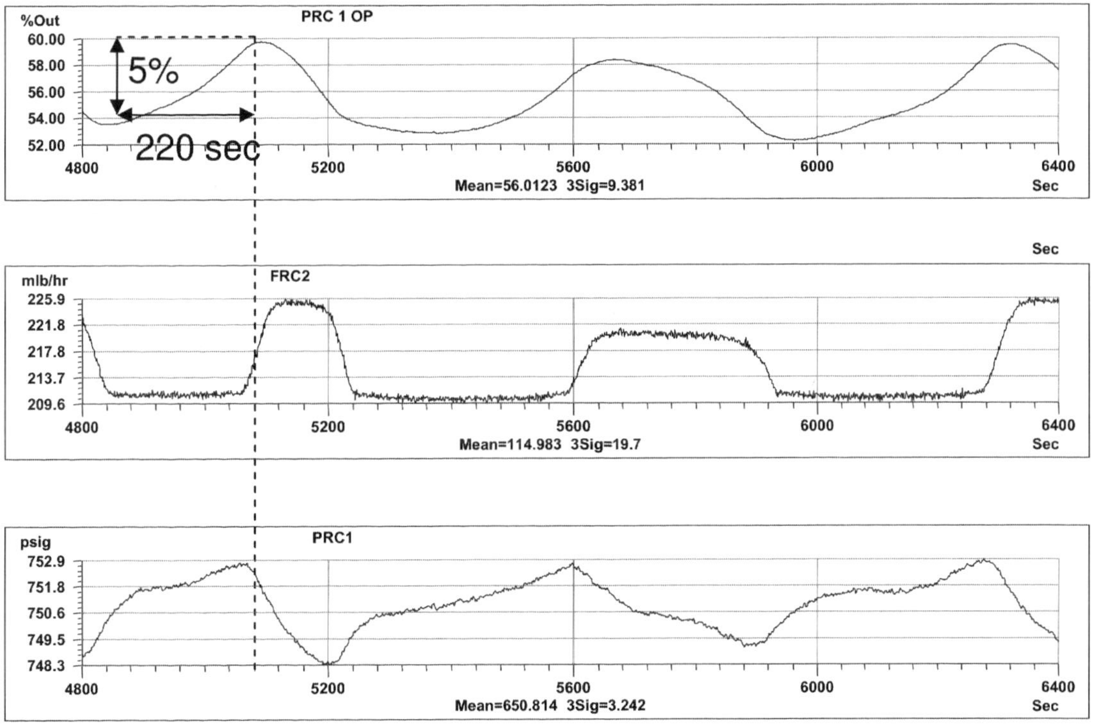


**Flow controller to TGB has 5% deadband; induces limit cycle in pressure**

**Correction: Fix TGB turbine governor/ steam valve**

**Tune controllers as system – not individually**

***Estimated value of increased flow to TGA - $3000/ day***

# Energy Losses Through Bypassing

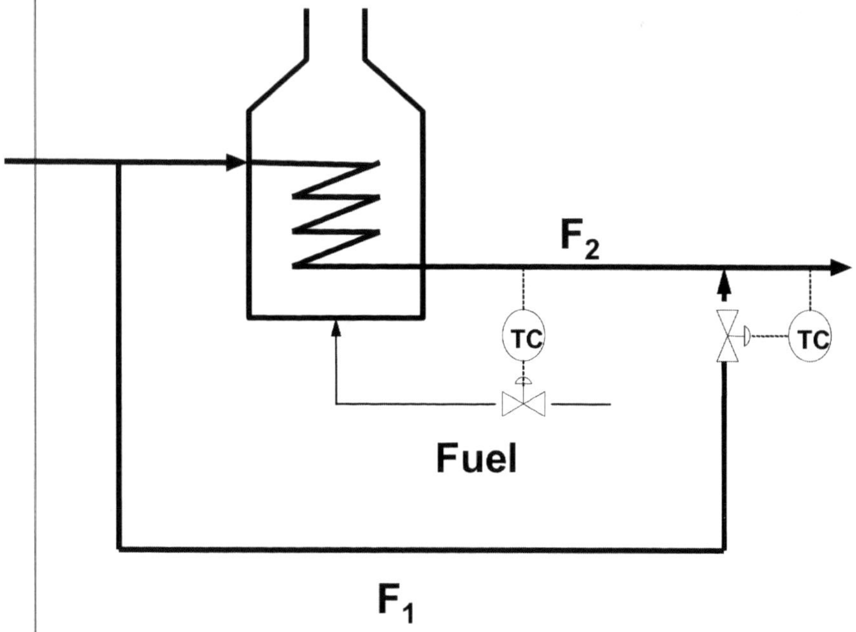


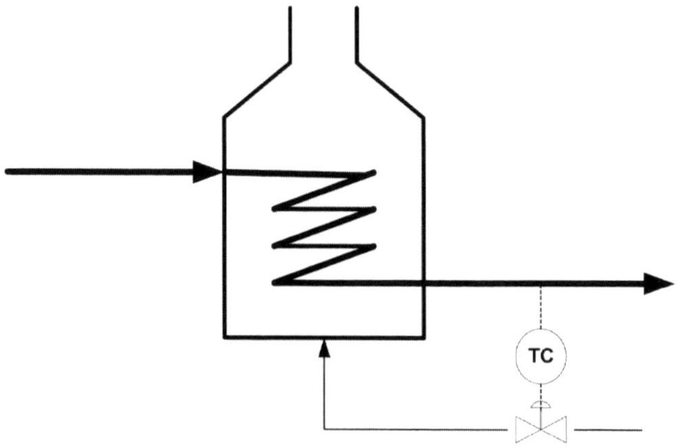


| $F_1/(F_1+F_2)$ | Heat Loss Increase -% |
|---|---|
| 0.14 | 3.2 |
| 0.25 | 6.8 |
| 0.4 | 14.3 |

Reference:
Shinskey;

<u>Energy Conservation Through Automation</u>

# Energy Savings – Equipment Level

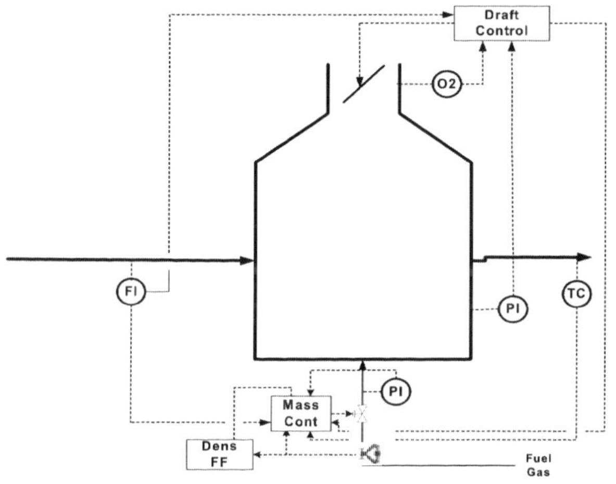


- •Improved Multi-Loop Control – Advanced Control
- •Improved Performance Monitoring
- •Improved Diagnostics

# Fired Heater Controls

# Combustion Control

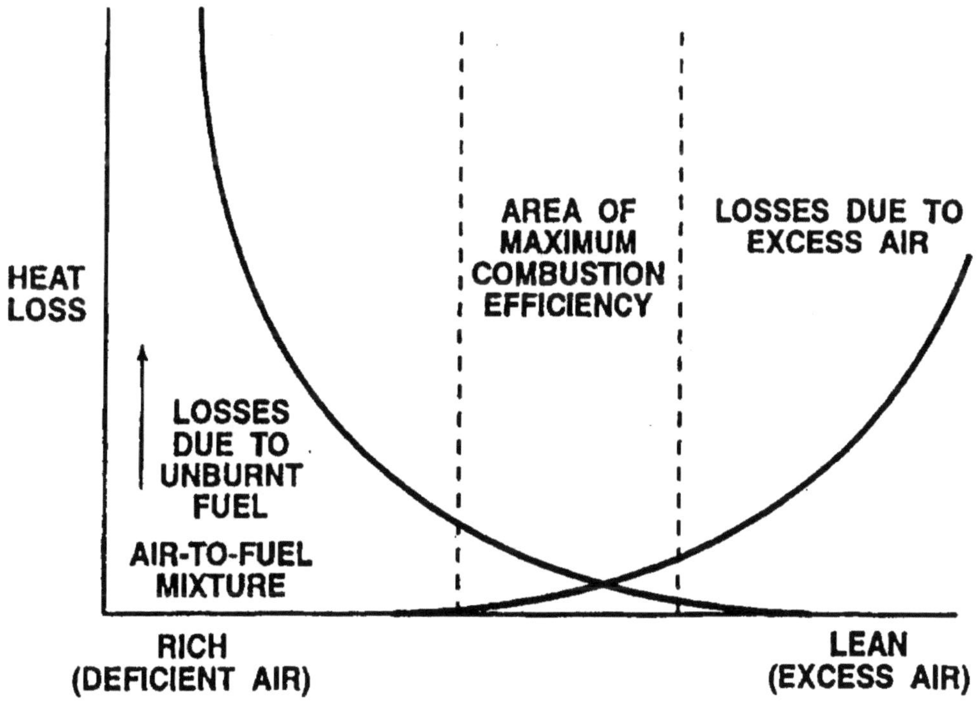

# Heater/ Boiler Combustion Control Savings

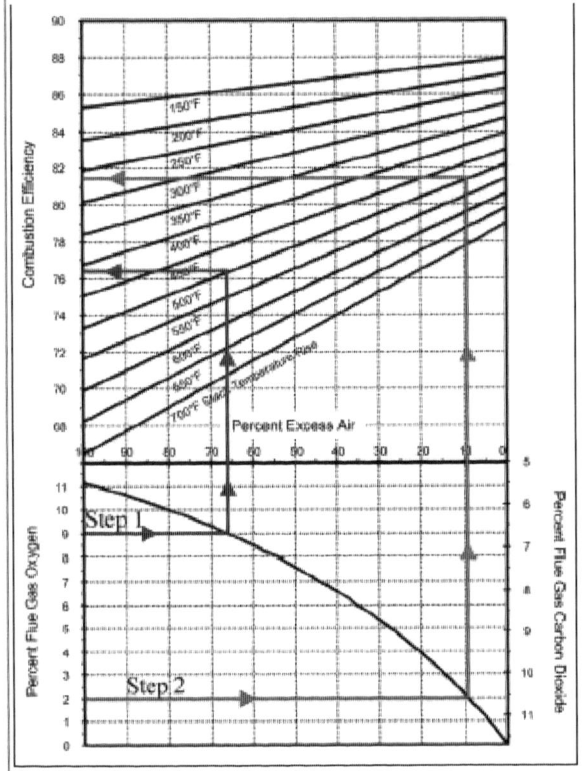


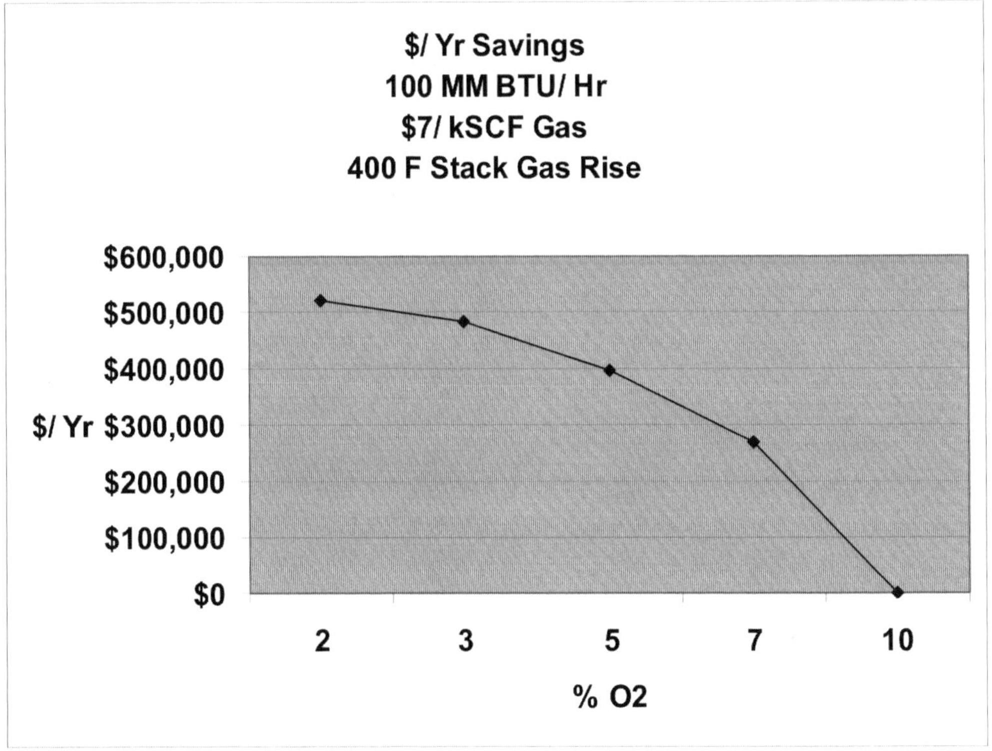

# Typical Heater APC Package

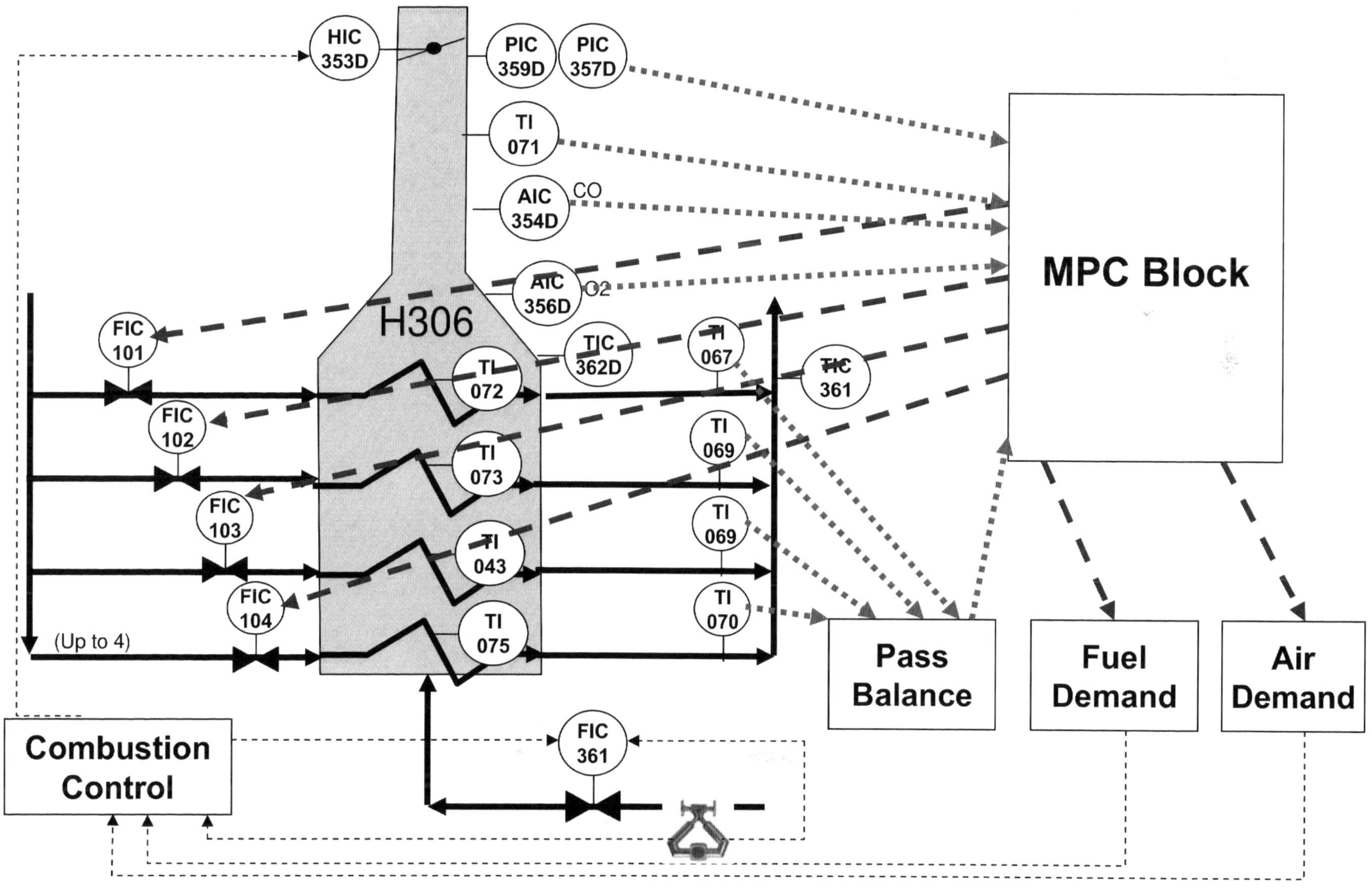

# Excuses For Not Improving Heater Controls

- Damper/ Air controls are not reliable
  - Answer: Add positioners to dampers, with feedback to control system; Analyze and fix controller problems
- Don’t have online analyzer/ can’t maintain them
  - Answer: Analyzers are cheaper and more reliable – particularly mass flow meters. With higher fuel costs, they are well justified.

# Boiler Controls

# Improved Boiler Controls
# Combustion Characterization

Figure 4 Desired $O_2$ Setpoint vs. Steam Load, $f_3(x)$, for Boiler 3 Burning Gas

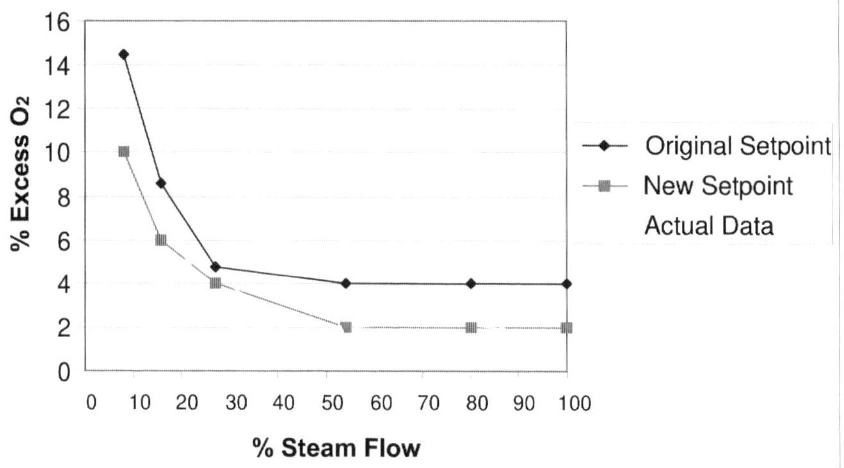


| | $f_3(x)$ for Boiler 3 for Gas | | |
|---|---|---|---|
| Full Scale | | Desired O2 Setpoint | |
| 250 | X | Old Y | New Y |
| Steam | Steam | Original | New |
| KPPH | % Steam | Setpoint | Setpoint |
| | | | |
| 20.00 | 8 | 14.5 | 10.00 |
| 40.00 | 16 | 8.6 | 6.00 |
| 67.50 | 27 | 4.75 | 4.00 |
| 135.00 | 54 | 4 | 2.00 |
| 200.00 | 80 | 4 | 2.00 |
| 250.00 | 100 | 4.00 | 2.00 |

**1% Fuel Savings; $228,000 per year ($8/ kSCF)**

# Wood fired boiler

# Wood fired boiler

- Board Mill
- Only one boiler for whole plant.
- Fuels = wood chips, wood waste, sawdust, and gas as a backup.
- Steam demand varies 60-120 Kpph (50%) every 6 to 12 minutes
- As found, boiler trips for drum level several times a day.

# Automation Results

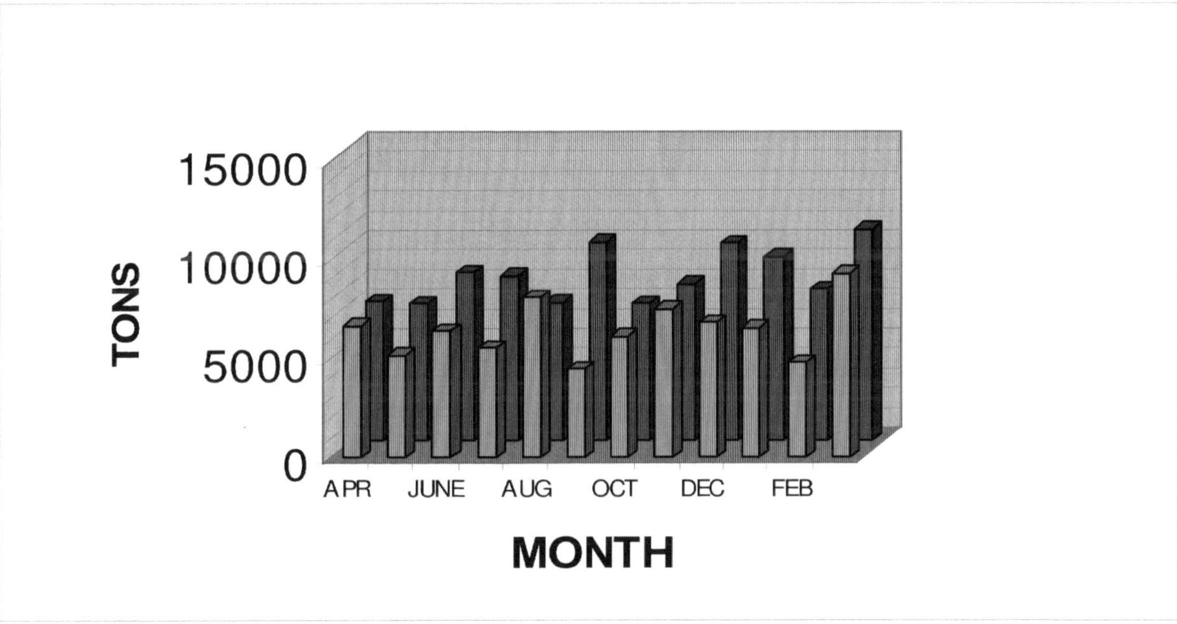


RESULTS:

- 21% less fuel consumed with new controls for annual savings of $223,000 (now >$500,000)
- boiler trips reduced from 3+/day to 1/week

# Energy Savings – Process Unit

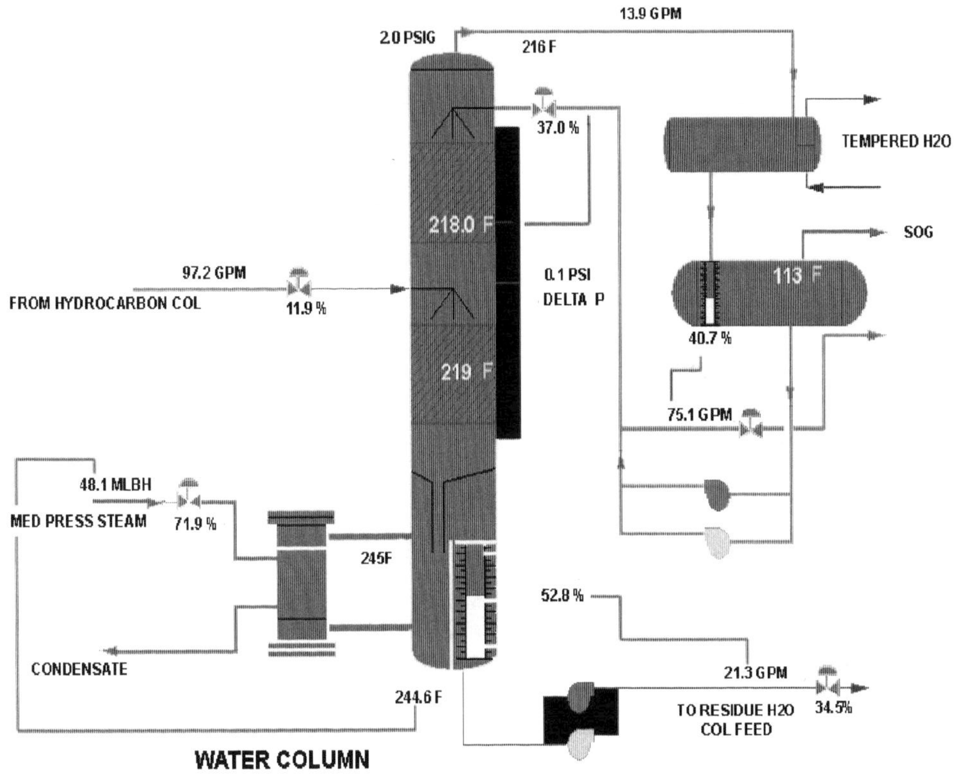


- Better Performance Monitoring and Control
- Advanced Control

# Distillation Controls

# Typical Distillation Column

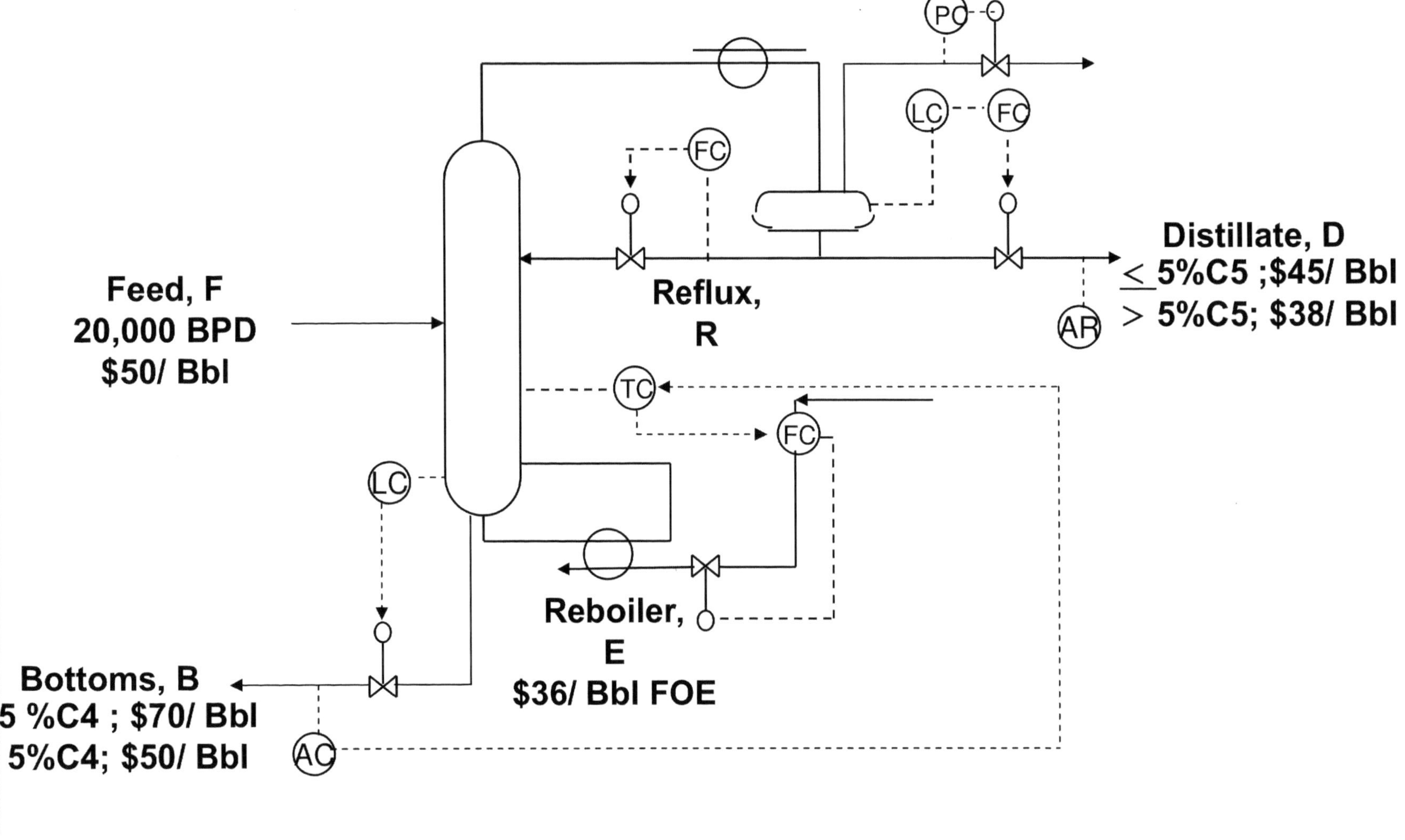

# Product Value

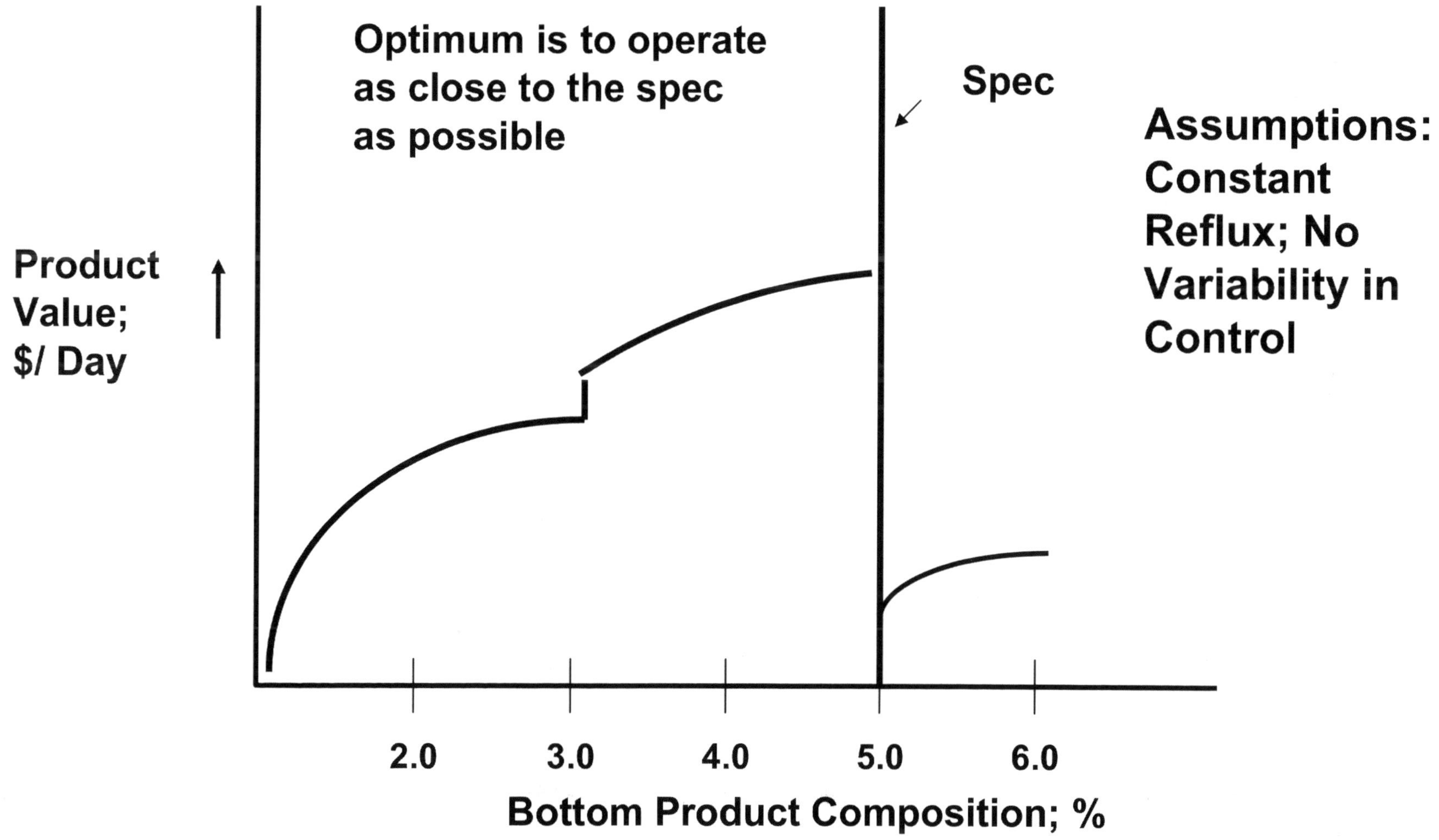

# Profitability

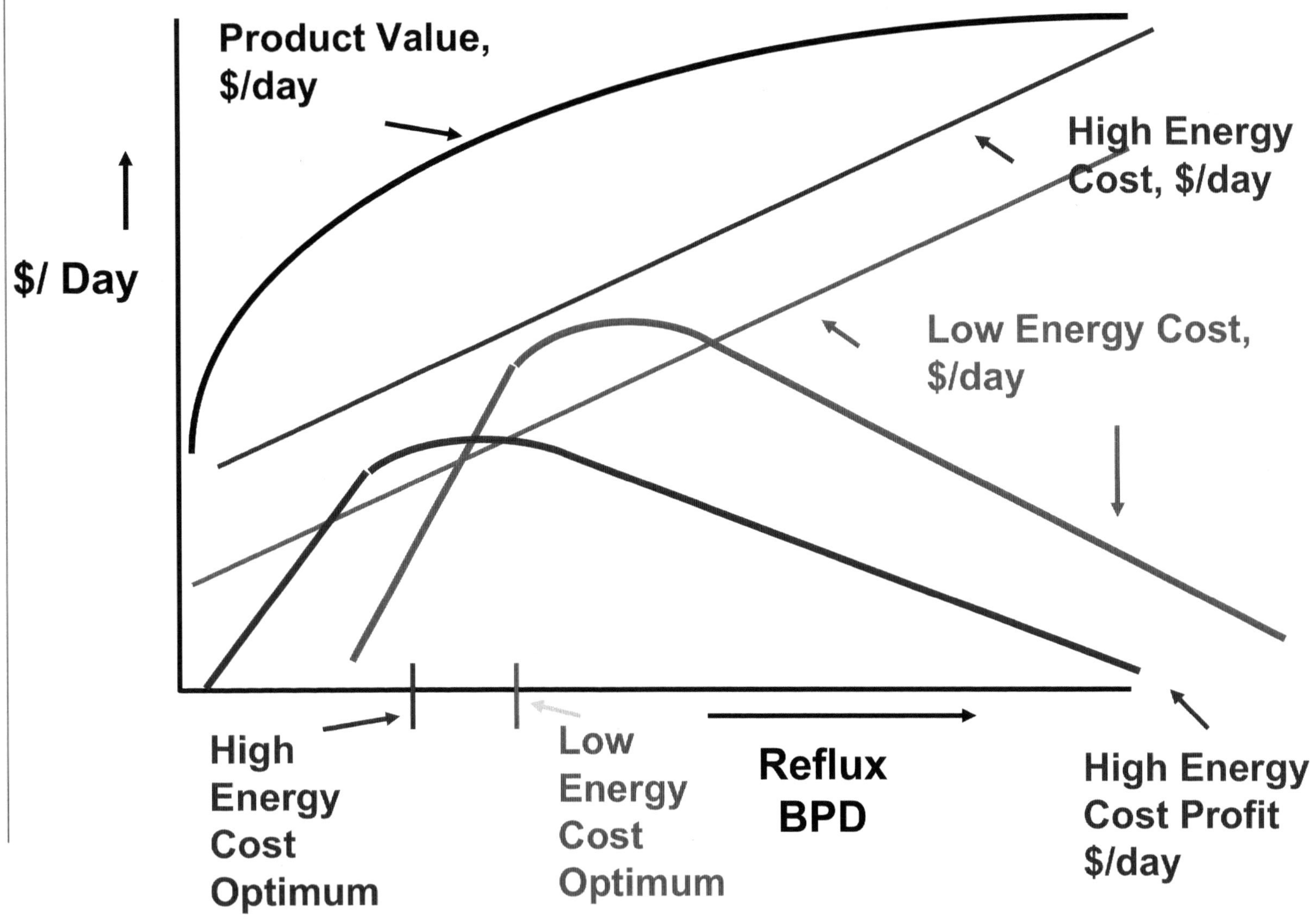

# Effect of Reduced Variability

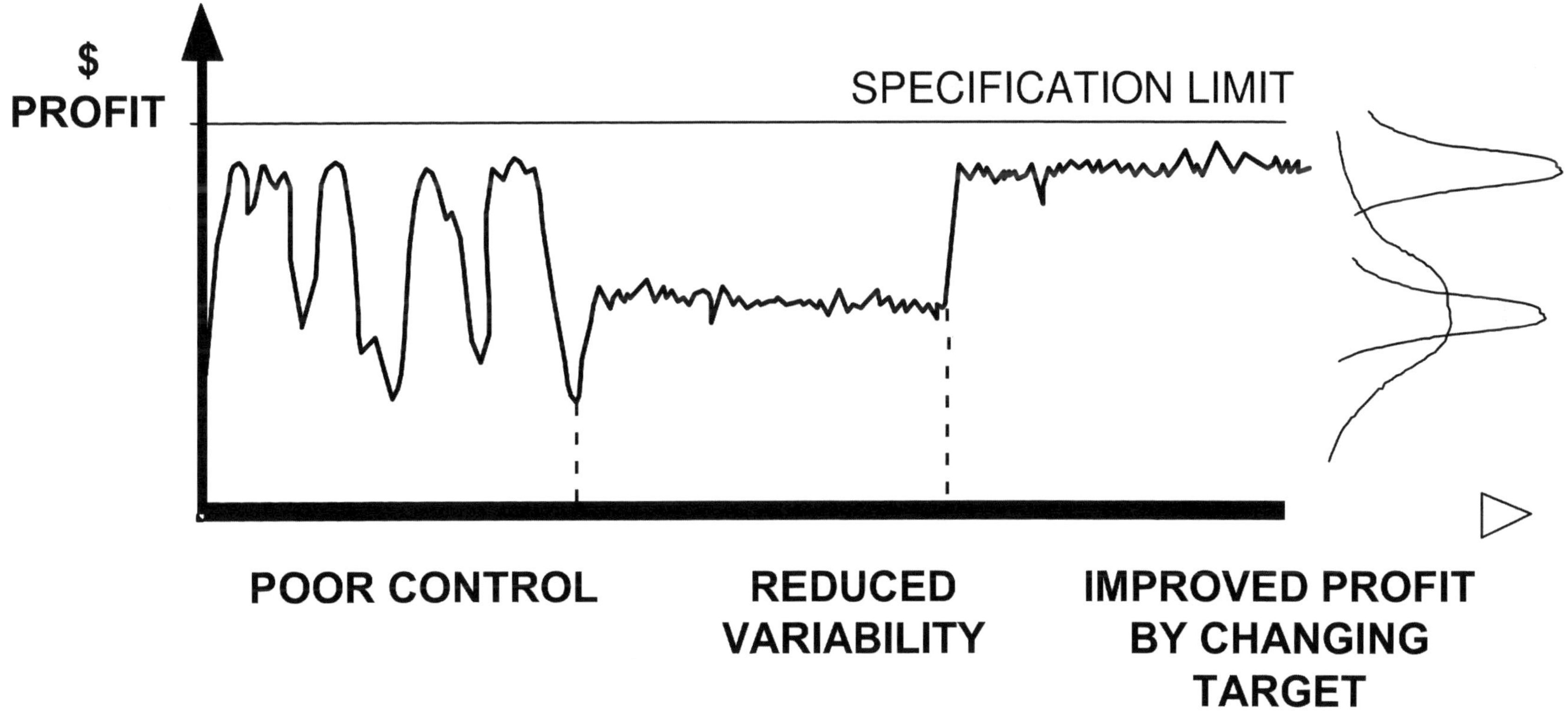

# Profitability with variability

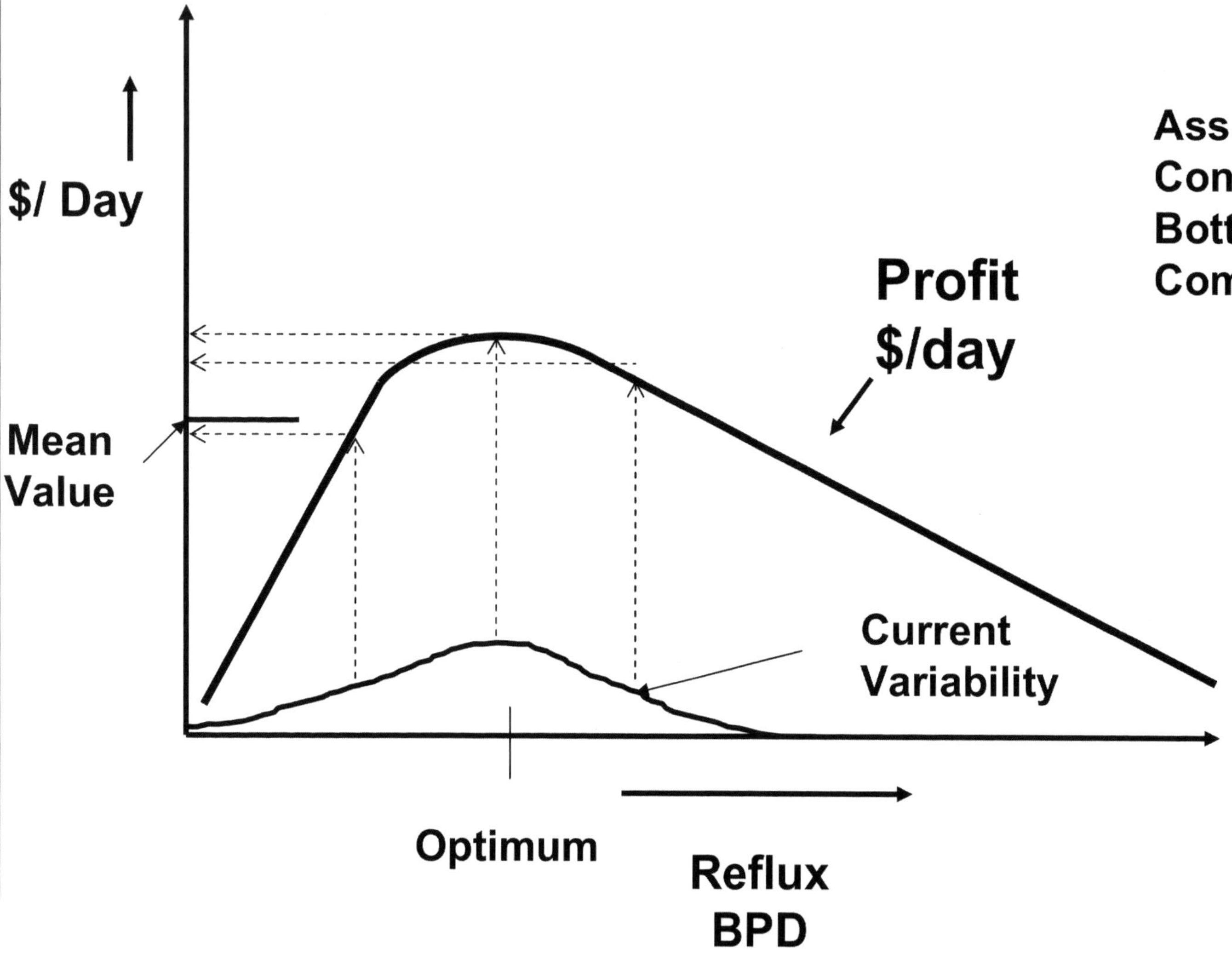

# Profitability with variability

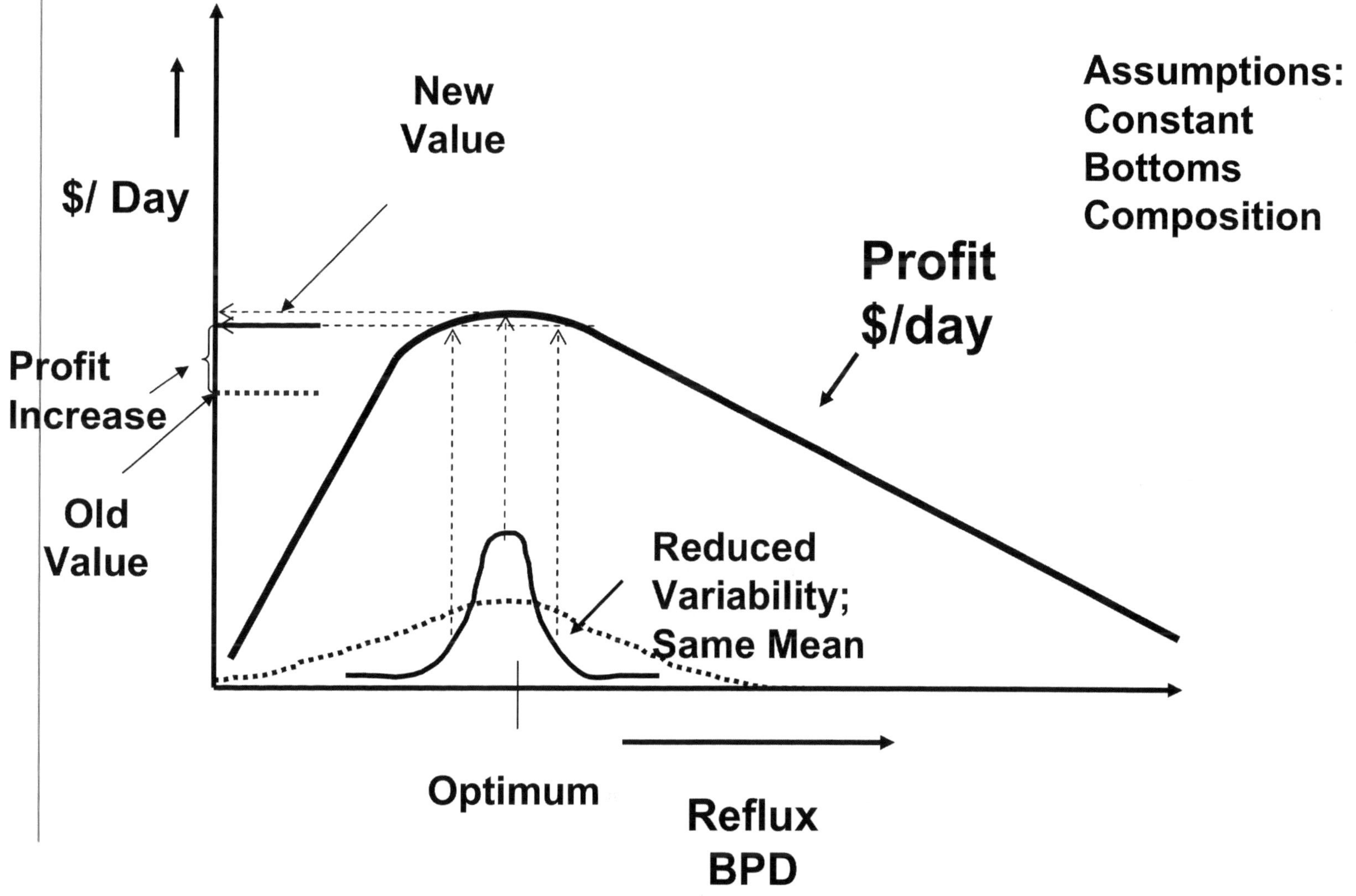

# Distillation Column Control Savings

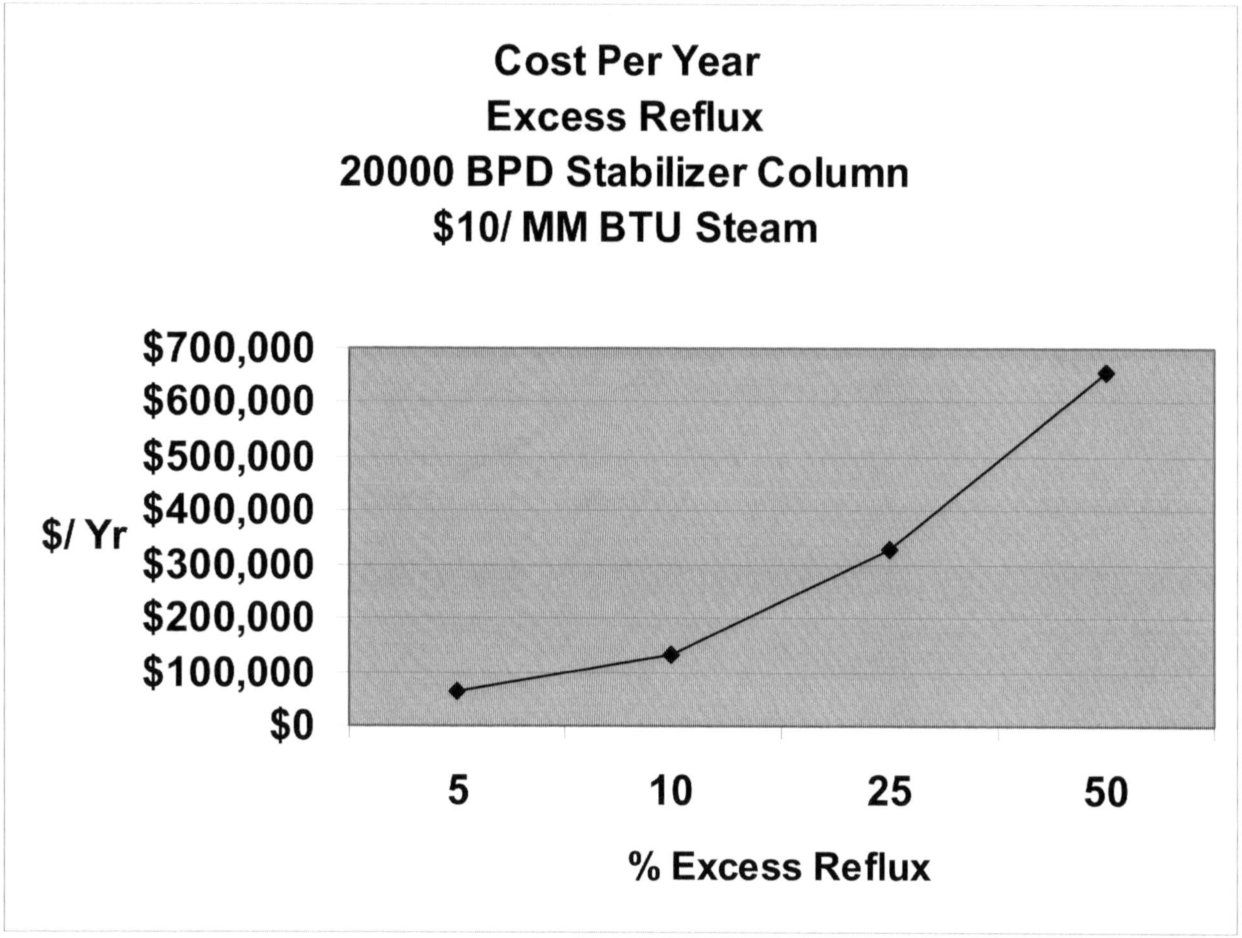

# Column Pressure Effect

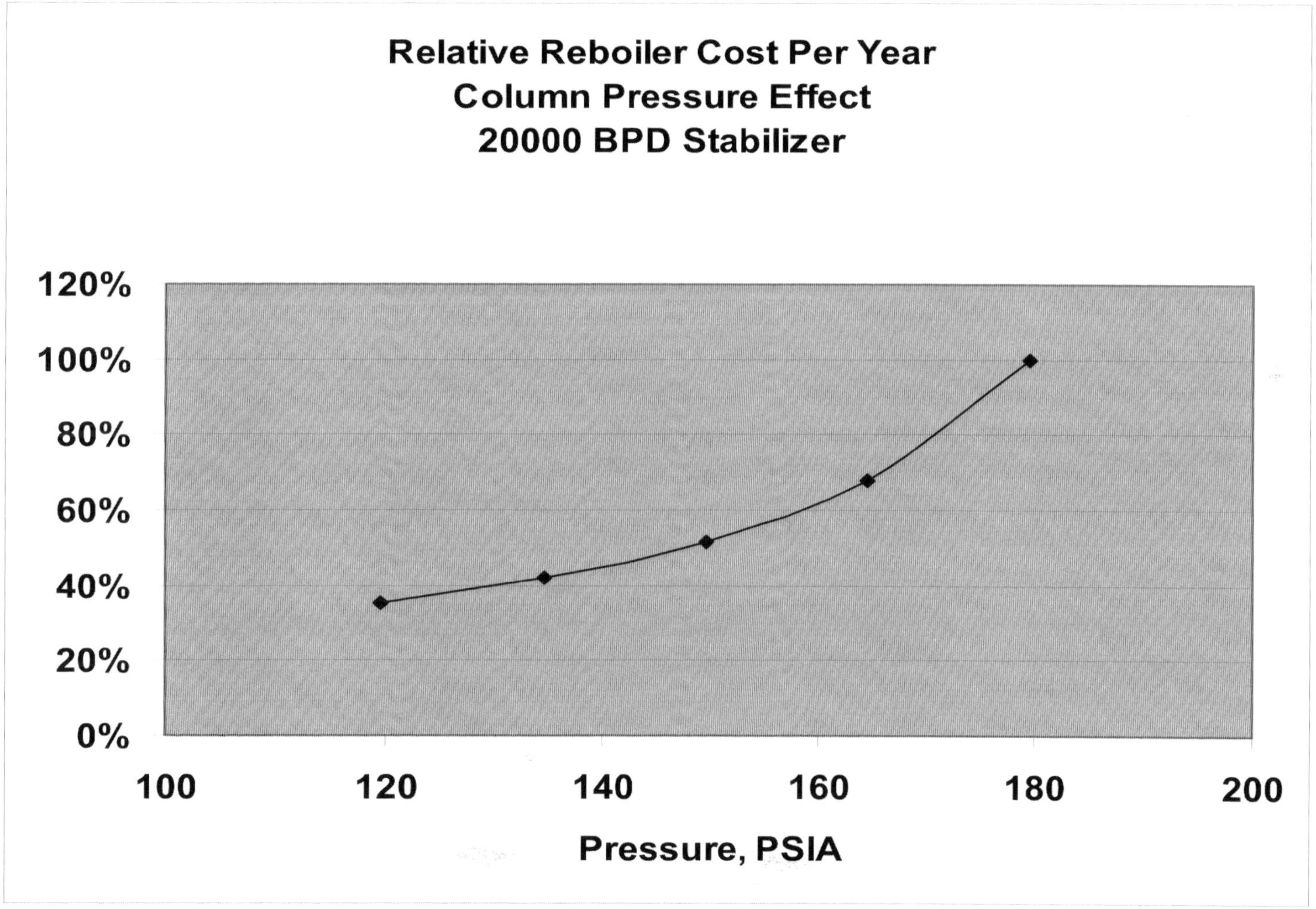


Basis: Constant Separation
Modeled With ChemSep
Peng Robinson Equation of State

# Energy Savings – Site Wide

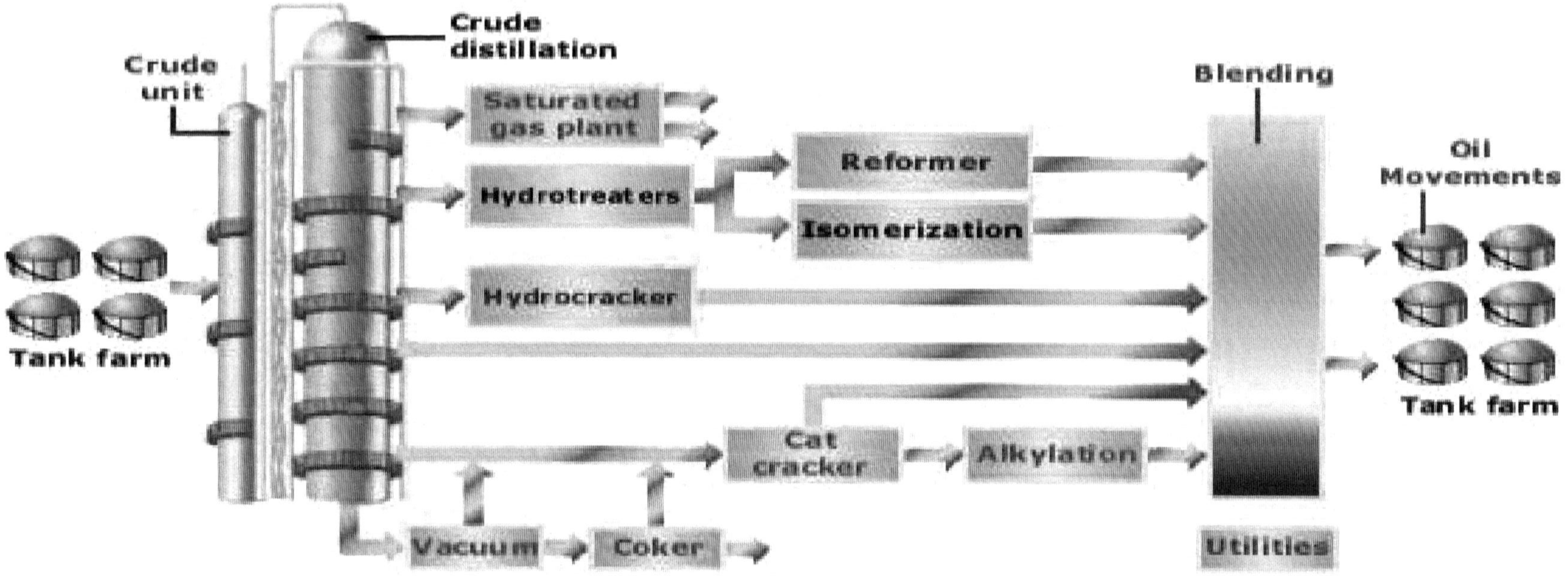


- Site Energy/ Utility Management
- Steam System Control
- Fuel System Control

# Energy Management and Optimization System

# Plant Utility Systems – Many Opportunities for Savings

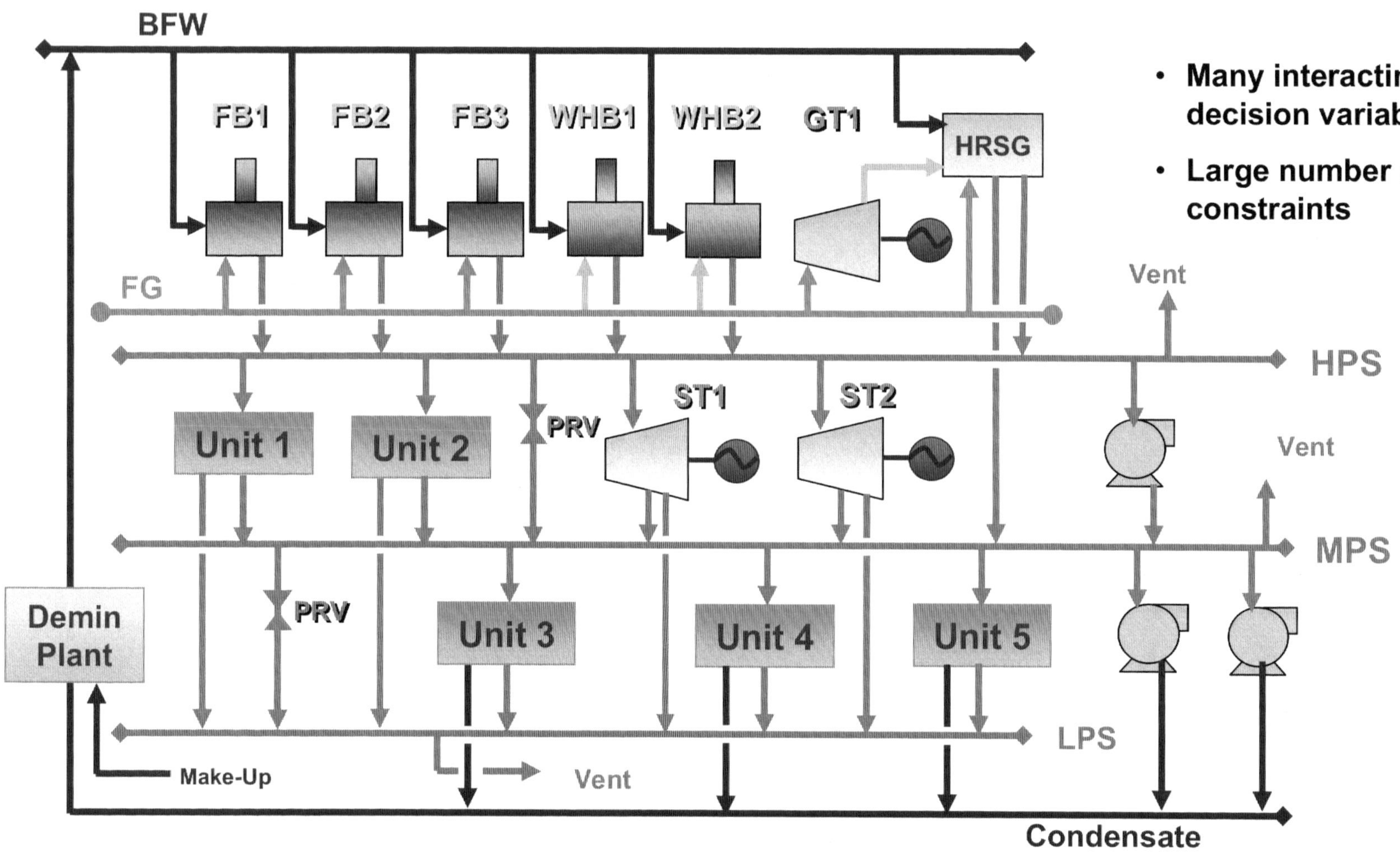


- Many interacting decision variables
- Large number of constraints

# Optimizer Decisions

- Which boiler(s) should I run? What load?
- How much electricity should I produce? Buy? Sell? Is it economic to run the steam turbine?
- Which fuel should I buy? How much?
- Should I be using more steam drives or more electric drives?
- When will efficiency gain from maintenance balance the cost of shut down for this equipment?
- How does my actual compare with plan corrected to standard conditions?

# Full Utility Optimization

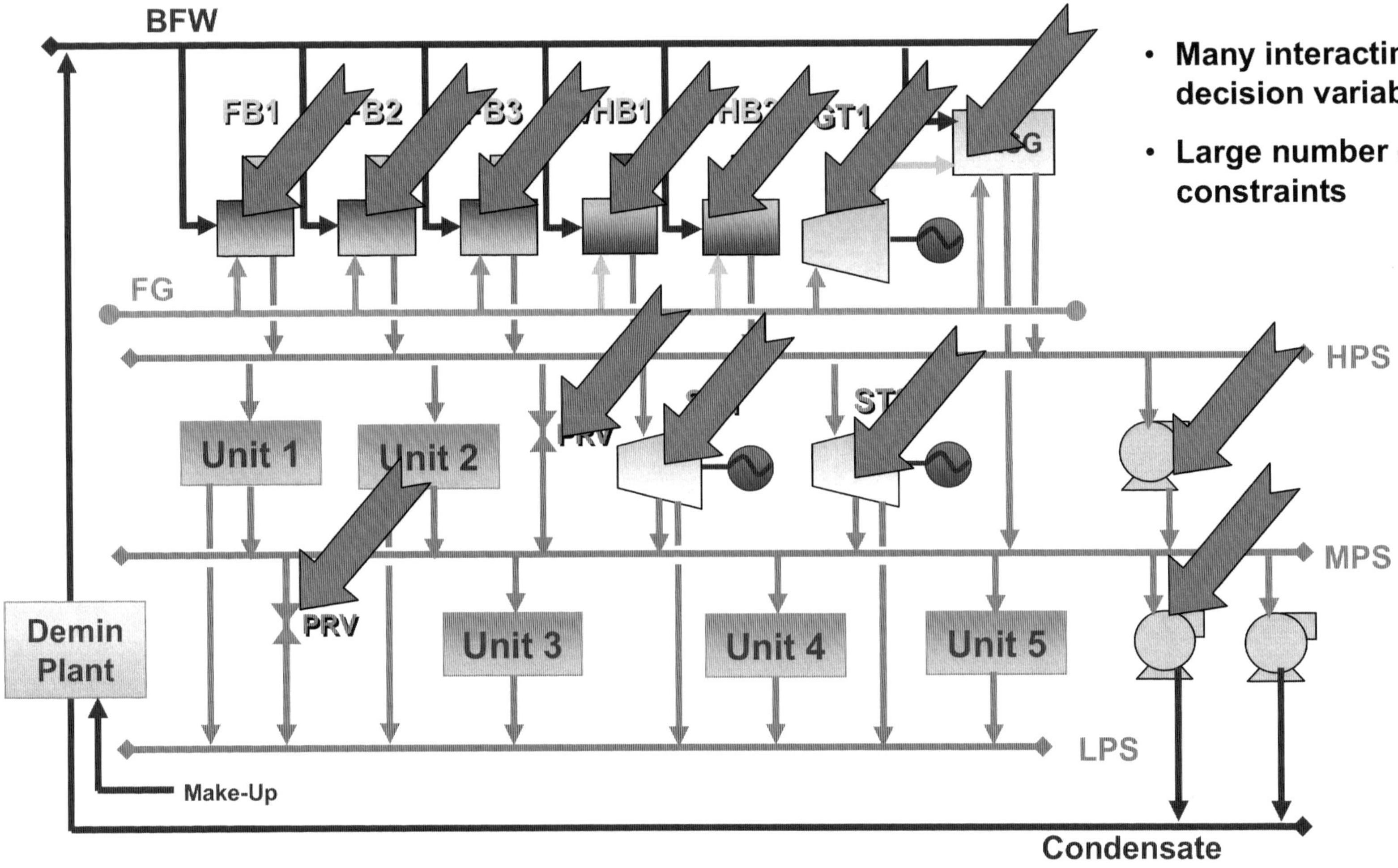


- Many interacting decision variables
- Large number of constraints

# Overall Energy Optimization Strategy

- Continuously Calculate Production Costs Over Load Range with Current Fuel Mix
- Incorporate Constraints on All Equipment
- Decisions Made Through Rule Based Logic
- Boiler Load Allocation
  - Distribute Steam Production Based on Cost and Constraints
- Turbine Load Allocation
  - Distribute Steam for Minimum Cost with Constraints
- Tie-Line Control
  - Control Electrical Purchase Based on Economic Decision and Constraints

# Boiler Load Allocation

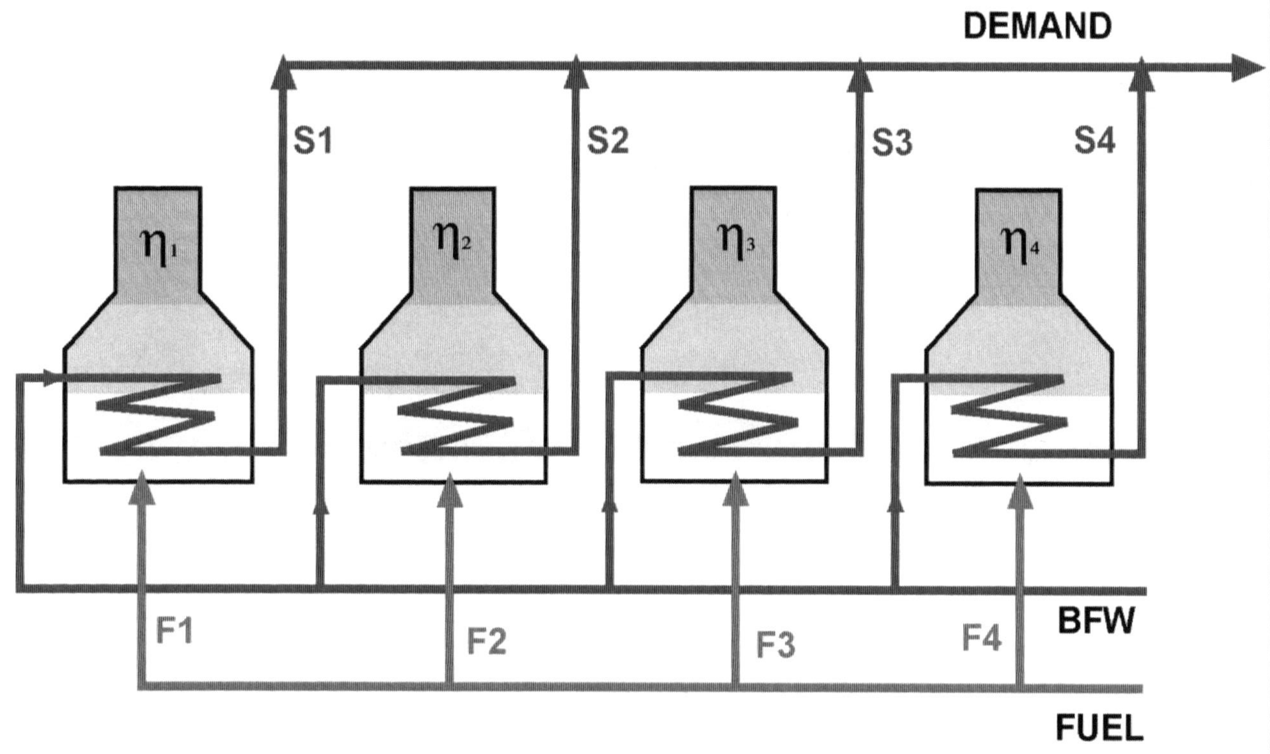

# Load Allocation

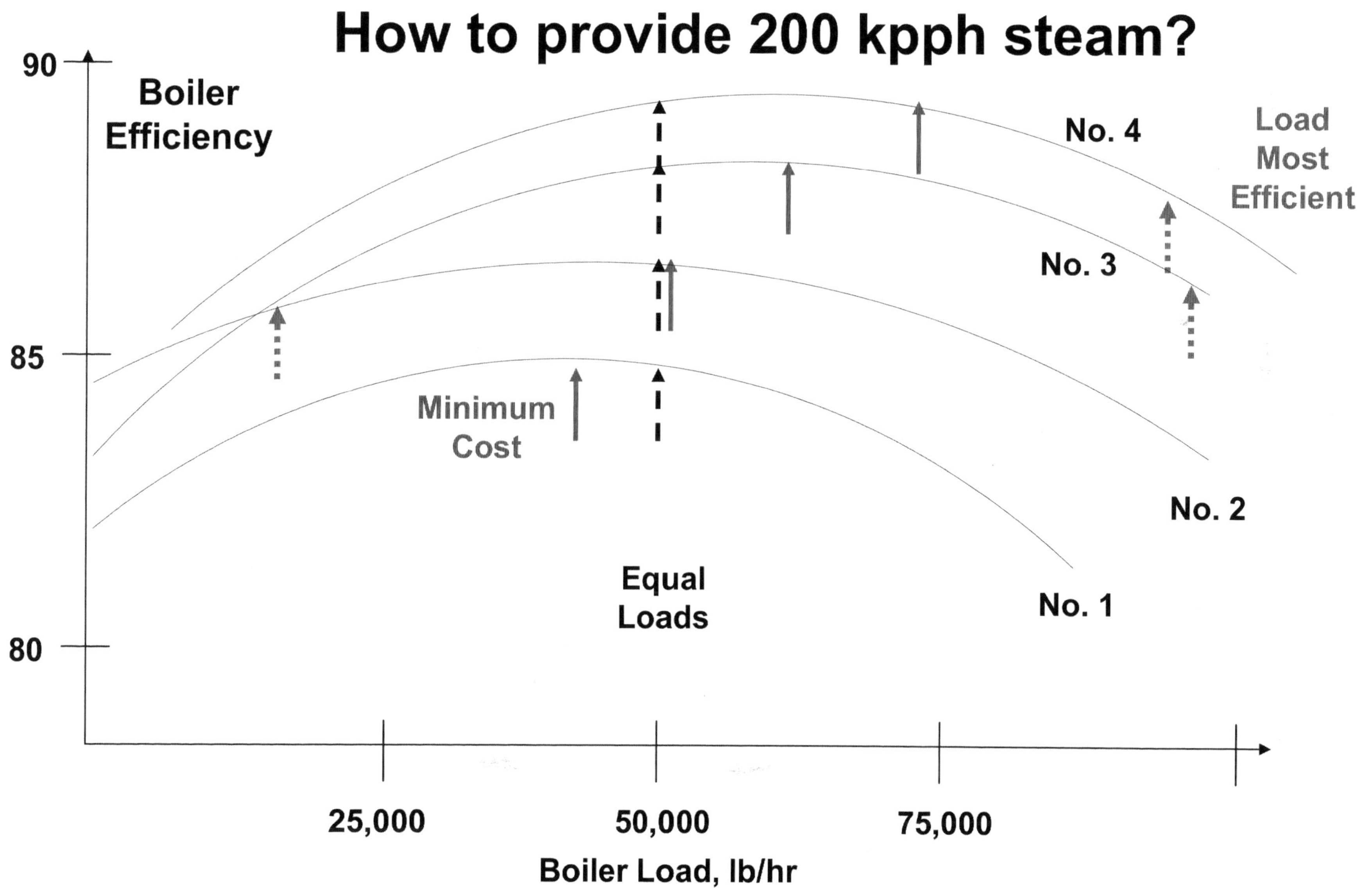

# Energy Savings via Site Energy Balance

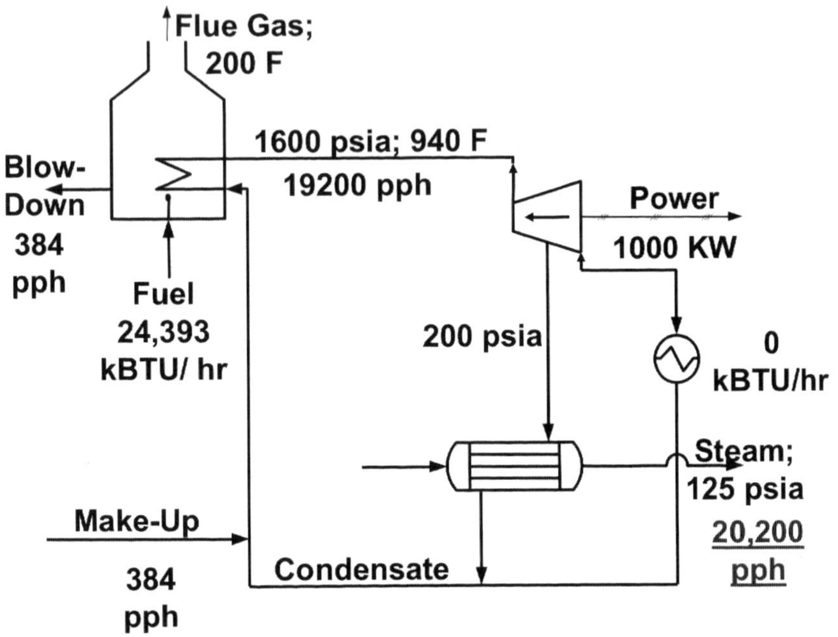


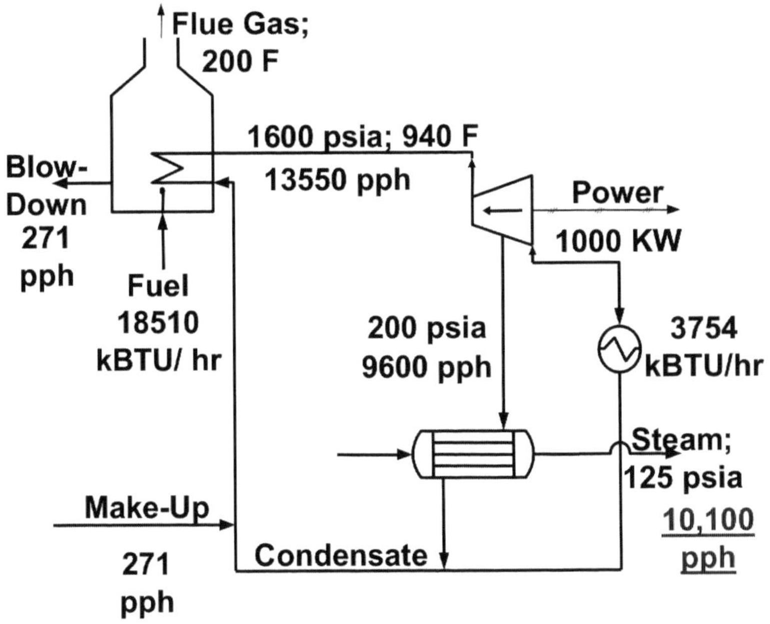


**Naïve calculation, value 125 psia steam reduction =**
**1000 x (1031 Btu/ lb ($\Delta H_v$) ÷ 0.7(eff))x \$7/MMBtu (Fuel) =**
**\$10.31 per klb**

**Actual site value 125 psia steam reduction =**
**\$4.08 per klb**

Reference: Kinney; Energy Conservation in Process Industries

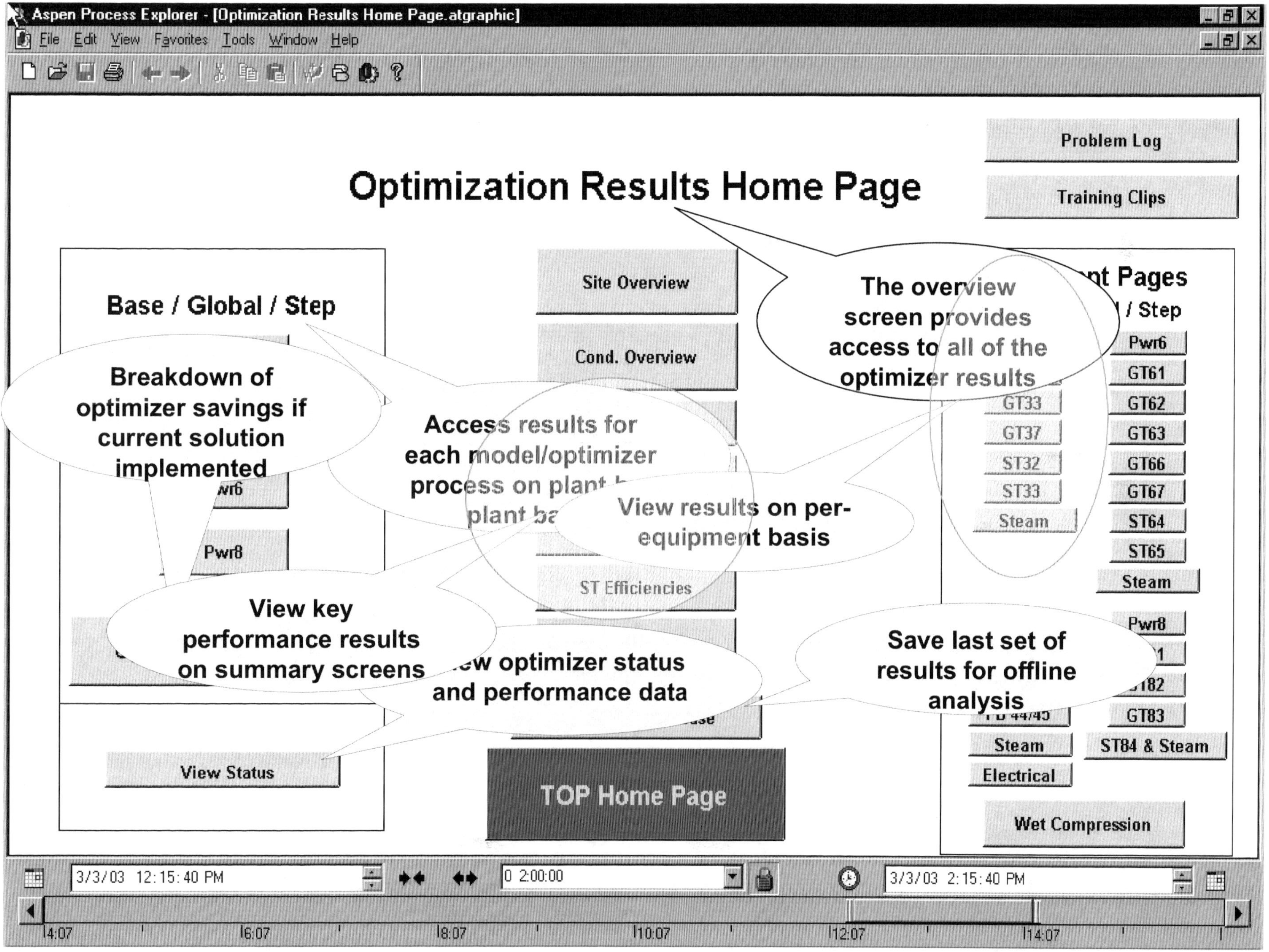
Aspen Process Explorer - [Optimization Results Home Page.atgraphic]
File Edit View Favorites Tools Window Help
Optimization Results Home Page
Problem Log
Training Clips
Base / Global / Step
Site Overview
Cond. Overview
ST Efficiencies
Pwr8
View Status
TOP Home Page
Pwr6
GT61
GT62
GT63
GT66
GT67
ST64
ST65
Steam
GT33
GT37
ST32
ST33
Steam
Pwr8
GT83
ST84 & Steam
Steam
Electrical
Wet Compression
The overview screen provides access to all of the optimizer results
Breakdown of optimizer savings if current solution implemented
View results on per-equipment basis
View key performance results on summary screens
and performance data
Save last set of results for offline analysis
3/3/03 12:15:40 PM
0 2:00:00
3/3/03 2:15:40 PM
4:07
6:07
8:07
10:07
12:07
14:07

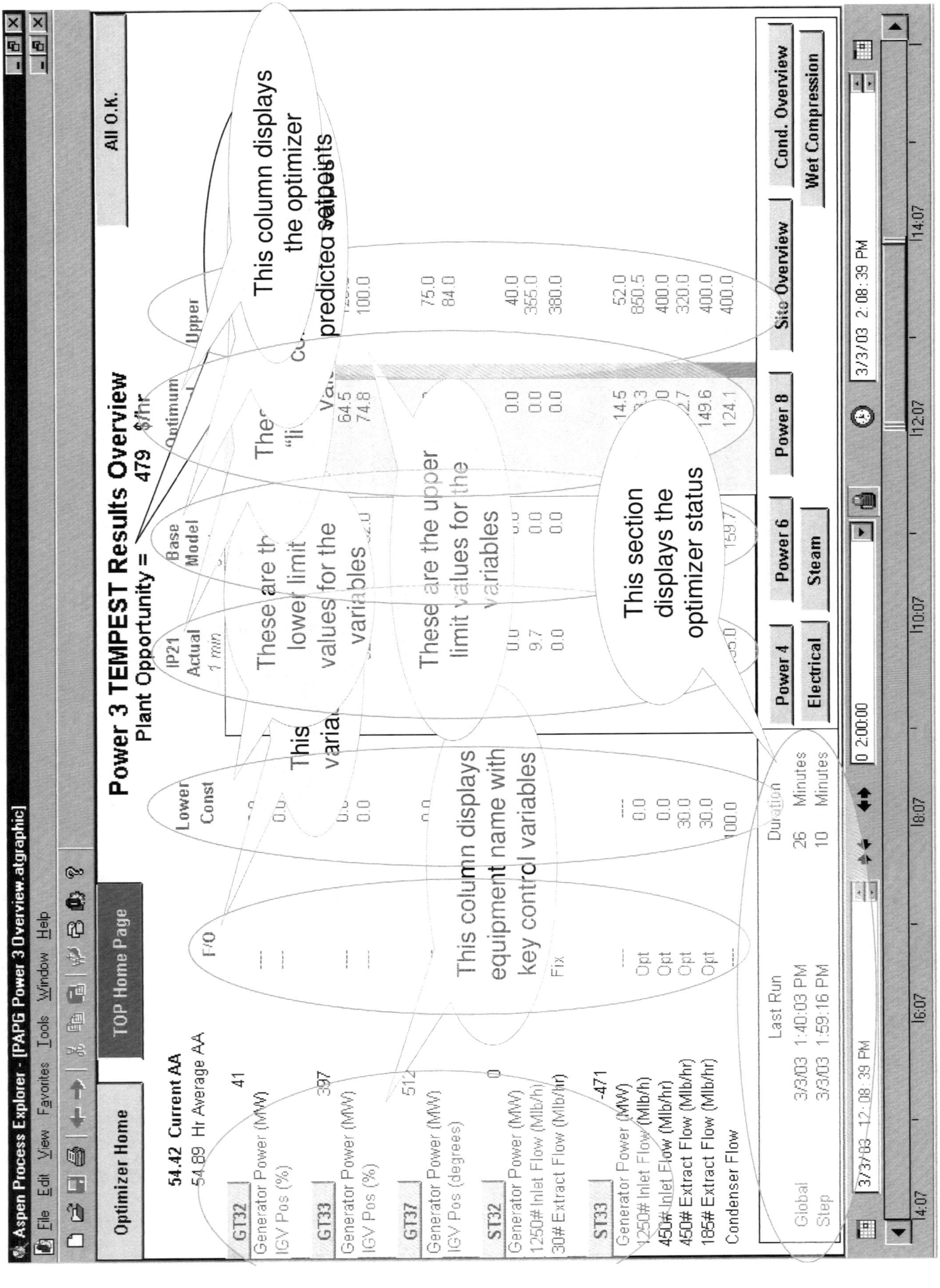
Aspen Process Explorer - [PAPG Power 3 Overview.atgraphic]
File Edit View Favorites Tools Window Help
Optimizer Home
TOP Home Page
All O.K.
Power 3 TEMPEST Results Overview
Plant Opportunity = 479 $/hr
54.42 Current AA
54.89 Hr Average AA
F/O
Lower Const
IP21 Actual 1 min
Base Model
Optimum
Upper
GT32
Generator Power (MW)
IGV Pos (%)
GT33
Generator Power (MW)
IGV Pos (%)
GT37
Generator Power (MW)
IGV Pos (degrees)
ST32
Generator Power (MW)
1250# Inlet Flow (Mlb/h)
30# Extract Flow (Mlb/hr)
ST33
Generator Power (MW)
1250# Inlet Flow (Mlb/h)
450# Inlet Flow (Mlb/hr)
450# Extract Flow (Mlb/hr)
185# Extract Flow (Mlb/hr)
Condenser Flow
Last Run
Duration
Global 3/3/03 1:40:03 PM 26 Minutes
Step 3/3/03 1:59:16 PM 10 Minutes
Power 4
Power 6
Power 8
Site Overview
Cond. Overview
Electrical
Steam
Wet Compression
3/3/03 12:08:39 PM
0 2:00:00
3/3/03 2:08:39 PM
This column displays the optimizer predicted setpoints
This column displays equipment name with key control variables
These are the lower limit values for the variables
These are the upper limit values for the variables
This section displays the optimizer status

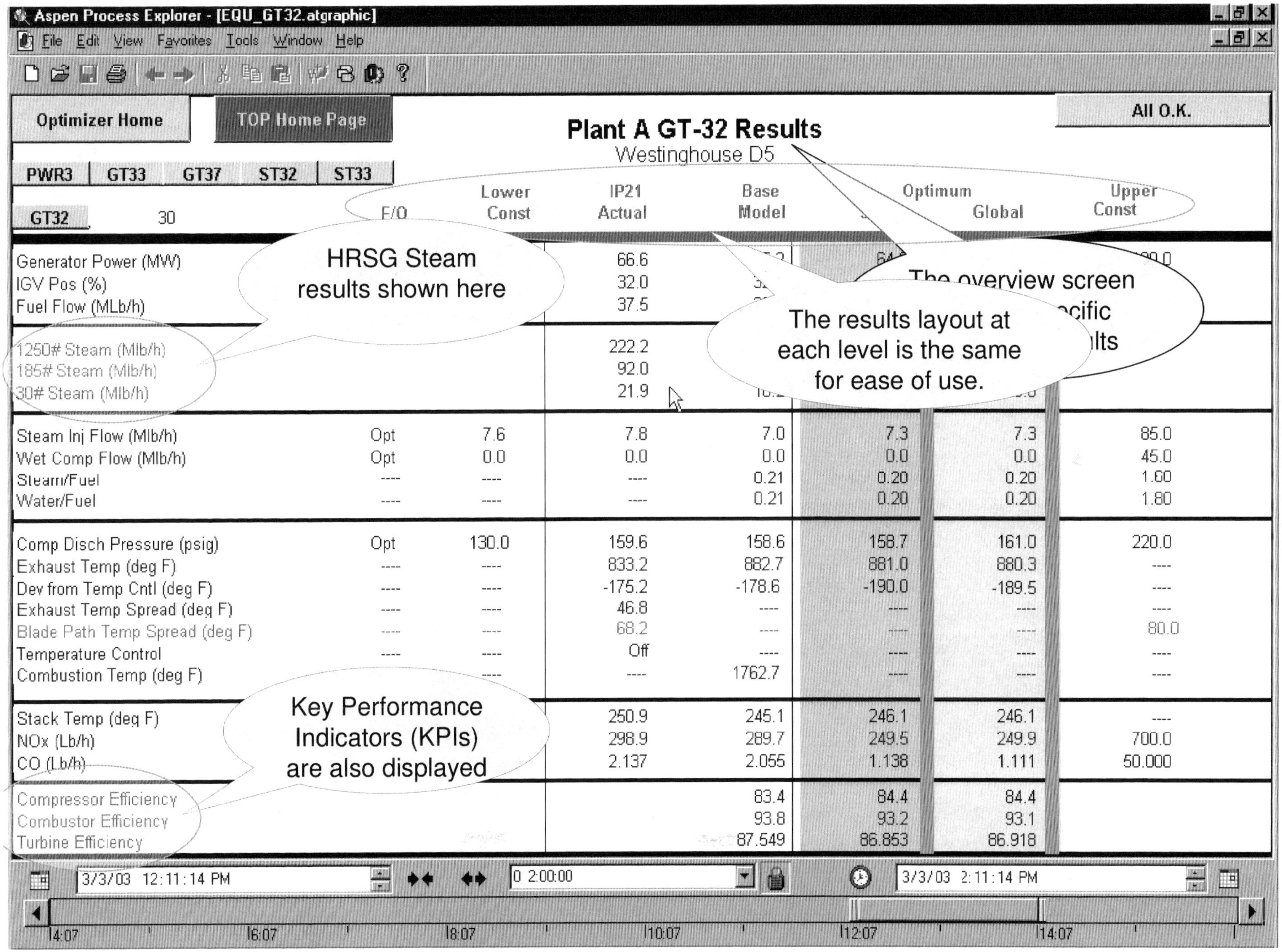
Aspen Process Explorer - [EQU_GT32.atgraphic]
File Edit View Favorites Tools Window Help
Optimizer Home
TOP Home Page
All O.K.
Plant A GT-32 Results
Westinghouse D5
PWR3
GT33
GT37
ST32
ST33
GT32
30
F/O
Lower Const
IP21 Actual
Base Model
Optimum
Global
Upper Const
HRSG Steam results shown here
The results layout at each level is the same for ease of use.
Key Performance Indicators (KPIs) are also displayed
Generator Power (MW)
IGV Pos (%)
Fuel Flow (MLb/h)
1250# Steam (Mlb/h)
185# Steam (Mlb/h)
30# Steam (Mlb/h)
Steam Inj Flow (Mlb/h)
Wet Comp Flow (Mlb/h)
Steam/Fuel
Water/Fuel
Comp Disch Pressure (psig)
Exhaust Temp (deg F)
Dev from Temp Cntl (deg F)
Exhaust Temp Spread (deg F)
Blade Path Temp Spread (deg F)
Temperature Control
Combustion Temp (deg F)
Stack Temp (deg F)
NOx (Lb/h)
CO (Lb/h)
Compressor Efficiency
Combustor Efficiency
Turbine Efficiency
3/3/03 12:11:14 PM
0 2:00:00
3/3/03 2:11:14 PM

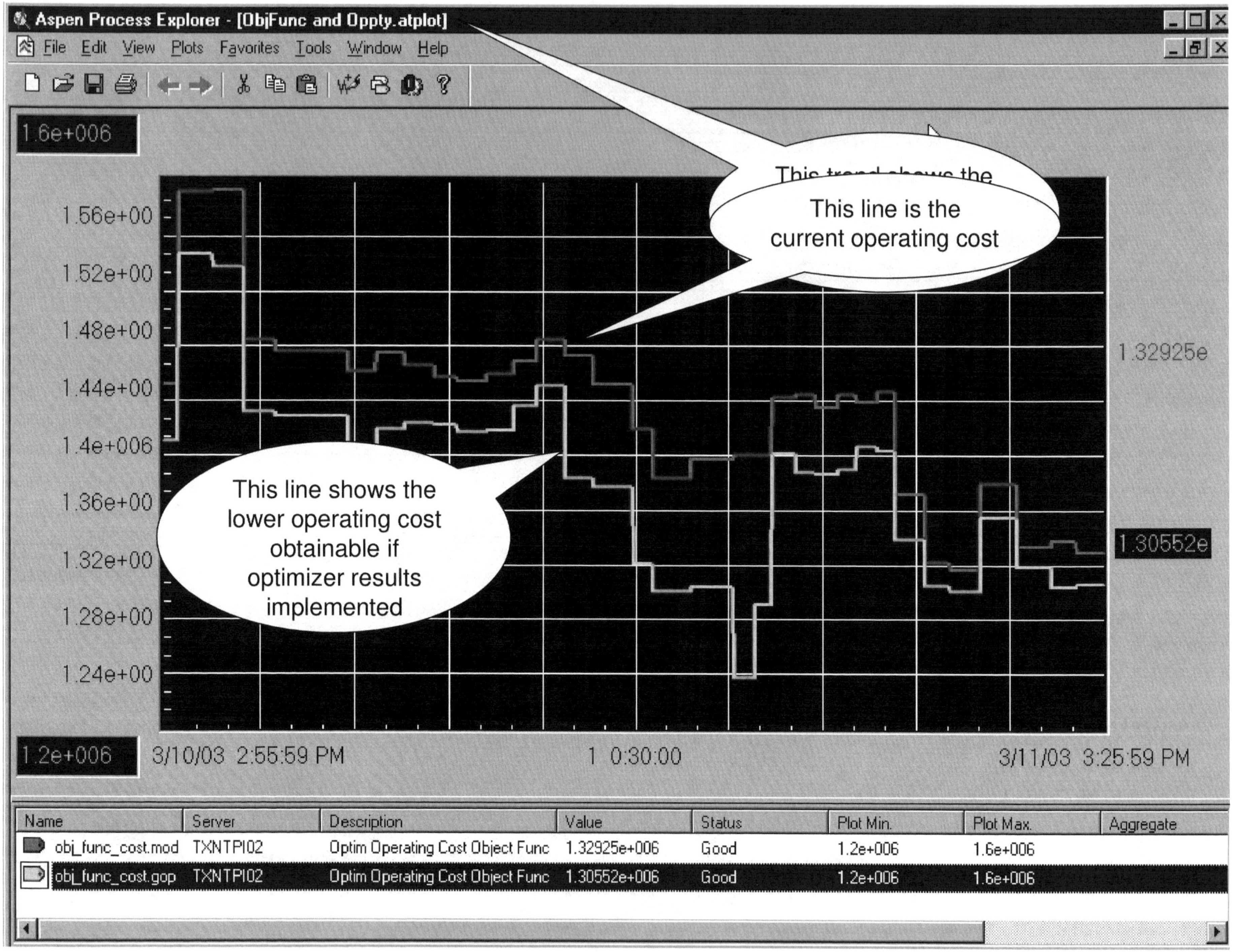
Aspen Process Explorer - [ObjFunc and Oppty.atplot]
File Edit View Plots Favorites Tools Window Help
1.6e+006
This trend shows the
This line is the current operating cost
This line shows the lower operating cost obtainable if optimizer results implemented
1.56e+00
1.52e+00
1.48e+00
1.44e+00
1.4e+006
1.36e+00
1.32e+00
1.28e+00
1.24e+00
1.32925e
1.30552e
1.2e+006
3/10/03 2:55:59 PM
1 0:30:00
3/11/03 3:25:59 PM
Name | Server | Description | Value | Status | Plot Min. | Plot Max. | Aggregate
obj_func_cost.mod TXNTPI02 Optim Operating Cost Object Func 1.32925e+006 Good 1.2e+006 1.6e+006
obj_func_cost.gop TXNTPI02 Optim Operating Cost Object Func 1.30552e+006 Good 1.2e+006 1.6e+006

# Typical Energy Management System Benefits

**1 – 3 % Overall site utility cost savings!**

# Example

# Utilities Example – Biomass Power Boiler

- Paper mill
- 160k PPH Fluidized-bed Boiler
- Fuels:
  - Sludge
  - Wood waste
  - Tires
  - Fuel gas
- Incentives:
  - Maximize use of cheap fuels (Tires & Wood)
  - Burn all the sludge to minimize land fill
  - Maximize steam production

# Solid Fuel Composition Control

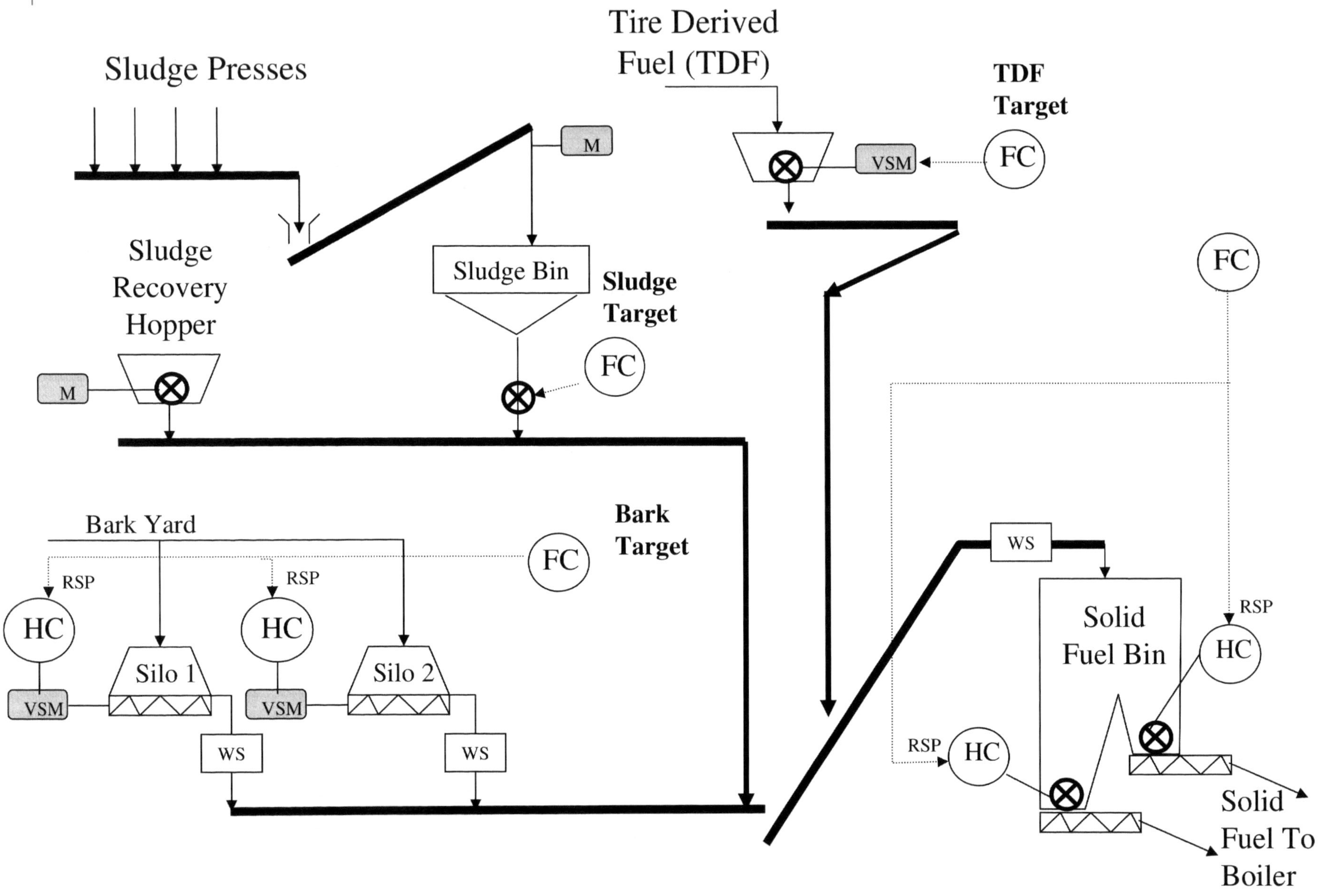

# Boiler Control

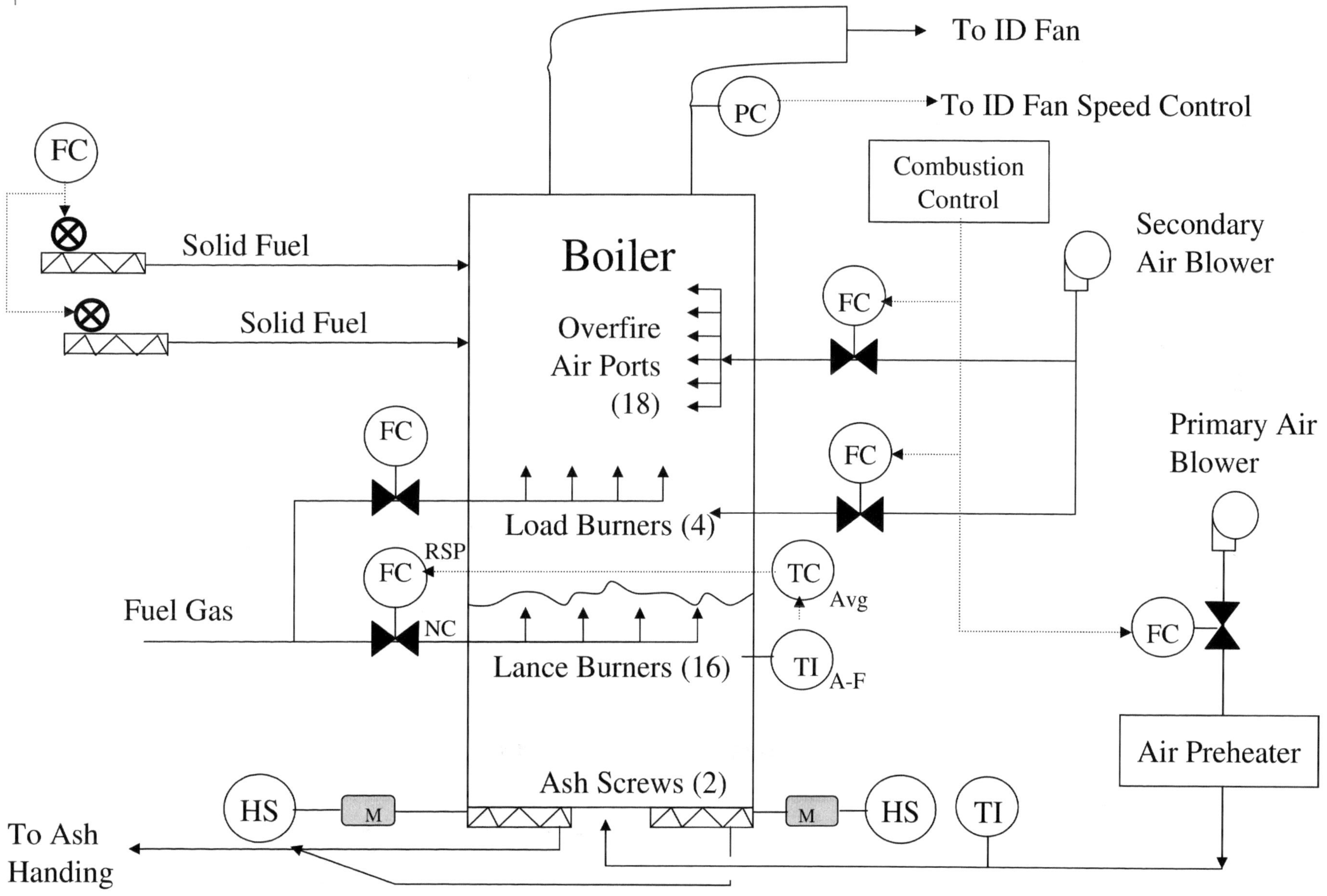

# Boiler Process Control Issues

- Varying water in sludge
- Long delay & lag times (20 – 60 minutes) to change fuel composition
- Fuel composition time constants are a function of fuel bin level
- Solid fuel composition in fuel bin is unknown
- Bed temperature constraints (max & min)
- Multiple operators controlling same unit
- Different operating philosophy used by each shift

# Solid Fuel Composition Control

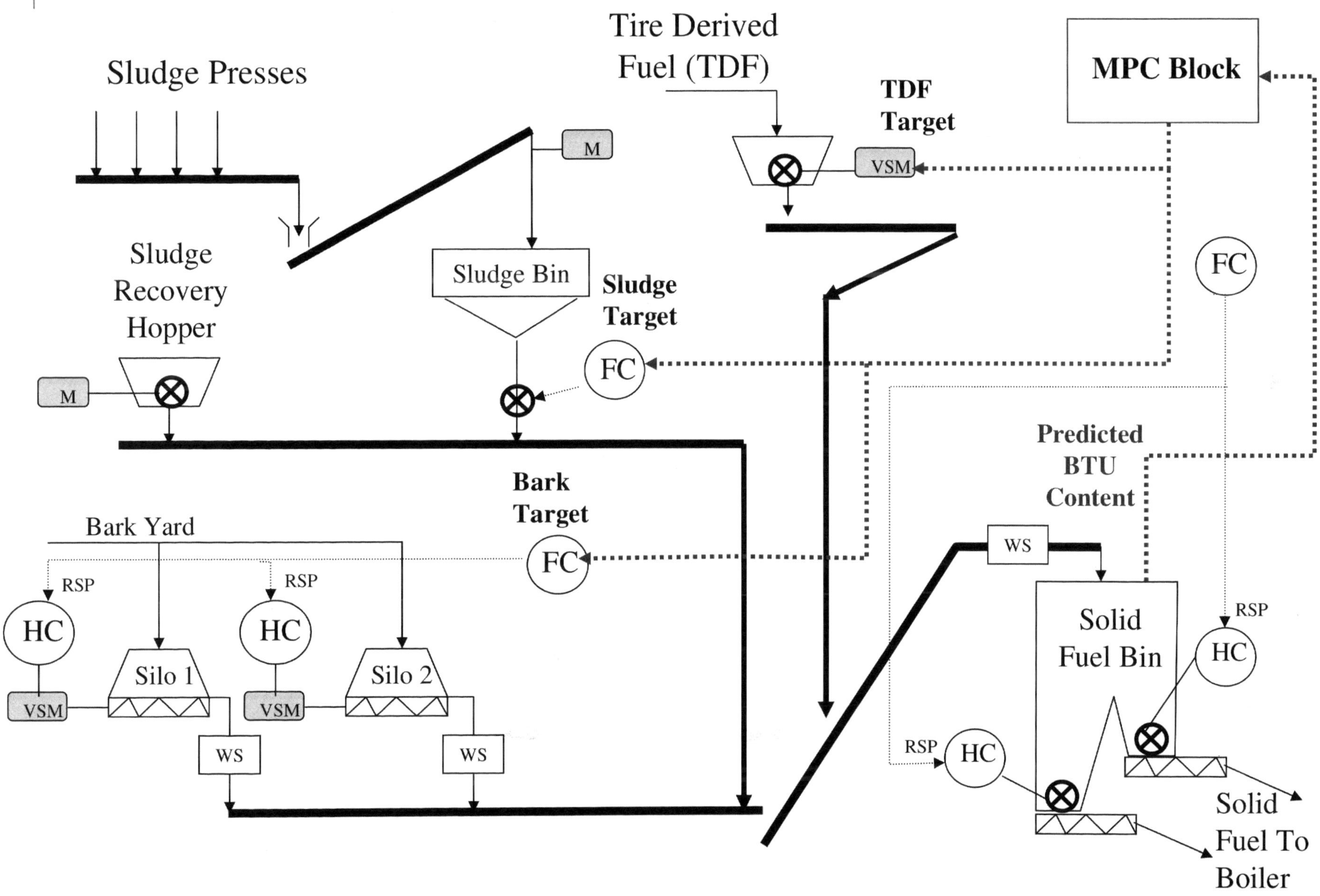

# Boiler Control

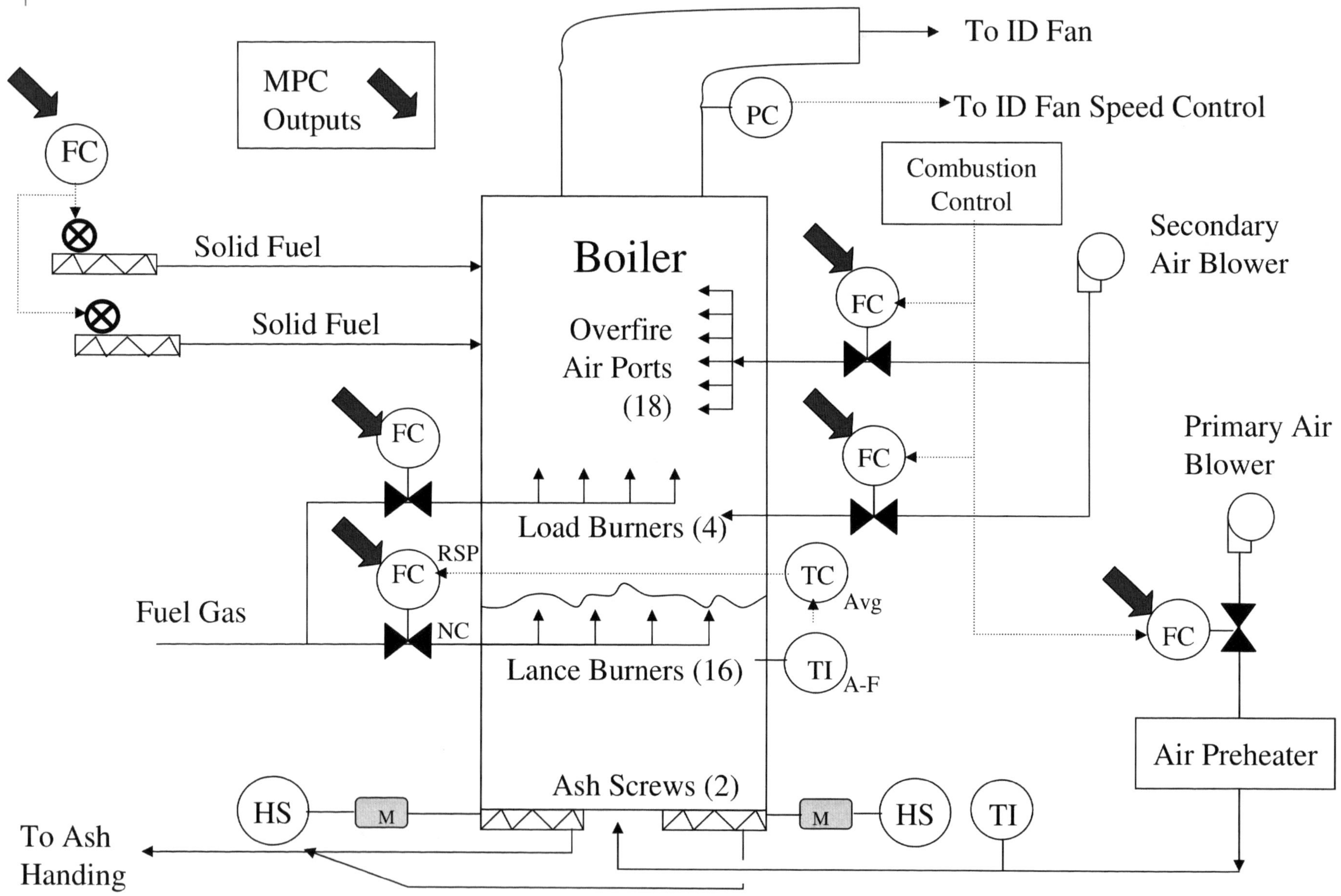

# Power Boiler Sludge

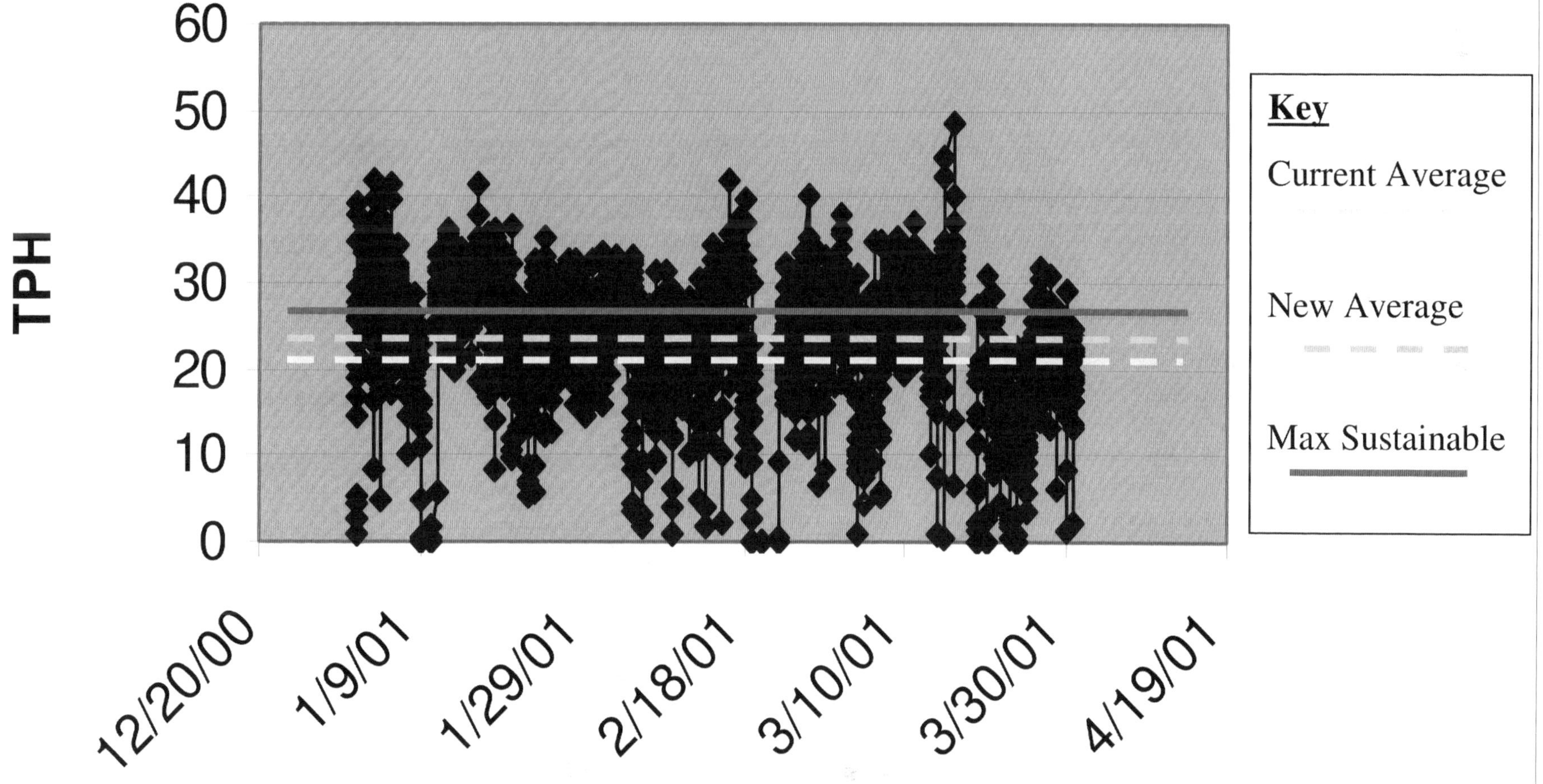

# Power Boiler Bark Fuels

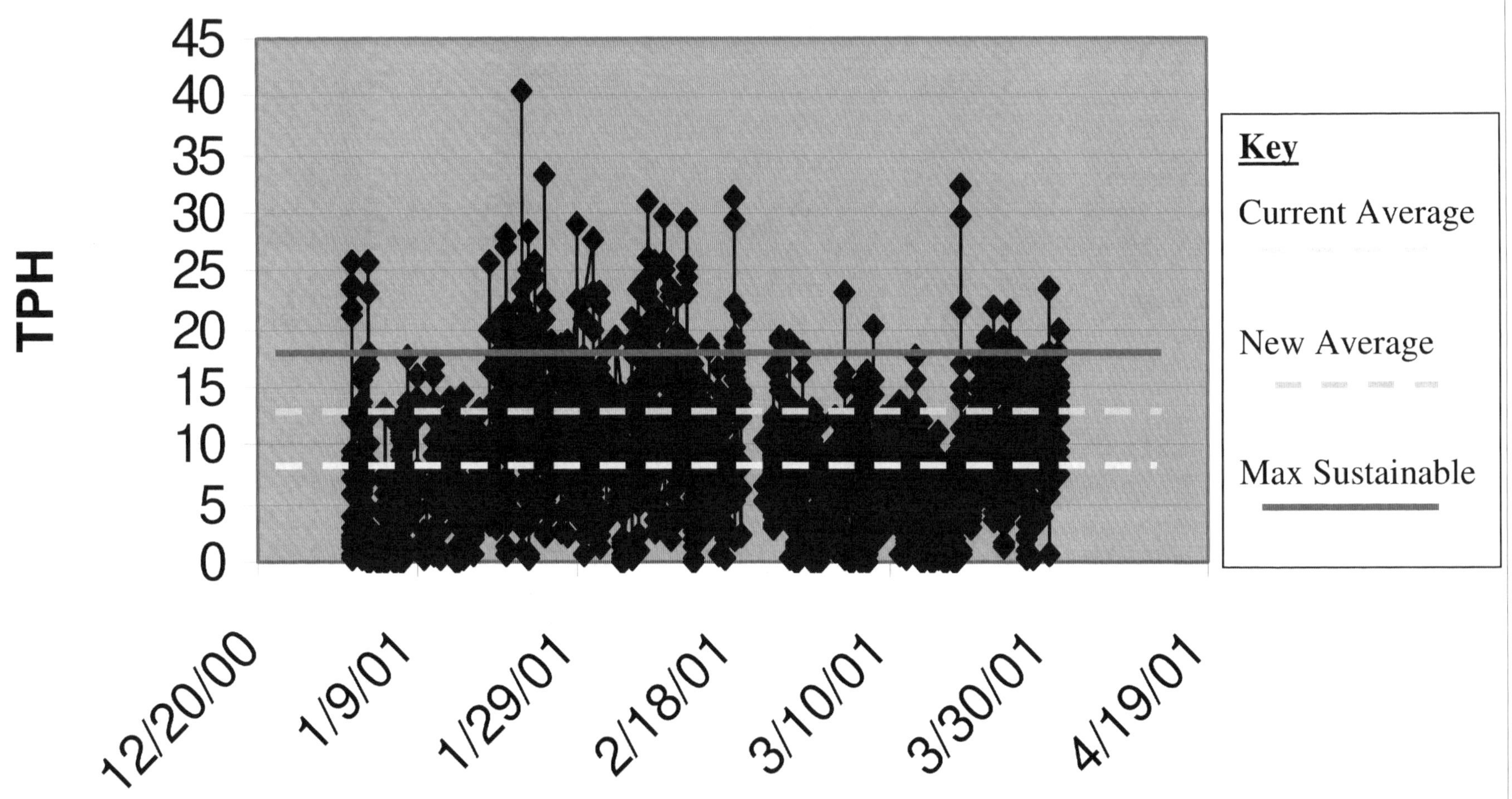

# Power Boiler APC Benefits

| | Difference in Hourly Costs & (Savings) |
|---|---|
| **Power Boiler** | |
| Natural Gas | $2.22 |
| Sludge Disposal | ($23.04) |
| Sludge Ash Disposal | $9.18 |
| TDF | $0.46 |
| TDF Ash Disposal | $0.00 |
| Waste Wood | $26.91 |
| W Wood Ash Disposal | $0.77 |
| Total | $16.50 |
| **Package Boilers** | |
| Displaced Natural Gas | ($98.42) |
| **Net Savings, $/Hr** | **($81.92)** |

**Total Savings**

$56k / mo

$672k / yr

**Project Justified:**

- Replacement of required pneumatic instruments
- DCS Hardware / Software
- APC Tools
- Turnkey Engineering Services

# Issues in Evaluating Plant Energy Usage

- Unit energy usage depends on production rate
- Unit energy usage variance dependent on production rate
- Need to correct to standard unit conditions

# Unit Energy Usage

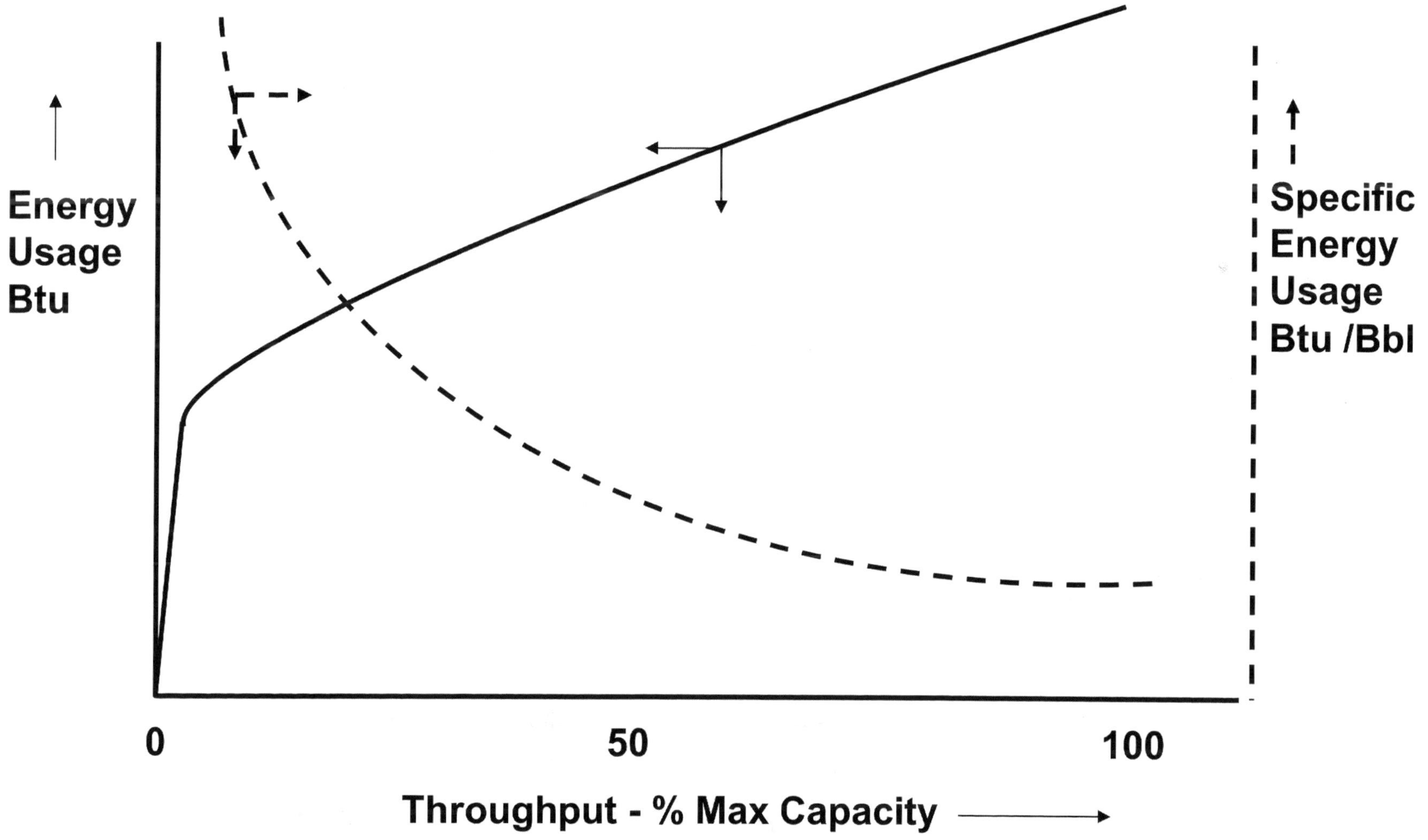

# Energy Usage - Example

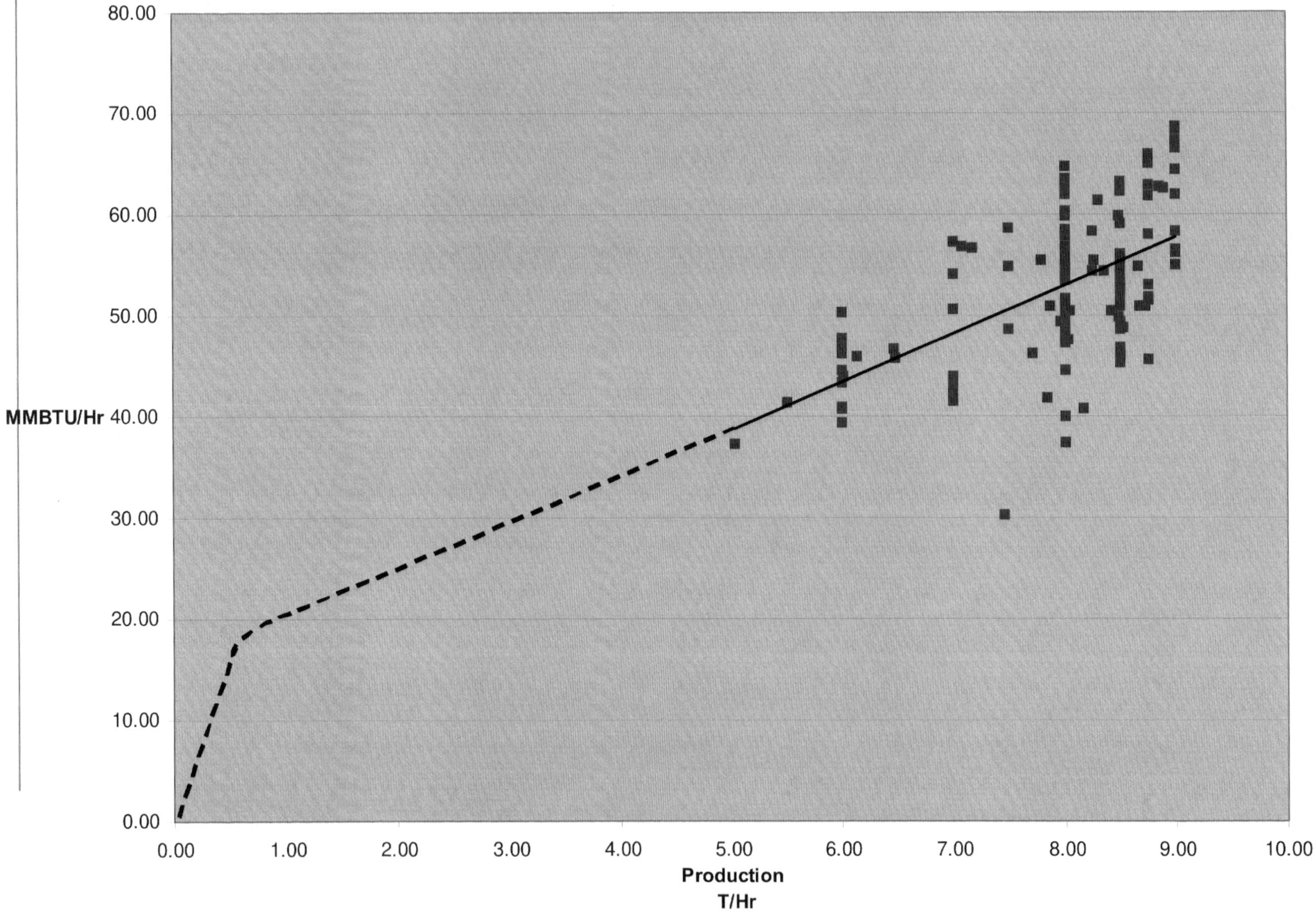

# Unit Energy Usage

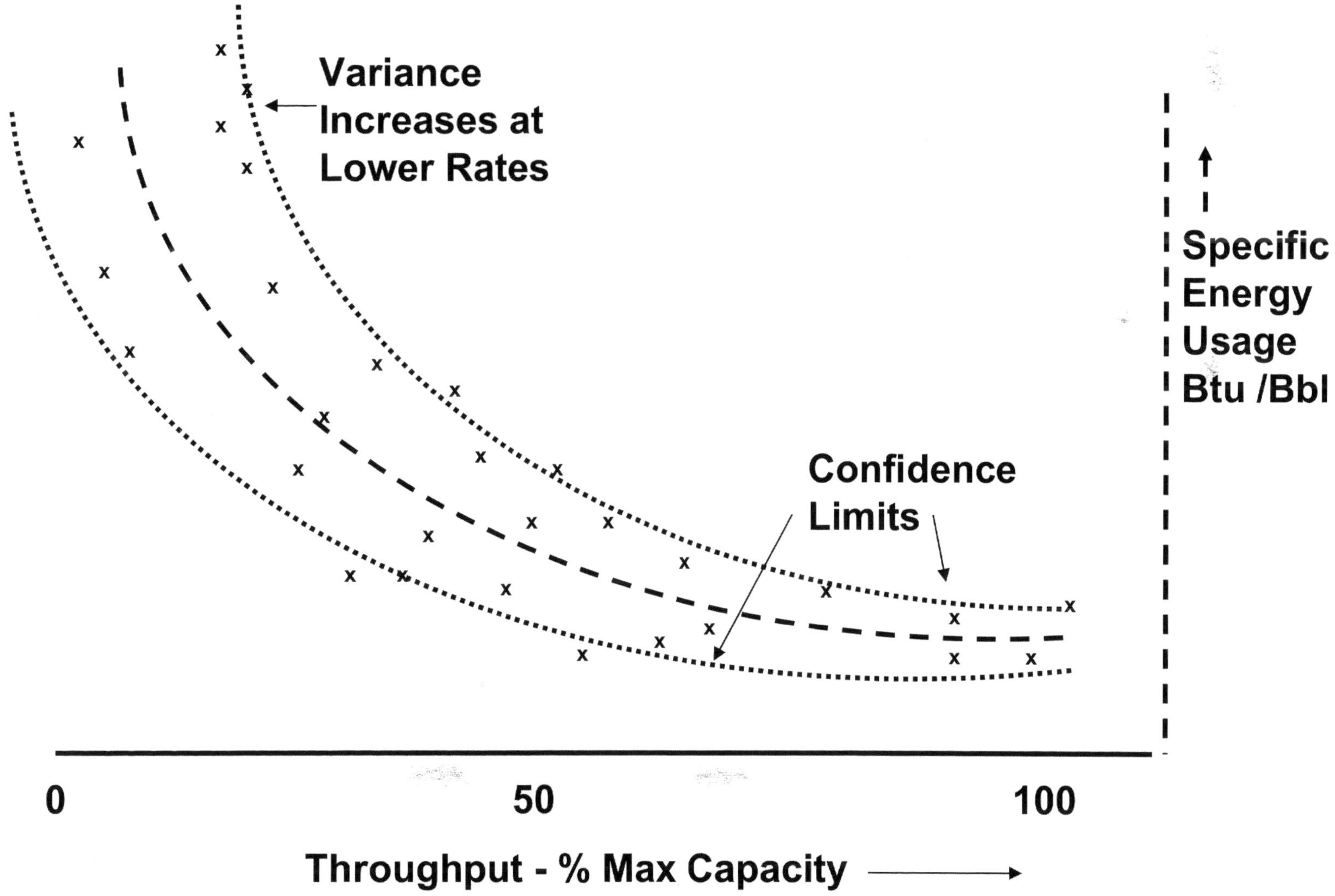

# Analyzing Energy Usage - Example

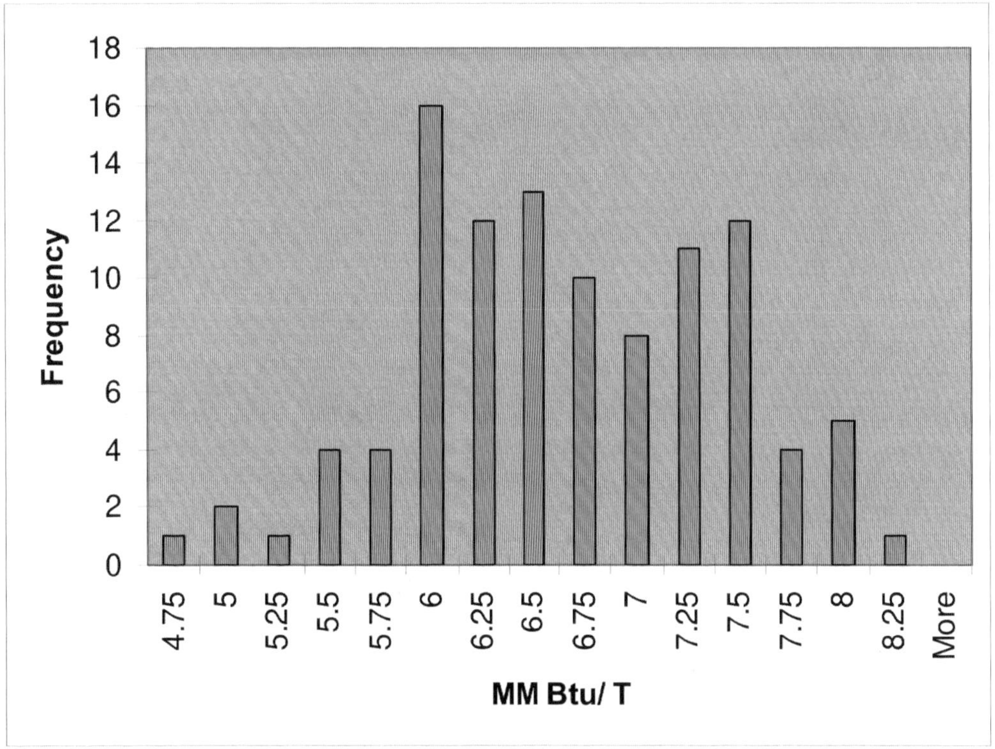

# Unit Energy Usage - Example

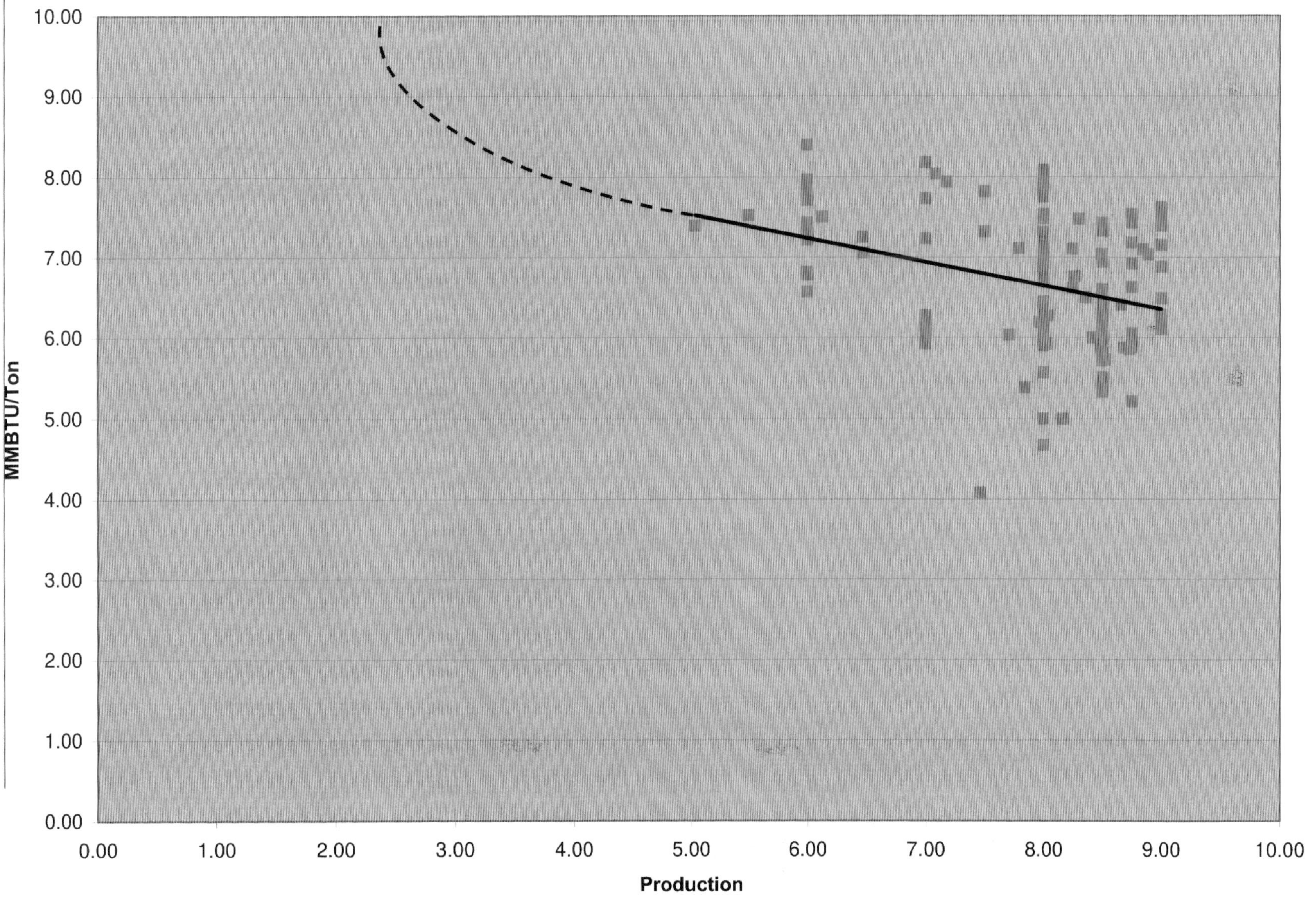

# Energy Performance Monitoring Problems

Example, Gas Turbine heat rate

will change as ambient temperature changes even though GT has not suffered any degradation

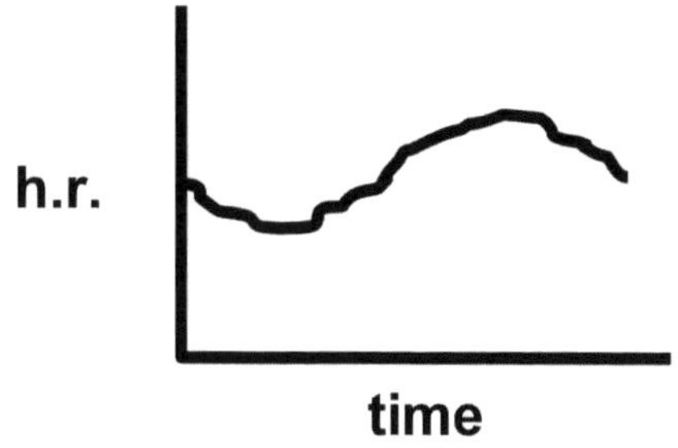


and as a result of changes in operating regime

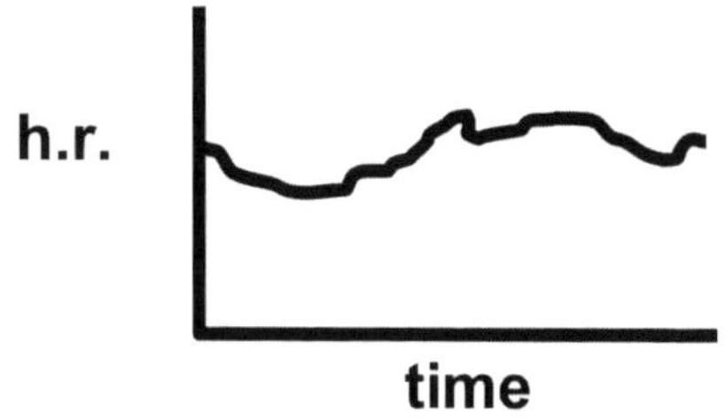


spot value calculations are affected by spurious measurements, drift and noise

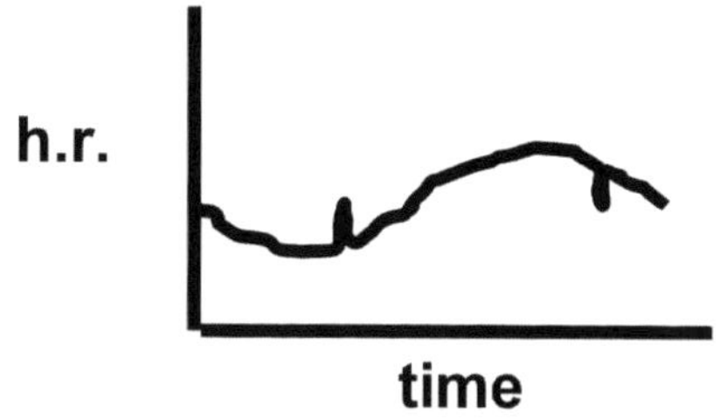

# Energy Monitoring - Relate to standard performance

**Actual Output**

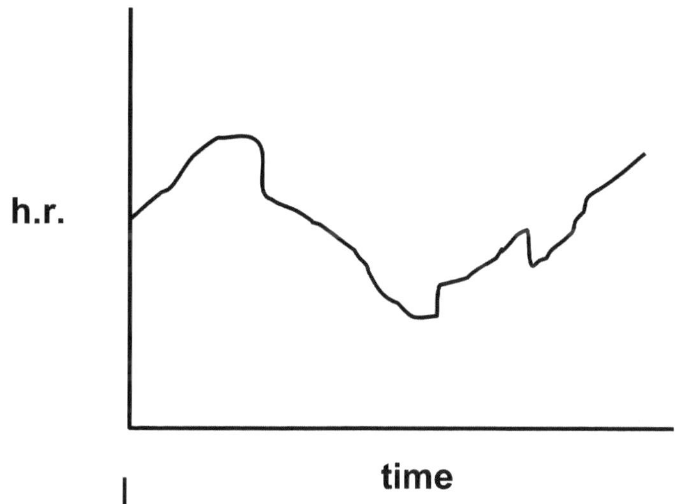


6.9MW

**Output related to standard performance**

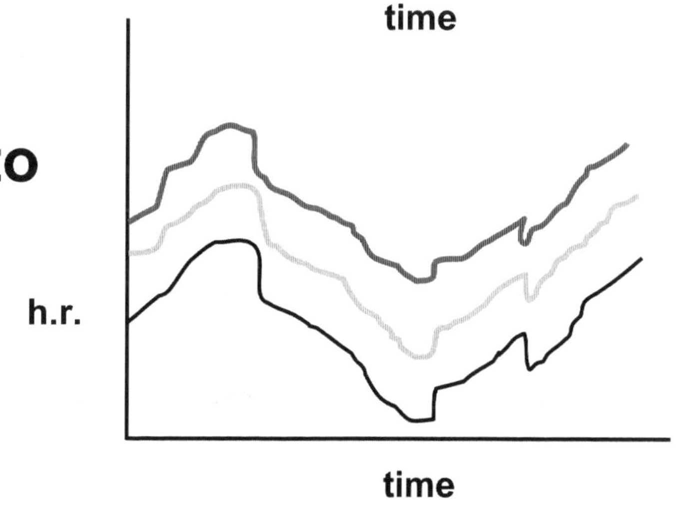


| | |
|---|---|
| Design | 8.4MW |
| Clean | 7.6MW |
| Actual | 6.9MW |

# Energy Monitoring - Relate to standard operating point

Performance based on current operation

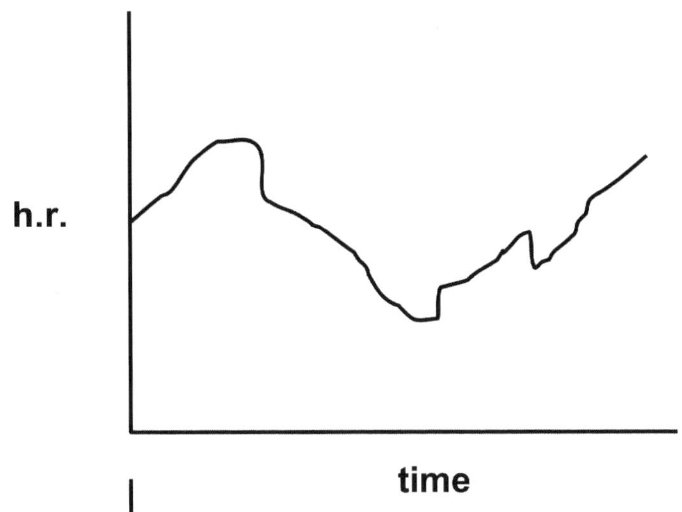


Performance related to standard operating point

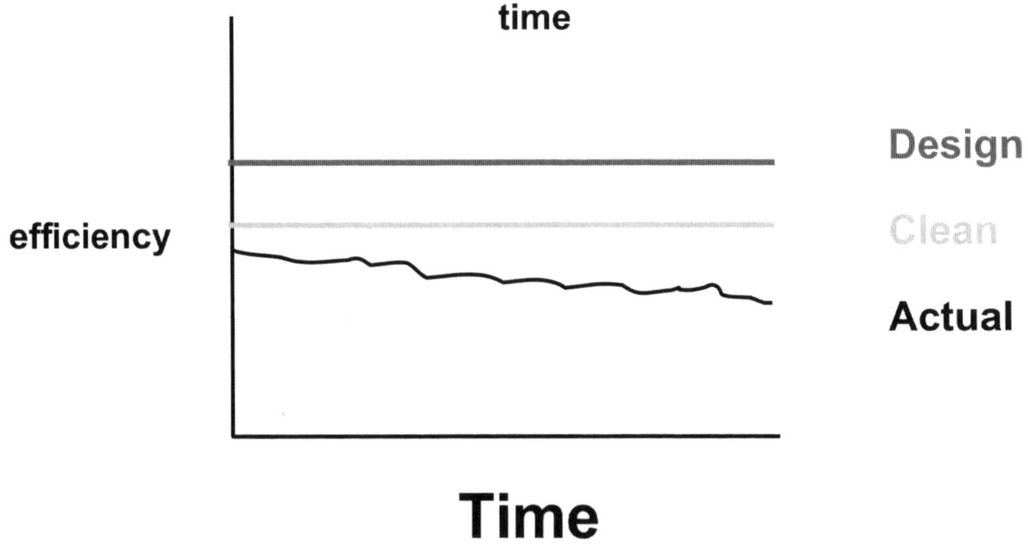

# Excuses for Doing Nothing

- Not enough manpower - Too busy doing other things
- Our plant is special – analysis based on other sites doesn’t apply
- We run our plant well already, there won’t be any big savings found
- Ostrich - (If we find something obvious, management will ask why we didn’t find it before)

# Summary

- Energy is the largest controllable cost in process operation – it's efficient production and use are keys to plant profitability
- Automation and Advanced Automation are keys to effective use and management of energy in the plant
- Implementation of a program to save energy requires a disciplined approach to evaluation and analysis

# Questions? Comments?

doug.white@emersonprocess.com

**More material on subject:**
**http://www.emersonprocess.com/solutions/services/aat**

# 202e Efficient Process Operations through Entropy Generation Optimization in Gas Fractionation Plants

Mahmoud Bahy Noureldin[1], **Zaher Barri**[2], **Mana Al-Owaidah**[1], Saleh Al-Qahtani[3], and Ahmed Aseeri[1]. (1) Consulting Services Department, Saudi Aramco, Dhahran, Saudi Arabia, (2) Yanbu Gas Plant, Sauid Aramco, Yanbu, Saudi Arabia, (3) consulting services, Sauid Aramco, Yanbu, Saudi Arabia

Gas industry is currently one of the most growing industrial sectors in the world. World demand for hydrocarbons is forecasted to grow rapidly during this decade and the one after. The growing demand for energy has initiated another wave of research focusing on new solutions for energy efficiency optimization. Gas fractionation plants are energy intensive processes; especially from the electricity consumption point of view. Entropy concept can be a systematic tool for generating ideas for process plants operations optimization.

This paper demonstrates through a typical gas fractionation plant; the benefits obtained from optimized process operations. It shows how simple process operations changes can result in sensible increase in the in-situ electricity generation and also sensible decrease of the electricity consumption. The paper shows how the entropy generation optimization concept can be conceptually utilized to systematically generate new intuition and suggest initial operations solutions for energy saving in gas fractionation plants. It also shows how the TEM_icons™ software can be used to select hot stream's dual minimum approach temperatures systematically. A typical gas fractionation utility plant combined heat and power model (CHP) is introduced as an efficient tool to evaluate the impact of operations modifications on gas fractionation plants utility bill.

Key Words: entropy generation; process integration; gas fractionation plant; TEM_icons™ ; CHP model

# Optimisation of Heat-integrated Distillation Schemes Based on Shortcut Analysis, Pinch Analysis and Rigorous Simulation

*Mansour Emtir* and Mansour Khalifa*
*Libyan Petroleum Institute, P.O. Box 6431 Tripoli, Libya*
**e-mail: memtir@yahoo.com*

## Abstract

Conventional and non-conventional distillation schemes for the separation of ternary mixture are investigated and optimized based on shortcut calculations, pinch analysis and rigorous simulation in order to achieve the highest saving values in energy and total annual cost (TAC) as the optimization objective function. The studied chemical system is benzene, toluene and m-xylene with feed composition of (25/50/25) and 99.9 mol % product purity. The optimization parameters are feed compositions. The state of feed condition is considered as the optimization parameters; feed at 20 °C, liquid at bubble point and vapor at dew point.

The results from shortcut and pinch analysis are found to be very close to rigorous simulation results regarding number of stages, reflux ratios and utilizing trim reboiler or trim condenser in studied distillation schemes. Both shortcut and rigorous optimization results indicate that heat-integrated schemes are consuming less energy compared to non-integrated distillation schemes, consequently TACs of heat-integrated schemes are attractive compared to non-integrated schemes. The state of feed conditions is playing imprtant role on the percentage of saving and direction of integration.

## Introduction

Distillation units are the most widely used technique for the separation of fluid mixtures in chemical and petrochemical industry. It is known that distillation is used for the separation of about 95% of all fluid separations in the chemical industry, and that around 3% of the total energy consumption in the world is used in distillation units (Hewitt et al 1999). The main disadvantage of the distillation is its high-energy requirement. As a result, new distillation sequences are emerging in order to reduce or improve the use of energy so that, there are several techniques which used to overcome this problem like integration of the distillation column with the overall processes which can give significant energy saving, e.g. Smith and Linnhoff (1988), Mizsey and Fonyo (1990), but these kinds of improvements can be limited. There are different configurations or distillation schemes that can be applied to get more energy saving, like integration of distillation columns with forward or backward heat integration, side-stripper, side-rectifier, fully thermally coupled distillation column (Petlyuk column) or dividing-wall column, heat-integrated of sloppy sequence, and double heat integration sequence. Energy-integrated distillation schemes give a great promise of energy savings up to about 70%. In addition to saving energy, which are accompanied by reduced environmental impact and site utility costs; there is also a possibility for reduction in capital costs.

Theoretical studies, e.g. Petlyuk et al. (1965), Stupin and Lockhart (1972), Fonyo et al. (1974), Stichlmair and Stemmer (1989), Annakou and Mizsey (1996), Dunnebier and Pantelides (1999), Emtir et al. (2001), Kolbe and Wenzel (2004) have shown that the column coupling configurations are capable of achieving typically 28-33 % of energy savings compared with the best conventional scheme. In addition, the coupling configuration can also be achieved with the so-called dividing-wall column (Kaibel 1987). By this arrangement, reduction in capital cost can be expected through the elimination of the prefractionator column shell (but not the column internals).

Emtir et al. (2003) compared five different energy integrated schemes, among them the forward and backward integrated prefractionator arrangement, with a non-integrated direct split sequence. The study compared the total annual costs (TAC) and the controllability of the different schemes. In terms of TAC they found that the backward integrated direct split configuration has the maximum savings of 37%. The integrated prefractionator arrangements have similar savings of 34% for the forward-integrated case and 33% for the backward-integrated case.

Skogestad et al (2005) had studied four pressure-staged distillation columns to see if multi-effect integration can be applied to any two columns in the sequence based on shortcut equations and *V*min-diagrams have been used for screening purposes to find the columns with the highest potential for energy savings. The results showed that when considering the existing number of stages available the ISF arrangement was the best, however when considering infinite number of stages the PF arrangement was the best.

## The scope of this work

This study is devoted towards separation of benzene/toluene/m-xylene (BTX) ternary mixture by continuous distillation at high product purity of 99.9 mol %. The feed composition to be separated consists of (25/50/25) mixture at atmospheric pressure and total flow rate of 100 kgmol/hr. The aim of this study is to investigate the effect of changing feed conditions on the saving potential of the heat-integrated distillation schemes. The tools which are implemented on the study to investigate the feasibility of energy and TAC savings are shortcut methods based on minimum vapour flow rate at infinite number of stages using shortcut equations, pinch analysis and rigorous simulation by utilizing HYSYS simulation for rigorous modelling with the following assumptions: UNIQUAC thermodynamic model is used, pressure drop across distillation columns is taken 5 Kpa, pumping is not considered in cost calculations, maximum internal flows are at 75% -80% of the flooding, and exchange minimum approach temperature (EMAT) = 10°C. In this study the total cost for each configuration is assumed to be the sum of utility costs (steam and cooling water) and equipment costs (purchase and installation). Detailed utility cost data are extracted from Emtir et al. (2001).

### Studied distillation schemes

Throughout this work A, B, and C denote the light, intermediate, and heavy components, respectively. The impurities are symmetrically distributed in the middle product stream. The studied distillation configuration are showing below: Figure 1; direct sequence without heat integration (DQ), direct sequence with forward heat integration (DQF), direct with backward heat integration (DQB), Figure 2; indirect sequence without heat integration (IQ), indirect sequence with forward heat integration (IQF), indirect with backward heat integration (IQB).

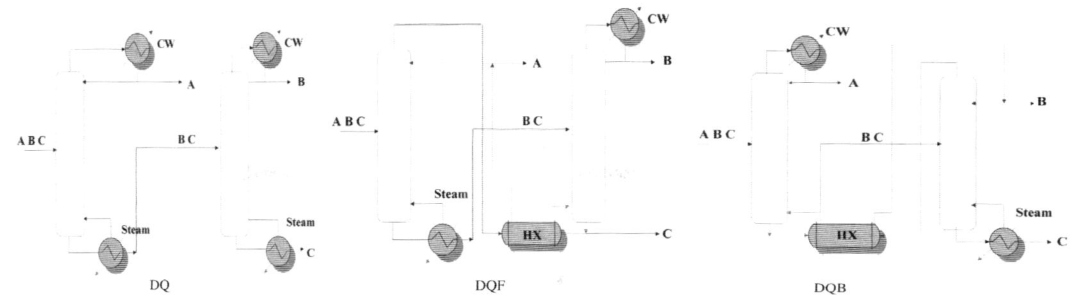


Figure 1. direct sequence with possible heat integration

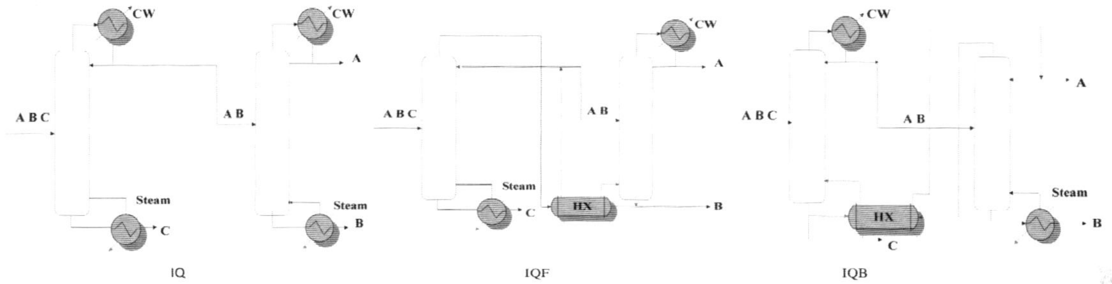


Figure 2. Indirect sequences with possible heat integration

## Rigorous simulation and optimization

The investigated schemes are simulated rigorously by HYSYS and design parameters are exported to Microsoft Excel where the final cost calculations for optimization are executed. Moreover, detailed column and heat exchanger costing are calculated using the default column and heat exchanger sizing. The sizing of distillation columns and heat transfer equipments requires the determination of flow rates, temperatures, pressures, and heat duties from the flow sheet of mass and energy balance, and these quantities can then be used to determine the capacities needed for the cost correlation. In addition, the concept of material pressure factor (MPF) is used to evaluate particular instances of equipment beyond a basic configuration. This concept is an empirical factor developed by Biegler et al (1997) as part of the costing process. For each column system, the pressure, number of trays and feed location are considered as the optimization variables and they are manipulated until the optimal design is found. Optimization variables can be more in case of additional feeds, draws or recycle streams are present. In every run, design parameters (optimization variables) are changed, specifications and optimality are checked. The process simulations are stopped when the global optimal system design is achieved.

## Shortcut analysis

In this work, minimum vapour loads are estimated based on shortcut method for conventional and heat-integrated distillation schemes at two different feed conditions, saturated liquid and saturated vapor. Underwood equations are used to find out minimum reflux ratio where reflux ratio was as set 1.1 times minimum reflux ratio, Fenske equations for minimum number of stages, Gilliland equqtion for theoretical number of stages and Kirkbride equation for feed stage location. FORTRAN program is written to solve all shortcut equations where the input data are feed, top and bottom flow rates and their compositions. The effect of pressure is taken into account for calculating average relative volatility of top and bottom streams for integrated and non-integrated configurations. The minimum vapor flow rate in every separation section can be calculated using

$$V_{min} = D(R_{min} + 1) \quad (1)$$

For saturated liquid feed, the vapor flow at the top is equal the vapour flow in the bottom but in saturated vapour feed vapour flow in the top is given as

$$V_T = V_B + (1 - q)F \quad (2)$$

Where q is zero for saturated vapour. The formulas listed in Table 1 have been used to determine the total minimum vapor flow rates of integrated and non-integrated distillation schemes.

Table 1. Total minimum vapor rates for integrated and non-integrated schemes

| | |
|---|---|
| Direct and indirect sequences without heat integration | $V_{min} = V_{min}^{COL1} + V_{min}^{COL2}$ |
| Prefractionator without heat integration | $V_{min} = V_{min}^{COL1} + max\{V_{min}^{COL2}, V_{min}^{COL3}\}$ |
| Direct with forward and backward | $V_{min} = max\{V_{min}^{COL1}, V_{min}^{COL2}\}$ |
| Prefractionator with forward and backward | $V_{min} = max\{V_{min}^{COL1}, V_{min}^{COL2}, V_{min}^{COL3}\}$ |

## Results and discussion

The shortcut results are summarized in Tables 2 & 3 for feeds at saturated liquid and saturated vapor conditions.

Table 2. Shortcut result of conventional and heat-integrated schemes at saturated liquid feed

| Sequance | | DQ | | IQ | | DQB | | DQF | |
|---|---|---|---|---|---|---|---|---|---|
| | | Column1 | Column2 | Column1 | Column2 | Column1 | Column2 | Column1 | Column2 |
| $N_{min}$ | | 16 | 18 | 18 | 15 | 16 | 20 | 21 | 18 |
| $R_{min}$ | | 2.13 | 1.24 | 0.816 | 1.91 | 2.13 | 1.38 | 2.944 | 1.24 |
| R | | 2.34 | 1.37 | 0.89 | 2.1 | 2.34 | 1.525 | 3.23 | 1.37 |
| N | | 40 | 48 | 48 | 39 | 40 | 51 | 52 | 48 |
| Feed Stage | | 19 | 24 | 26 | 19 | 19 | 26 | 23 | 24 |
| Vmin | kgmol/hr | 78.25 | 112 | 136.2 | 72.75 | 78.25 | 119 | 98.6 | 112 |
| $V_{min\ total}$ | kgmol/hr | 190.25 | | 208.95 | | 119 | | 112 | |
| $V_{min\ svaving}$ | % | 0.00 | | -9.83 | | 37.45 | | 41.13 | |

For conventional direct and indirect sequence, its clear that IQ scheme has been improved from -9.83 % to 17.81 % with respect to the base case (DQ), where as in case of heat-integrated cases the maximum savings is achieved in case of DQF with 41.13 % at saturated liquid feed. The higher savings in minimum vapor load for heat-integrated cases at satured liquid feed is attributed to the energy added inside the system are utilized effecintly by recycling the heat between the integrated columns.

Table 3. Shortcut results for conventional and heat-integrated schemes at saturated **vapor**

| Sequance | | DQ | | IQ | | DQB | | DQF | |
|---|---|---|---|---|---|---|---|---|---|
| | | Column1 | Column2 | Column1 | Column2 | Column1 | Column2 | Column1 | Column2 |
| $N_{min}$ | | 16 | 18 | 18 | 15 | 16 | 20 | 21 | 18 |
| $R_{min}$ | | 5.0633 | 1.28 | 1.178 | 1.91 | 5.06 | 1.38 | 6.236 | 1.24 |
| R | | 5.56 | 1.37 | 1.296 | 2.1 | 5.56 | 1.525 | 6.859 | 1.37 |
| N | | 38 | 48 | 46 | 39 | 38 | 51 | 51 | 48 |
| Feed Stage | | 18 | 24 | 26 | 19 | 18 | 26 | 22 | 24 |
| Vmin | kgmol/hr | 51.5825 | 114 | 63.35 | 72.75 | 51.5 | 119 | 80.9 | 112 |
| $V_{min\ total}$ | kgmol/hr | 165.5825 | | 136.1 | | 119 | | 112 | |
| $V_{min\ svaving}$ | % | 0.00 | | 17.81 | | 28.13 | | 32.32 | |

Pinch analysis has been conducted for the heat-integrated cases by utilizing SPRINT software, results are showing in Figures 3 & 4 for case of DQB. It is obvious from pinch analysis that trim condenser is needed in case of DQB, steam temertaure level can be decided also.

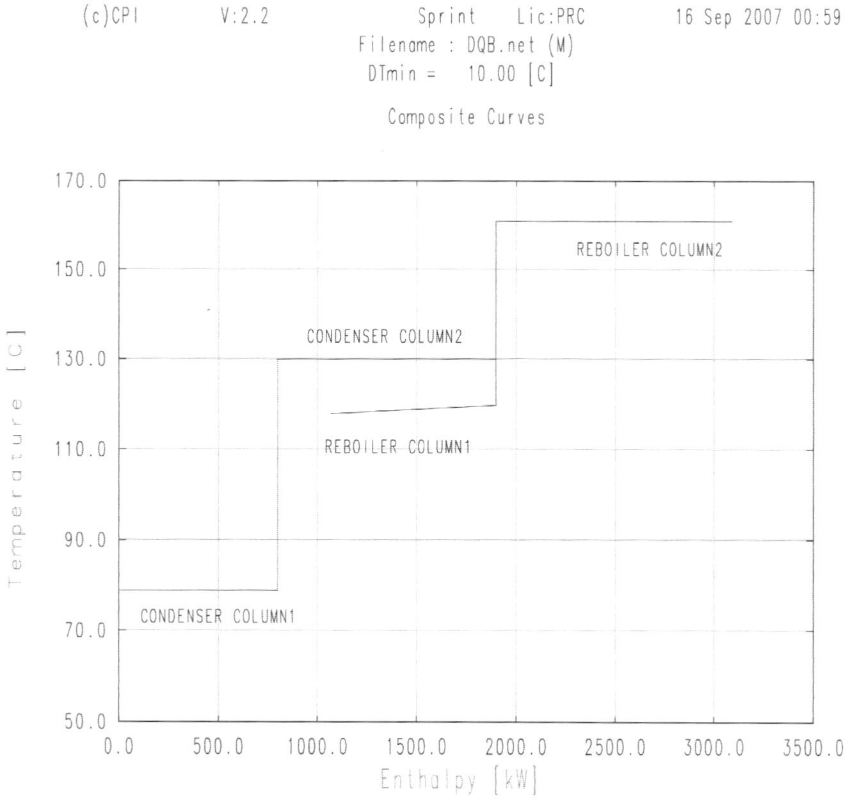


Figure 3. Composite curve for DQB case

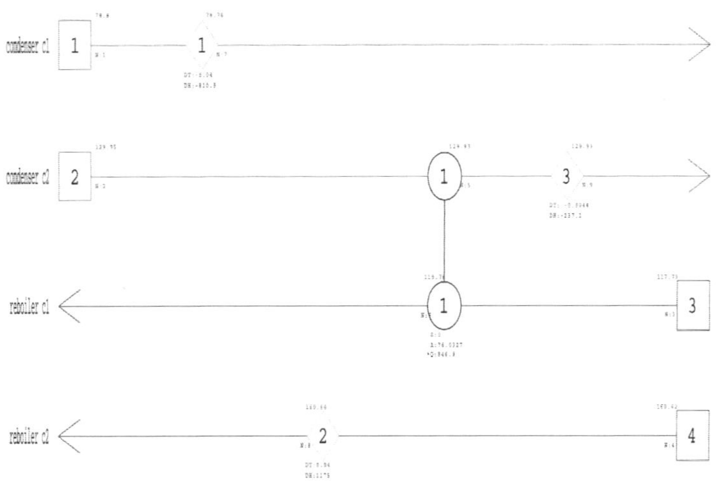


Figure 4. Matching between hot and cold streams

Rrigorous optimization results are showing similar trend in case of savings in both energy and TAC. DQB shows maximum saving values of 44.44 % in energy with 25.4 % of TAC saving. IQ, IQF and IQB integrated schemes indicates inferior savings in both of energy and TAC which is attributed to the wide gap in temperatures between heat source and heat sink and use of high pressure steam in distillation system.

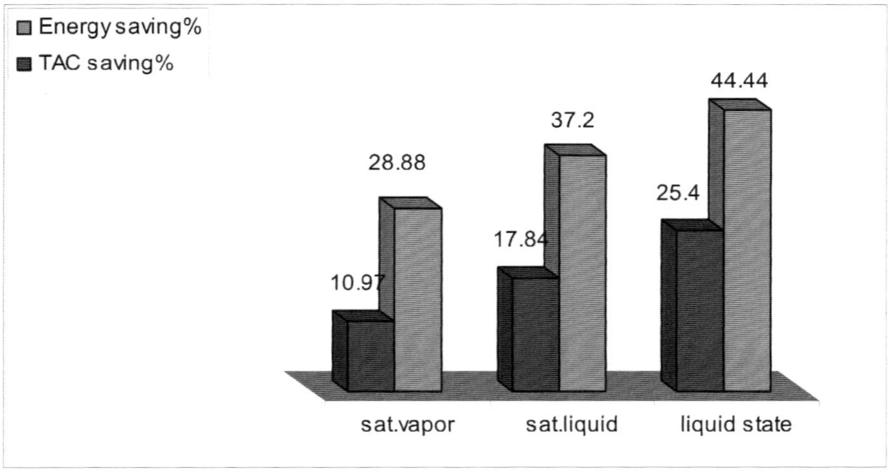


Figure 5. Effect of feed conditions on DQB distillation scheme

Considering the effect of different feed conditions on DQB savings in energy and TAC are shown in Figure 5 and the results can be summarized as follows:

- Direct sequence with backward heat integration gives maximum saving values 44.44 % in energy, and in addition to 25.4 % of TAC saving with respect to direct sequence without heat integration.
- IQB distillation scheme are showing the highest energy saving of 51.42 % but lower values in TAC saving (not shown).

## Conclusions

The state of feed conditions plays important role on the ranking and evaluation of distillation schemes from heat integration point of view. DQB heat-integrated scheme is showing higher energy saving for feed entering at liquid state, this due to the recycling of heat added in the system to change the phase of the feed inside the distillation scheme, where as in case of feed entering at vapor phase the recycled heat between the columns is reduced which will effect the saving of the heat-integrated scheme directly. In some case although the saving in energy is high, but due to increasing in preussures of the integrated columns which will lead to utilization of high pressure steam causing its TAC saving to drop down. Shortcut and pinch analysis methods gives priliminary indication regarding energy saving in distillation schemes, rigourous simulation and TAC saving comparison is showing the most rigourous method of selection between heat-integrated schemes.

### References


Annakou, O. and P. Mizsey, 1996, Ind. Engng. Chem., 35, 1877.

Dimian, A.C., Eds., 2003, Integrated Design and Simulation of Chemical Processes. Elsevier, Amsterdam.

Dunnebier, G. and C.C. Pantelides, 1999, Ind. Eng. Chem. Res., 38, 162.

Emtir, M., E. Rev, and Z. Fonyo, 2001,. Applied Therm. Eng., 21, 1299-1317.

Emtir, M., Mizsey, P., E. Rev, and Z. Fonyo, 2003,.Chem. Biochem. Eng. Q., 17(1), 31-42.

Fonyo, Z., J. Szabo, and P. Foldes, 1974, Acta Chimica Academia, 82, 235.

Hewitt, G., Quarini, J. and Morell, M., 1999, Chem. Eng, 21 Oct.

Kaibel, G., 1987,.Chem. Engng. Technol., 10, 92.

Kolbe, B. and S. Wenzel, 2004, Chemical Engineering and Processing, 43, 339.

Mizsey, P. and Z. Fonyo, 1990, Computers and Chemical Engineering, 14, 1303.

Petlyuk, F. B., V. M. Platonov, and D. M. Slavinskii, 1965, Int. Chm. Engng., 5, 561.

Smith, R. and B. Linnhoff, 1988, Trans. IChemE., 66, 195.

Stichlmair, J. and A. Stemmer, 1989, Chemical Engineering Technology, 12, 163.

Stupin, W. J. and F. J. Lockhart, 1972, Chemical Engineering Progress, 68, 71.

Biegler, L. T., Grossmann, I. E. and Westerberg, A. W., Eds., 1997, Systematic Methods for Chemical Process Design, Prentice-Hall, NJ.

# Innovative Approach to Solve CCR Regenerator Pinning Problem

Rabea M. Al-Saggaf, Sajeesh Padmanabhan, Hamzah Z. Abuduraihem and Neelay Bhattacharya.
Saudi Aramco-Saudi Arabia

## Abstract

Saudi Aramco Yanbu' Refinery & Process & Control System Department engineers worked jointly to troubleshoot and resolve the successive plugging of the Continuous Catalyst Regeneration (CCR) regeneration tower. The CCR section of the Yanbu' Refinery (YR) Platformer experienced successive catalyst blowouts leading to a regeneration section shutdown. This caused the Platformer section to operate at reduced feed rate and severity.

It was concluded that the root cause for this problem was the over sizing of the regeneration gas blower. The available option to reduce pinning was to lower the regeneration gas flow. Installation of the restriction orifice on the regeneration gas blower suction eliminated the pinning problem and normal CCR Platformer operation resumed.

## Introduction

The YR Platformer unit was revamped in June 2006 from a fixed bed unit to a Continuous Catalyst Regeneration Unit. Coked spent catalyst from the Platforming reactors is continuously sent to the CCR Regeneration Tower where the coke is burnt off and the spent catalyst is regenerated in four steps:

1) Coke Burning
2) Oxychlorination – for dispersing the catalyst metals and adjusting the catalyst chloride.
3) Catalyst Drying
4) Reduction – for changing the catalyst metals to the reduced state. Finally, the regenerated catalyst is circulated back to the first Platforming reactor.

As the catalyst is in a continuous movement, over a period of time catalyst fines plugs the Regenerator screen. The CCR has to be shutdown and the screen removed for cleaning every 12 months. Figure (1) illustrates a schematic of the CCR flow and equipment.

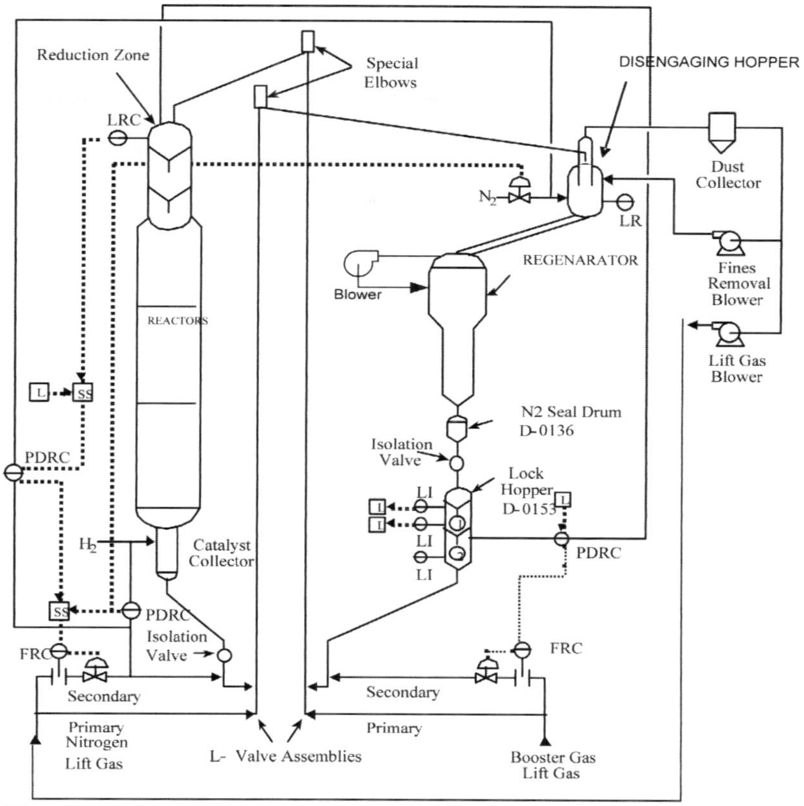


**Figure (1): Schematic of Cyclemax Regenerator**

## INCIDENT BACKGROUND

Since the first startup in June 2006, the CCR Regeneration Tower has been operating with a partially plugged screen, as a result of catalyst fines, generated in the system from the containment loss in the Platformer reactors, due to a weld failure in the center pipe of the first reactor. This effectively reduced the Regeneration gas flow without affecting the CCR operability. It was recommended by the licensor to clean the screen at the earliest opportunity. The CCR was shutdown for a period of 5 days to carry out the cleaning of the regeneration tower screen.

The unit was restarted in black burn mode. The startup was normal and the regeneration tower temperature profile had its peak at the second TI in the Burn Zone. The operation of the regenerator for the next 36 hours was absolutely normal with regenerated catalyst carbon at 0.095 wt% and spent catalyst carbon at 4.3 wt%.

A blowout occurred between the disengaging hopper and the regeneration tower, which caused the disengaging hopper level to drop from 52% to 47%. The Disengaging pressure dropped and the Regeneration Tower pressure increased momentarily. Figure (2) below shows the pressure fluctuation and level drop at the time the blow out occurred.

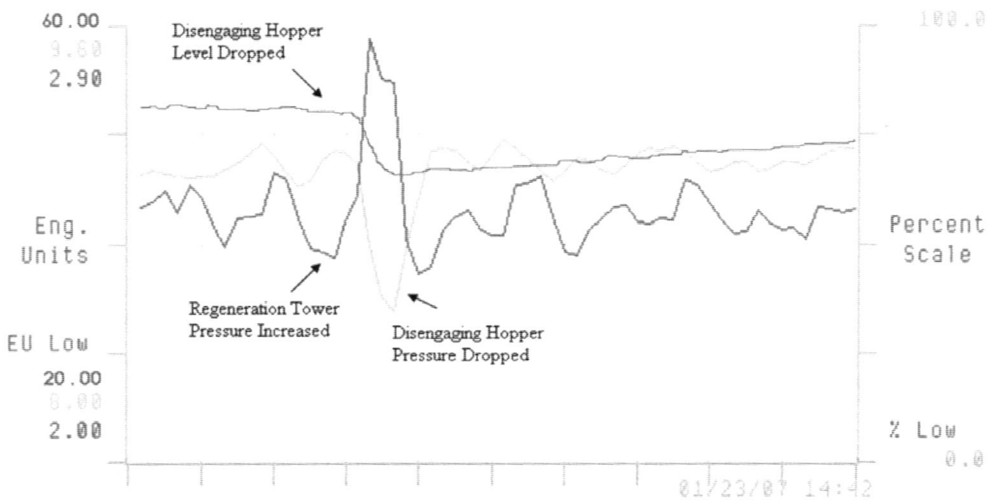


**Figure (2): Pressure and level profiles during the incident**

Due to the blowout, a lot of fines and chips were generated that plugged the Regenerator screen. A shift in burn zone temperature profile was experienced and the peak temperatures shifted to fifth TI, which is at the bottom of the burn zone (refer to Table 1).

Maintaining a proper burn profile in this section is extremely important for safe operation of the unit. If the burn profile shifts down, un-regenerated catalyst will enter the chlorination zone, potentially causing catalyst to agglomerate and damage the Regenerator internals.

**Table (1): Regeneration Tower Normal and Incident Temperature Profiles**

| TI | Licenser Specified Range | Normal Profile | Shifted Profile |
|---|---|---|---|
| TI -1 | 479-510°C | 380°C | 250°C |
| TI -2 | 493-593°C | 563°C | 465°C |
| TI -3 | 493-593°C | 557°C | 475°C |
| TI -4 | 493-560°C | 503°C | 511°C |
| TI -5 | 491-504°C | 492°C | 543°C |
| TI -6 | 491-504°C | 487°C | 479°C |
| TI -7 | 491-504°C | 485°C | 472°C |
| TI -8 | 479-499°C | 485°C | 465°C |

Intensive discussions were held between Refinery engineers and licensor experts on this subject and it was concluded that it is very difficult to pin-point the root cause of the blowout. An attempt was made, as suggested by the licensor, to stop the Regeneration blower and carry out cold circulation of the catalyst to clear the screen. Accordingly the regeneration tower was cooled, as per procedure, at 50°C/Hr. when the blower was switched to low speed at 350°C, a blowout was again observed.

A similar phenomenon was observed when the blower was switched from low speed to high speed during the reheat. There was no improvement in unit performance. Therefore, the regeneration section was again shutdown as a result of repetitive blowout occurring between the disengaging hopper and the regeneration tower. The CCR Platformer throughput was lowered from the design of 40 to 30 MBD and the operating severity reduced from 100.0 to 95.0 to control coke lay down.

## ANALYSIS & FINDINGS

It is very unusual for a blowout to take place in a CCR because it is normally catalyst full. During normal operation the catalyst flows down due to gravity. The Disengaging Hopper operates at approximately 9.0 $kg/cm^2$ (g) and the Regeneration Tower operates at approximately 2.5 $kg/cm^2$ (g). This huge pressure differential is taken by the catalyst in the long transfer pipes between the Disengaging Hopper and the Regeneration Tower. If due to any reason, a void is created in the Regeneration Tower, the huge differential pressure will force the catalyst down, causing a pressure fluctuation and generating fines and dust. The fines will block the Regenerator screen, causing the temperature profile to slip down.

The problem was analyzed with the following facts:

1. One probable cause could be something blocking the catalyst transfer pipe. But this would be a one-time occurrence that would be automatically cleared after the blowout. But since the blowout reoccurred, it was obvious that blowout was not due to plugging of the transfer pipes.
2. The only difference between the two startups was the condition of the screen. Since initial startup the unit was operating with a plugged screen. The recent start-up was the first time with a clean screen.
3. The blowout was clearly related to the blower operation because the blowout had happened on two occasions when the blower speed was switched.

A detailed analysis of the blower data was carried out. During the unit start-up, a test run was carried out to evaluate the performance of the blower. The head delivered was equivalent to 29.53" $H_2O$ versus 30.94" $H_2O$ (design), based on the performance curve of the blower; this motor was delivering 13,300 ACFM versus 12,986 ACFM (design), as indicated in Figure (3).

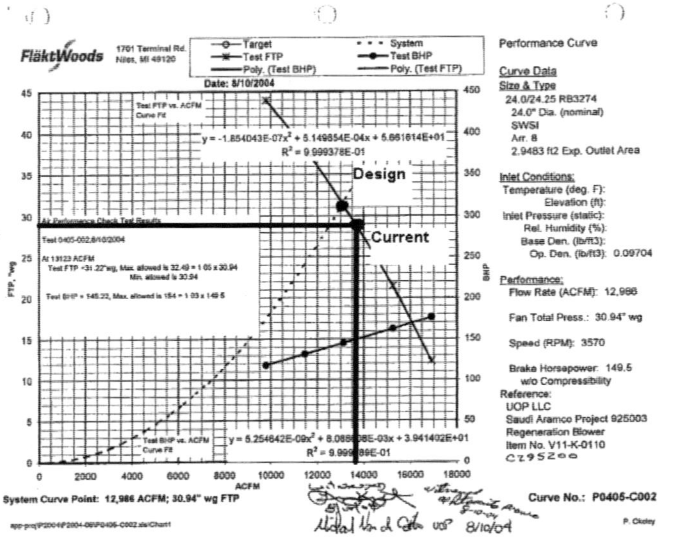


**Figure (3): Performance of blower v/s design conditions**

It was concluded that the blower flow was 2.4% above design, even with a partially plugged screen, indicating that the actual degree of over design was higher.

Once the screen was cleaned, the flow increased, causing the catalyst to pin in the Regeneration Tower. Since the Lock Hopper is removing catalyst, a void was created at some location in the Regeneration Tower. When the differential pressure between the Disengaging Hopper and the Regeneration Tower was high enough, it forced the catalyst to fill the void creating a blowout. This would also explain why the blowout occurred when the blower speed was changed. The first time when the blower was switched from high speed to low speed the catalyst unpinned, causing an immediate blowout. The second time the blower was switched from low-speed to high-speed, causing the catalyst to pin. The blowout did not occur immediately. During subsequent heat-up, the catalyst slumped, creating a void below the pinned area, which caused a blowout.

Discussions were held by the team with the licensor, and the licensor was convinced that the blower could be a likely cause for the blowout. They agreed to the following:

1. The licensor would size a restriction orifice to reduce the Regeneration Gas flow by 15-20% to avoid pinning. YR would keep it ready for installation in the Regeneration tower gas outlet line.
2. Conduct inspection of the disengaging hopper and regeneration tower screen.
3. If inspection did not reveal any obvious reason to explain the blowouts, then install the restriction orifice in the Regeneration gas outlet line and observe the unit performance on restart.

## Field Inspection

The regeneration tower and the Disengaging Hopper were open for inspection. While unloading the catalyst, the last few drums from the Disengaging Hopper contained fresh catalyst. YR had added 12 drums of fresh catalyst to the Disengaging hopper during the recent reload. Since the catalyst had been circulated for 500 cycles, there should not have been any fresh catalyst remaining in the Disengaging Hopper. This clearly indicated that the catalyst was not moving as a result of localized catalyst pinning on the regeneration tower screen. No clumps or debris were found between the disengaging hopper and the regeneration tower. The Regeneration screen was heavily plugged. There was no sign of integrity loss or severe damage to the screen.

It was concluded that the root cause for this problem was the over-sizing of the regeneration gas blower.

## Recommendations

The available option to reduce pinning was to lower the Regeneration gas flow. The least cost-effective option was to install a restriction orifice in the Regeneration gas blower suction line, to reduce the Regeneration gas flow which would increase the pinning margin. Discussions were held with the unit licensor on the most optimum sizing for the restriction orifice. It was concluded that a 15% reduction in the total regeneration gas flow would improve the pinning margin to 25%.

The licensor designed a restriction orifice with a bore size of 340mm, which was fabricated by YR and installed in the Regeneration gas outlet line. The reduced burn zone capacity, due to reduction in Regeneration gas flow, would be compensated by an increase in burn zone oxygen content, from the normal operating values of 0.85 mol to 0.95 mol%, to 0.95 to 1.1 mol%.

## CCR Restart

The regeneration section was restarted after the installation of the restriction orifice. The start-up was smooth without any pressure fluctuations or signs of blowout. No major impact or deficiencies were observed in the Regeneration blower operation. The temperature profiles across the Regeneration tower were back to normal as indicated in Figure (4) below.

To test the capacity of the Regeneration Tower, a coke burn test was conducted. The circulation rate was slowly increased to 95% of design and the oxygen concentration was slowly reduced from 1.0 mole% to 0.8 mole%. Laboratory results showed less than 0.07 wt% carbon, indicating complete regeneration. The evaluation of the test results indicated that there was no major reduction in the unit capacity.

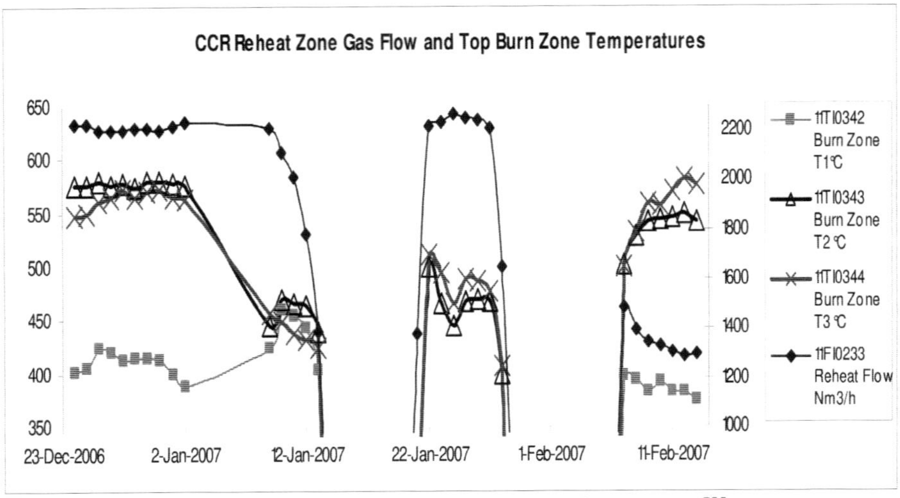


**Figure (4): Regenerator temperature profiles**

## CONCLUSION

In conclusion, the blowouts were caused by pinning that resulted from the over-design of the blower. The installation of the restriction orifice on the Regeneration gas blower suction line eliminated the pinning problem. Normal CCR Platformer operation was resumed.

## 208b Detailed Process Knowledge about Refinery Process Operations Using Radioisotope Technologies

**Lowell Pless**, Tracerco, 4106 New West Drive, Pasadena, TX 77507

Detailed Process Knowledge about Refinery Process Operations Using Radioisotope Technologies

For understanding refining processes Engineers have historically relied upon plant process measurements such as flow rates, temperatures, pressures, and model or simulation results for data with which to solve troubleshooting or re-design problems. An array of on-line diagnostic services applying radioisotopes are available that can provide additional data offering real-time information on how pieces of process equipment are actually operating.

One common application is gamma scanning. This technique is primarily applied, but not limited to, distillation or separation columns. Gamma scans provide a density profile of the internal process of operating distillation columns and other process vessels. The density profile can be used to diagnose the hydraulic operating conditions of mass transfer devices such as damage to internals, flooding, degree of entrainment or weeping, liquid levels on trays and distributors, liquid distribution through packed beds, etc. The presenter will show case studies where scanning revealed vital process information that helped solve an operating problem, or helped make a revamp/re-design successful.

A derivative from gamma scanning, CAT-scans, will be demonstrated. The CAT-Scan technique is a specialized horizontal gamma scan used to generate a topographic profile of the internal cross-sectional density of process equipment. This profile is useful for the detailed study of liquid flow distribution through packed columns as well as having applications on fluidized catalytic reactors such as FCC unit risers.

Radioactive tracers are used on operating processes to acquire real-world measurements of process variables that otherwise could only be calculated or estimated. Process variables that can be directly measured with radioactive tracers include volumetric flow rates, flow distribution, residence times, mixing characteristics, etc. Case studies of tracer applications applied specifically to refinery operations such as entrainment studies, flow distribution, and leak tests will be discussed.

## 208d Refinery Furnace Fouling

**Paul E. Eaton**, Technical Marketing, Champion Technologies Inc., 19827 Sunbridge Lane, Houston, TX 77094 and Murray Gray, Chemical & Materials, University of Alberta, T6G 2G6, Edmonton, AB T6g 2G6, Canada.

The problem of fouling is treated from three points of view; laboratory, plant, and economic. A laboratory apparatus, which simulates furnaces operating in high fouling circumstances, is discussed and test results are given for real situations. A fouling problem, which occurred in an actual operating furnace, is evaluated with respect to operating variables and chemical anti-foulants. Good lab-plant correlation is observed. The economic aspects of fouling are presented in terms of cleaning costs, lost profits, and energy consumption. An economic model is proposed, containing the key operating costs.

# 208e Impact of Hydrotreater's Ammonium Chloride Salt Deposition on Refinery Operations

**Nelmo F. Fernandes Sr.**, REGAP - Gabriel Passos Refinery, PETROBRAS, Rod. Fernao Dias km 427, BR 381, MG, Betim, Brazil

Fouling and consequent underdeposit corrosion caused by ammonium salts, especially ammonium chloride, have serious impact on the reliability of operation of various process units. In hydrotreating units salt deposition on heat exchanger tubes causes a decrease in heat transfer efficiency, decrease hydrotreating efficiency, increased pressure drops, and corrosion. This paper will discuss the causes of ammonium chloride fouling,methods to help prevent and/or mitigate the fouling, and provide a case history demonstrating the effects of ammonium chloride formations in one refinery operation.

## 208f Opportunities to Minimize Flaring

**Omid Zadakbar**, Ali Vatani, and Kianoosh Karimpour. Faculty of Chemical Engineering, University of Tehran, Enghelab St., Tehran, Iran

ABSTRACT

Recently, the need for energy conservation and environmental protection has been gaining increased importance. Due to huge amount of gas flaring in oil and gas facilities we tried to find alternative methods to reduce, recover and reuse flare gases. Therefore, comprehensive process evaluation of plants especially units that produce flare gases, comprehensive monitoring of flow and composition of flare gases, investigation of existing flare systems and finding alternative choices to reuse flare gases have been done in 11 petroleum refineries, natural gas refineries and petrochemical plants. Based on experimental results, many alternative methods have been advised. Evaluation of flare tips, flare pilot and flare ignition system, purge gas and blanketing gas replacement, number of flare stacks in use and equipment redundancy and also installation of waste heat refrigeration unit, ejector vapor recovery system, flare gas recovery unit and local flare gas recovery system for FCC unit and delayed cocking unit have been recommended to minimize flaring.

In this paper we explain the expected results, operation guidelines and economics of each method to develop a global standard to reduce flaring and venting of hydrocarbon gas that is produced in oil and gas refineries. The correct selection of technology will take the unique challenges of the oil and gas processing industry into account.

NALCO

# Guidelines for Using Overhead Simulation Models to Assess Overhead Corrosion Potential

S.A. Lordo[a] C.J. Claesen[b],
[a]Nalco, Energy Services Division, 7705 Highway 90A Sugar Land, Texas, USA
[b] Nalco, Energy Services Division, Mechelsesteenweg 290, 2550 Kontich Belgium

## Abstract

As more refineries are expanding their crude selection it is important to look at more than the assay information. Understanding the operability issues of the new crude is paramount. One important aspect is the impact on corrosion in the fractionator top and overhead systems of refinery crude distillation units.

Overhead simulation models can be used successfully to assess the impacts new crude may have on the effect of opportunity/challenge crudes have on the corrosion potential in these systems. Models can be also used as a proactive/diagnostic tool to define some of the changes required to maintain or improve the reliability of the system.

The key to getting the maximum benefit from overhead simulation models is the understanding of their limitations and user guidelines.

This paper discusses some of the key guidelines and limitations of overhead simulation programs. Some of the topics covered include data quality, overhead sampling techniques and interpretation of the results with respect to actual refinery experience. Three field examples where the Nalco PATHFINDER® overhead simulation program provided key diagnostic information are reviewed.

Keywords : Overhead, Corrosion, Simulation, Software, Model

## Introduction

Opportunity crude processing is rapidly becoming an issue in the majority of refineries. In the past many Refiner's did not rely on opportunity or challenge crudes as part of their crude diets. With crude pricing >$90 per barrel and the projections that crude quality will continue to decline, more refiners have begun processing more opportunity crudes.[1]

Reduction in crude quality and changes in crude fractionator operating conditions such as top pressure, top temperature, crude source, crude temperature, can have a significant influence on the conditions in other parts of the unit. One condition that can change is the corrosiveness of the water that is condensing in a fractionator overhead (OVHD) system due to changes in the content of mineral acid species (i.e., Hydrochloric, $H_2SO_3$, etc.), organic acids, and weak acids, such as $CO_2$, and $H_2S$. Another change can be the formation temperature of ammonium chloride or amine chloride salts. Some of these changes can be significant and result in corrosion. In the past, the strategy often has been a wait-and-see approach when changes of this nature occurred and often this led to drastic unexpected corrosion and loss of plant integrity.

Simulation programs based on water condensation profiles and ionic equilibria can be used to evaluate the potential impact of these changes on a circuit like a crude distillation unit (CDU) OVHD. Simulation programs can also be used as part of the economic assessment process used to develop a successful corrosion mitigation strategy.

Some of the mitigation strategies that can be considered are:

- Changing operating conditions (often not an option because of economic penalties)
- Changing unit overhead configurations (large capital expense)
- Changing overhead metallurgy (expensive, often requires a shutdown)
- Changing overhead chemical additive programs and application strategy (relatively simple and easy to do)

Computer simulation models that were developed for designing and troubleshooting fractionation units often are missing the capability to handle the complex nature of a condensing water system and its changing chemistry. This is beyond the scope of these models, as the detailed sets of information required to simulate ionic interactions are not readily available or do not exist beyond a simple understanding. Hence, over the past 20 years there has been an effort in Industry to develop the information needed to close this gap.

Several computer simulation models have become available that address some of the existing gaps in the understanding of ionic equilibria and interactions. Some of these models are commercially available and some are provided through consultants or companies offering anti-corrosion chemicals and services. These models provide some of the information needed to better identify the potential impacts that changes in operational conditions and in the concentrations of acidic and basic ionic species have on a water condensing system as typically found in CDU OVHD circuits. These models can provide better understanding of:

- What could have caused corrosion related failure on piping, exchangers, etc.
- What could be currently happening in the overhead circuit?
- What could happen when operational changes occur?
- Where could amine- or ammonium-chloride salt formation occur in the overhead circuit?

Experience has shown that some of these simulation models are able to provide reasonably good results. The results generated from simulation models can only be as good as the understanding of the system and the quality of the input data. A good understanding of the guidelines for using these types of simulation models and their limitations is paramount in utilizing them to their fullest capabilities.

The objective of this paper is to outline a range of guidelines that should be considered when planning to utilise a simulation model and some of the limitations of various model types. In addition, three examples are presented where simulation models have been used to analyze a crude tower overhead system.

## Types of Overhead Computer Simulation Models

Typically these computer simulations are used to determine crude tower overhead parameters, such as,

- Water dew point temperature and its location in a circuit
- Vapor-Liquid distribution of ionic species (partitioning) as water condenses
- Interaction between the acids and bases present
- Kinetics of amine/ammonia chloride salting
- Initial water dew point ionic conditions (pH, concentrations, etc.)

The types of programs available can be classified into two broad categories:

**DETAILED USE**
**GENERAL USE**

**DETAILED USE** models often require extensive data collection and samples that are not routinely available and hence they are not used for routine monitoring. These models are typically used to troubleshoot a problem, or to aid in defining the impacts of long-term strategic operational changes. Because of the data and sample requirements, these models take a long time (weeks to months) to develop and are not suitable or practical for tracking trends or making rapid decisions.

It is recognized that the ideal point to control the pH of a CDU OVHD is at the water dew point[2]. A **DETAILED USE** model is not suitable for routine dew point pH control due to the amount of time that is required to produce useful output. The pH at dew point would only be available weeks or months after the data and sample collection.

**GENERAL USE** models use readily available data and samples to model overhead systems for frequent routine monitoring and trending. These models allow a refiner to make short-term adjustments or decisions, for example, in response to a change in crude slate, involving the introduction of an opportunity crude. Opportunity crudes have brought a host of challenges to industry, such as, increased loading of organic acids or increases in chloride loading caused by operating the upstream desalter outside of its original design criteria (i.e., different crude density, viscosity, conductivity, etc.).

**GENERAL USE** models are also useful in controlling pH within the dew point regime, as they can be run on a frequent basis (i.e., daily, weekly). Some of these models generate trend data that can highlight problems (increasing incidents of low condensing dew point pH, increasing amine chloride salting temperature).

Both types of models can provide information to make more informed decisions.

**Guidelines for Using Ionic Overhead Simulation Models**

Before using these types of computer modelling programs, one needs to define the objectives. Some typical objectives are:

- Determine the impact on the OVHD pH control program if there was an increase in OVHD acids (HCl, $H_2SO_3$, organic acids, etc.)
- Confirm if shock condensation could be a cause for a particular exchanger failure
- Determine if the current chemical additive program is still providing optimal performance or if it should be modified to compensate for changes in process conditions, crude slates, etc.
- Determine an optimal neutralizer program (with respect to cost performance, type of neutralizer used, chloride salting potential, etc.)

The desired objectives will determine which model should be selected.

When selecting a model some of the key attributes to evaluate are discussed below.

- Understand the model mechanics. In other words, one should understand the basis of the calculations and what methods are utilized. The model's basic calculations should be based on established and recognized thermodynamic principles, specifically in determining water and hydrocarbon dew points. A simple model may utilize steam tables and consider no hydrocarbon condensation to determine the water partial pressure and dew point in an overhead system. More advanced models will utilize recognized methods, such as the Peng-Robinson method, as part of a rigorous dew point calculation. For example, in a particular system operating at a pressure of 14.5 psig (1 barg) a model not considering hydrocarbon condensation may give a water dew point of 163°F (73°C), while one using a more rigorous method for two-phase condensation would yield a water dew point of 201°F (94°C).

- Understand the basics of the ionic interaction portion of the model. Unlike heat transfer modelling, there is no standard method in handling ionic balancing. Some models utilize a kinetic approach, others a thermodynamic approach, while some consider aspects of both approaches[3,4]. This is one of the areas that differ between the various models. It could be misleading to use only one of the approaches, kinetic or thermodynamic, as they may lead to different conclusions.

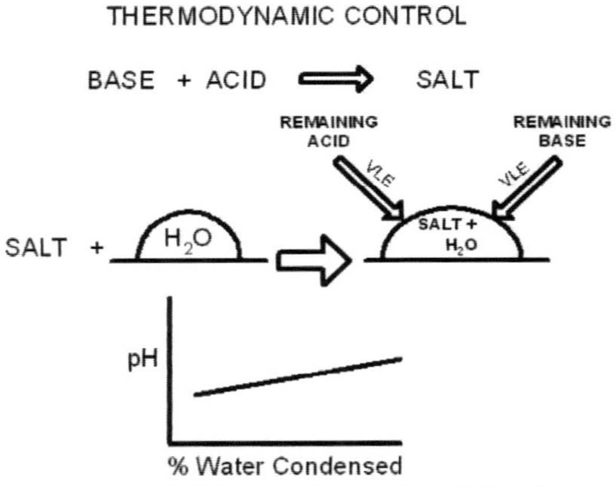

**Figure 1: Thermodynamic pH Profile**

In a thermodynamic equilibrium based model it is assumed that cations and anions have enough time to completely react, hence, the strongest bases react with the strongest acids. For example, at equilibrium, as much chloride salt will be formed in the vapor phase as possible. After all of the chloride salt is formed, there may be small amounts of residual acid or base left un-reacted in the vapor phase. pH profile calculations, in this case, assume that all of the chloride salt that can form will dissolve in the liquid water and the residual acids or bases will participate the water phase at a rate determined by its vapor-liquid equilibrium constant (VLE) and the amount of water condensed (Figure 1). Assuming that there was enough amine to react with all of the acid, the amine salt dissolved in the water behaves as a buffer, and the resulting pH profile tends to be relatively flat.

Assuming a thermodynamic equilibrium regime, amine/ammonia salts dissolved in the water will be formed with respect to their ionic strengths. In other words, strong bases will react with strong acids until one of them is depleted and then the next strongest reacts and so on (Figure 2).

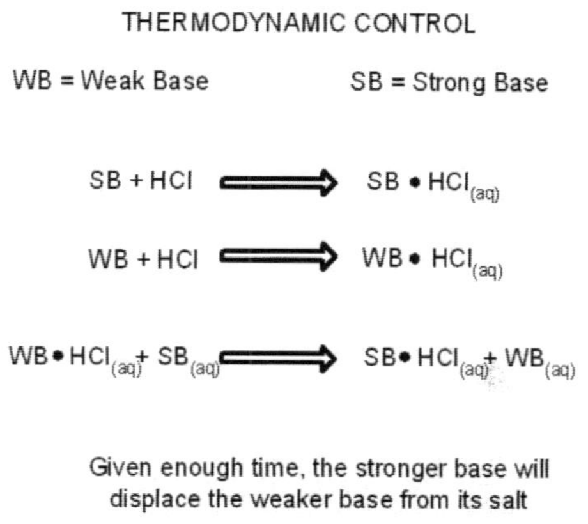


**Figure 2: Thermodynamic Salt Formation**

In a kinetic based model, the concentrations of the various species are the determining factor regarding which reactions take place and the relative concentrations of the reactants will determine how rapidly the reaction takes place. In the example used for the thermodynamic equilibrium based model (Figure 1), the kinetic model assumes that NO reaction in the vapor phase has occurred. The rate that acids and bases present in the vapor phase enter the water is determined by their individual concentrations and VLE constants and the percentage of water condensed (Figure 3).

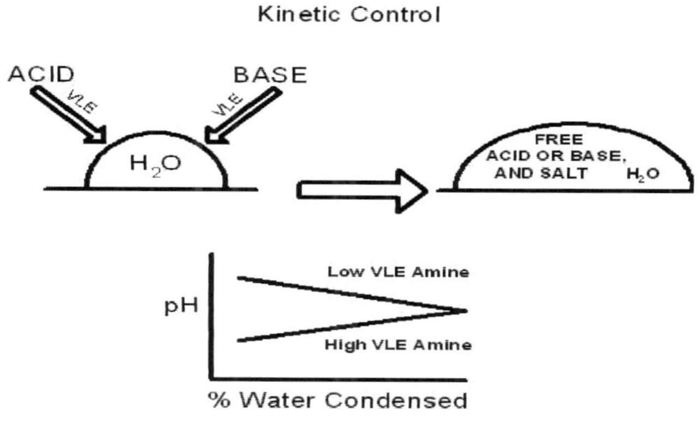


**Figure 3: Kinetic pH Profile**

Some bases, such as ammonia, have a very low partitioning into the hot, first condensed water droplets so these droplets may contain only acids and have a very low pH. Some amines partition into the first condensate as rapidly or more

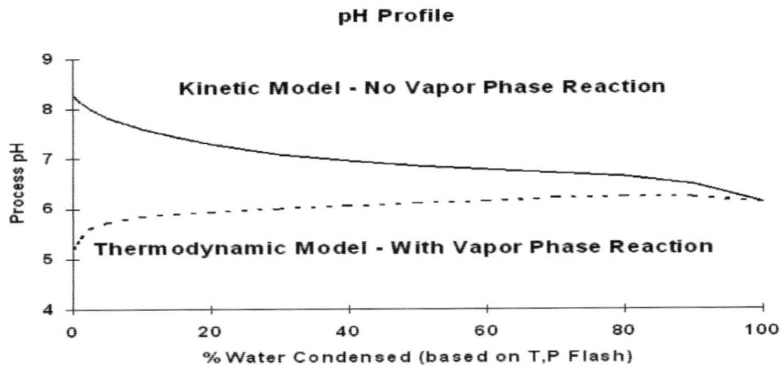


**Figure 4: Example of Condensing System pH Profile**

rapidly than hydrogen chloride does. When such amines are present the initial condensates can show elevated pH values that gradually decrease as more acid dissolves in the condensing water (Figure 4).

Under kinetic control, amine/ammonia salts formation is based on the molar concentration of the reactants and their individual formation constants (Figure 5).

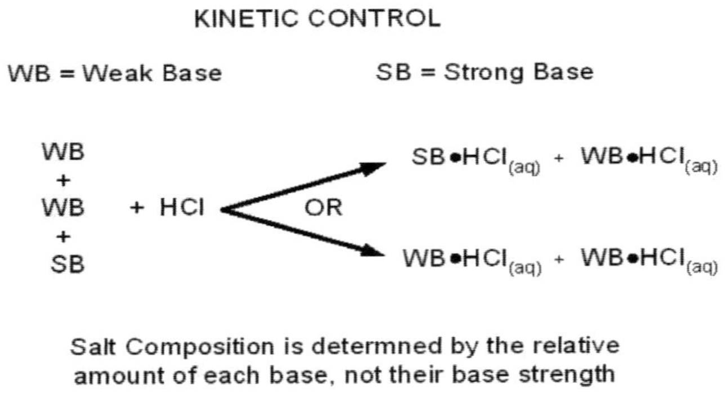


**Figure 5: Kinetic Salt Formation**

In reality what is occurring is somewhere between these two interaction mechanisms as both are occurring at the same time.

- Understand the calculation basis used to determine the amine/ammonia chloride salt formation temperatures. While the ammonium chloride salt formation temperature can be found in the literature[5], most of the other amine chloride salt formation temperatures are not documented or published. These data have been developed by the models' developers and are proprietary.

- Understand what data is required and how input data is addressed. For some models, rigorous analytical information on water samples is required for each model run, while others use the typical field analytical data on the samples taken. Some models can handle the various ASTM methods for defining the distillation profiles of the hydrocarbon portion while others utilize pure component data. This can create a source of variation between models developed for the same unit.

- Understand the terminology being used by the model supplier. Initial water dew point definitions do differ. Some models use an academic definition, for example, defining the first water drop as when a very small amount, 0.0000001 mole fraction (in a typical CDU OVHD this would be 2-3 grams) of free liquid water forms. Different precision can be put into these models and thereby change the calculated water dew point temperature and corresponding pH.

  Although academically correct, a high level of accuracy/precision does not necessarily reflect what is impacting an overhead circuit. In dynamic systems, such as those found in CDU OVHD's, controlling the pH of a higher percentage of water condensed may be more representative of the system[4]. This would also compensate for the inherent inaccuracies in the data gathered.

- Few models are capable of estimating the pH of a condensing system along the condensation profile (Figure 6). This capability shows how well a neutralizer program is performing with regards to controlling the pH as water condenses in the system. This is valuable in selecting the appropriate neutralizer for a particular system. Figure 6 shows the pH profiles of two amines and ammonia. In this example one can see that ammonia (NH3) and morpholine (MORPH) start at a low pH and increase, morpholine increased rapidly after 15% water condensation ammonia only after 80% of the water has condensed. The monoethanolamine (MEA) initial pH starts very high and decreases.

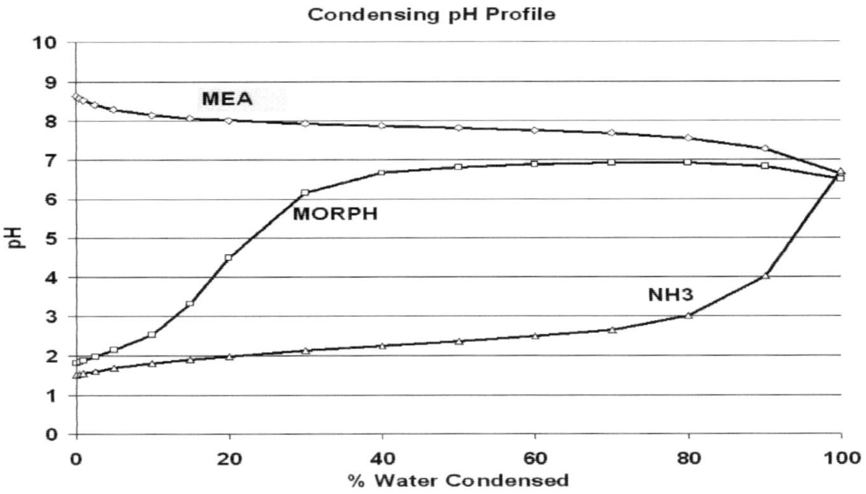


**Figure 6: Condensing Profile of Ammonia and Two Neutralizing Amines**

- All models will require some form of analytical support to determine the species present in the system. Care needs to be taken when sampling as improper sample handling may introduce error. Water and hydrocarbon sample collection can range from the simple glass or plastic bottle to nitrogen purged sample bombs.

  Analyses of the samples for the various anion and cation species are generally done using established and acceptable methods. Depending on how the sample is taken, stored and processed some of the issues to be aware of are:

  - oxidation of some species (e.g., $H_2S$ and sulfite to elemental S and thiosulfate, etc.)
  - loss of volatile un-reacted components, such as excess $NH_3$, $CO_2$, $H_2S$,etc.
  - pH of the water taken during sampling will change as the chemistry of the water changes. (This is a consideration when using routine water samples.)

- Operating data (e.g., flows, temperatures, pressures) that is used to drive the model is important and should be taken at the time that the samples are taken. When using a Detailed Use model, the proper time lag should be applied to the data collection, so the water and hydrocarbon samples correspond directly with the operating data. Unit fluctuation impact on data quality can be minimised by sampling during 3-4 hours of fairly stable operation.

Following the basic guidelines above will help the refiner to produce the data at the quality level that is required for installing and running the computer simulation model. Depending on how the models are being used the variations in results from different models may or may not be significant.

**Limitations and Opportunities for Misinterpretation**

All ionic equilibria-based models used in Industry have limitations. Understanding these limitations are important so the interpretation and significance of the results can be taken within context to avoid making unnecessary changes.

The major limitation common to both types of models is the assumption that the system is in thermodynamic equilibrium. The systems being modelled are dynamic condensing systems that may not reach thermodynamic equilibrium.

Results from models that do not use a rigorous T-P flash calculation to determine the condensation profile will differ from the models that do use a more rigorous calculation method. Not only will the dew points differ, both for hydrocarbon and water, but formation temperatures of ammonium chloride, ammonium bisulfide, amine-chloride salt and other compounds controlled by partial pressure[2] will differ.

Another limitation is the assumption that a neutralising amine when injected immediately flashes and is perfectly mixed. Whitehead and Ramachandran[6] point out that overhead naphtha will coat water droplets as heat is removed from the vapors in the process of warming and evaporating the water in the droplet. This hydrocarbon coating creates a physical barrier to the evaporation of more water, slowing the evaporation process. This also applies to the injection of a liquid amine into an overhead circuit. The currently available models do not account for the diffusion limitation of the various overhead components (i.e., hydrocarbon, amine, ammonia, hydrogen sulfide), within the residence time of the system.

All models are based on ionic species identified in water and gas samples taken from the unit. However, there can be instances where not all of the ions reach the sample point or the sample point misses the ions. For example, chloride salts could have already formed and deposited in equipment and piping ahead of the sample point(s), thus reducing the amount of chloride salt forming constituents measured in the water sample. In this case, the chloride salt formation temperature calculated would actually be higher due to the concentrations of the chloride salt components being understated. This in turn would also affect the calculated water dew point. This is a recognized issue and hence most modellers will signal an alert when the salt formation temperature is within a specified range of the tower overhead or water dew point temperature. Typically these temperature ranges are ±5-25°F (±3-14°C) of the calculated reference temperature. This alert point varies with the model supplier.

Some models have included the impact that various ionic species have on the initial water dew point. These impacts are usually based on binary mixtures. The impacts of multi-component systems (e.g., $H_2S$+HCl+$NH_3$+Free Amine+Amine Salt) have not been developed.

The VLE behaviour of partitioning species changes with the pH of the water condensed, and the incremental partitioning calculations required for truly correct phase partitioning must take into account the variations in liquid pH caused by the concentrations of ionic species already dissolved in the water.

The academic purity of the calculations is also suspect due to the condensed hydrocarbon phase that will exist above the water layer. The first condensed liquids that form in the overhead are the high-boiling hydrocarbon components. Any water droplets that subsequently form will sink below the hydrocarbon-vapor interface.

In most cases, there will be around 70% of the hydrocarbon mass in liquid form, before any free water begins to appear. Diffusion through the hydrocarbon barrier is required for the acidic and basic species to partition into the water. There are thus two partitioning effects in practice. The first is from vapor to hydrocarbon liquid, and the second is from hydrocarbon liquid to the aqueous liquid phase, with a physical hydrocarbon diffusion process in between.

The calculations from these models usually provide numerical answers. It is important to note that a probability factor needs to be considered. For example, when a model indicates the salt formation temperature of an amine chloride salt is 20 degrees above the process temperature one cannot assume that this chloride salt is forming there on an absolute basis. Rather, the correct way of interpreting the answer is that one has a high probability for the salt to be forming. All results from a model must be compared to the system being modelled to ensure what is being described by the model is in alignment with what is observed in the system.

Given that computer simulation models have these and other limitations, it is imperative that the model be validated with the actual process. Two validation questions for these types of models are typically used. These are:

- Is the calculated accumulator pH equal to the measured pH
- Does the model reflect what is being observed on the unit (i.e., increased corrosion, pressure drop increase)?

Understanding these inherent limitations will give the refiner the capability to better interpret the output from these models.

**Practical Application of Computer Simulation Modelling**

Included in this paper are three examples of using computer simulation models in refinery applications.

**EXAMPLE 1 – IDENTIFYING THE NEED FOR CHANGE IN OPERATING CONDITIONS.**

A vacuum tower top pumparound was showing symptoms of fouling, as was the packing in the associated tower section. A review of the system used the Nalco PATHFINDER overhead simulation model to determine the potential for ammonium chloride fouling.

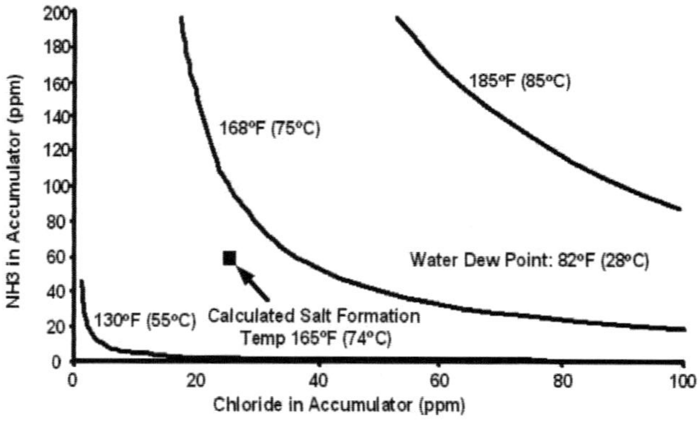


**Figure 7: $NH_4Cl$ salt formation isotherms**

Although the simulation model is typically used for modelling overhead systems, it was used to model the top pumparound of the tower to get a directional indication of what might be the cause of the fouling.

Modelling showed that under the existing tower top conditions there was a high potential for the formation of ammonium chloride ($NH_4Cl$). Tower top temperature was 113°F (45°C) while the calculated salt formation temperature was 165°F (74°C). Figure 7 graphically shows the $NH_4Cl$ formation temperature with respect to temperature isotherms.

Overhead water analysis was used to estimate the amount of chloride, ammonia, $H_2S$ and other species that could be present in any water formed in the top pumparound due to cold tubeside temperatures. The hydrocarbon properties (i.e., gravity, distillation, etc) used were those of the hydrocarbon in the top pump around.

It was recommended to increase the temperature of the overhead and top tower section sufficiently to move away from the probability of ammonium chloride formation. The typical recommendation is to maintain the system temperature 14°C above the salt formation temperature. The objective was to raise the temperature high enough that $NH_4Cl$ would not precipitate until it was outside the tower and into the overhead condensers, where condensing water was present. After raising the temperature the top pumparound, fouling stopped.

When the temperature was raised, the amount of chloride measured in the overhead was seen to increase from <10 ppm to >40 ppm, thus validating the prediction made by the model (Figure 8). The observed increase in chloride also reinforces one of the limitations discussed above.

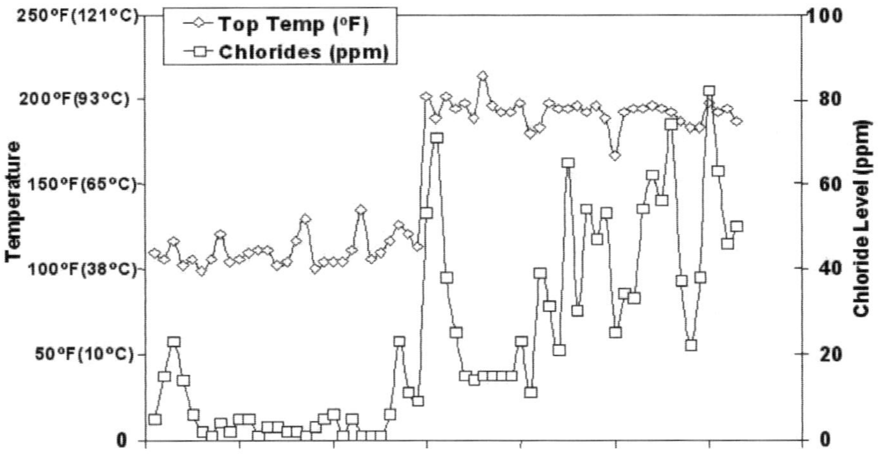


**Figure 8: Actual Tower Overhead Temperature and Overhead Accumulator Drum Chloride Concentration**

## EXAMPLE 2 – USING SIMULATION MODELS TO DETECT ONGOING SYSTEM CHANGES

A refinery began processing a new opportunity West African crude on one of their crude units. The crude unit does not have an OVHD water wash and had shown issues with amine/ammonium chloride salt formation in the past. Unit operations during the initial processing showed no measurable changes in desalter performance, however the OVHD chlorides began to increase. Chlorides in the OVHD increased from nominal 60 ppm to near 110 ppm. In addition, the OVHD accumulator water iron content increased from nominally 0.2 ppm to 0.8 ppm. Final pH of the system was maintained within the normal operating range of 6.8 – 7.2. The observed changes in the OVHD were attributed to the new crude.

One of the monitoring tools used to gauge the corrosion potential of the OVHD on a routine basis was the Nalco PATHFINDER overhead simulation model. PATHFINDER is run weekly. PATHFINDER started to show an increase in salt formation potential as the refiner increased the percentage of the new crude in the crude slate. Figure 9 shows the observed increase of both the amine-chloride and $NH_4Cl$ salt formation temperatures relative to water dew point and tower OVHD temperature.

Although the increased formation potentials were not large, the trend showed that higher percentages could potentially push the salt formation temperatures into even more problematic regimes. The refiner used the model results to restrict the amount of the new crude process on this particular unit and shift the balance of the opportunity crude to another crude unit that was more capable of handling increased chloride loading without losing the economic advantage of running the new crude.

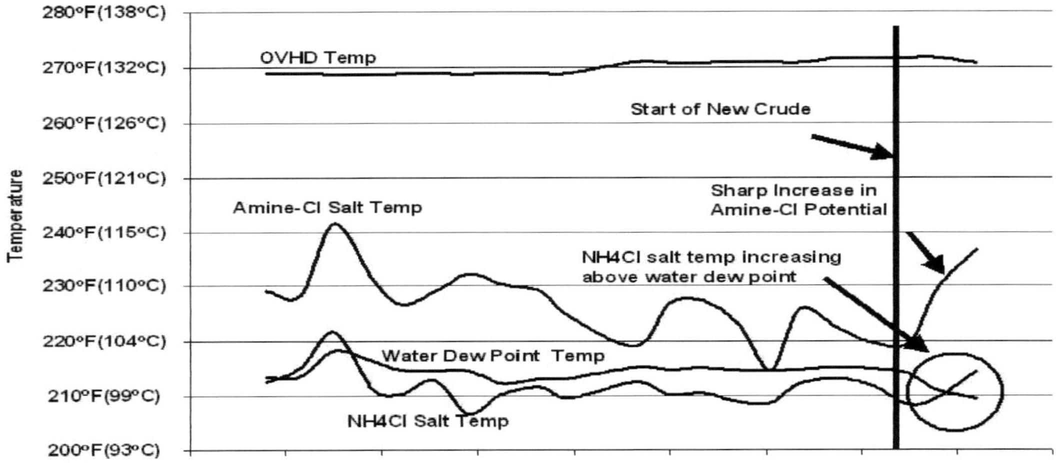


**Figure 9: Ovhd Trend Data**

## EXAMPLE 3 – USING SIMULATION MODELS TO ASSIST ROOT CAUSE FAILURE ANALYSIS

A refinery operating a CDU with a double drum OVHD design was experiencing frequent failures in the coldest part of the first stage OVHD exchangers. The metallurgy of the exchanger shell and tube components was carbon steel and the run time to failure had decreased to six months. Electrical Resistance corrosion probes located at the outlet of the exchangers showed significant variations in corrosion rates with peaks exceeding 100 mpy.

A root cause failure analysis was conducted and data was gathered to run the Nalco PATHFINDER overhead simulation program. Simulations showed that, depending

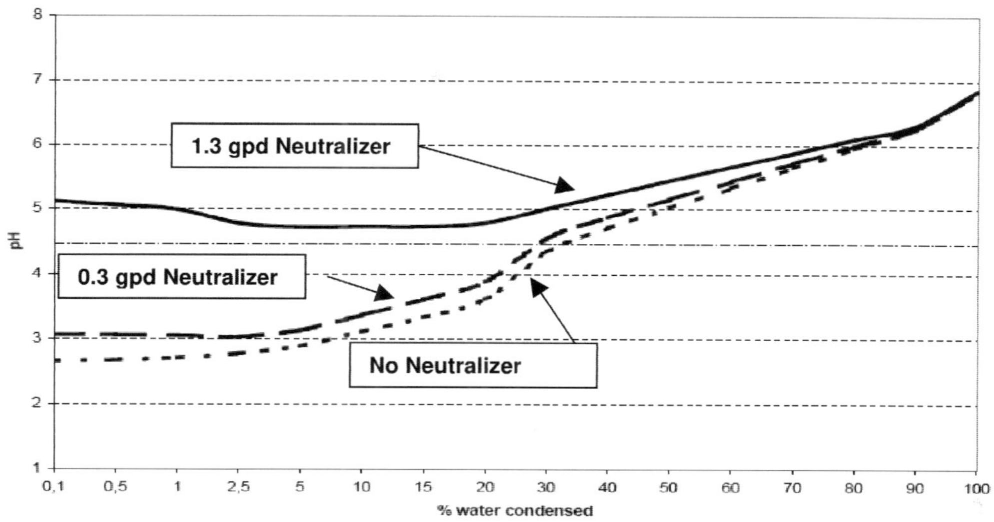


**Figure 10: Neutralizer Dosage and pH**

on the mode of operation; the pressure in the overhead was changing up to 14.5 psi (1 bar) giving a variation in water dew point between 223°F and 243°F (106°C and 117°C). Due to a revamp the temperature of the cold crude that provided the cooling medium on the tube side of the coldest exchangers had decreased and sometimes was below 104°F (40°C). The conclusion was that the skin temperature of the coldest tubes would be cool enough to cause localized shock condensation.

Fearing amine-chloride salt formation the refinery had been injecting no or only very low levels of neutralizer into the first stage overhead. The pH profile of the condensing water was simulated and the results for one set of conditions are shown in Figure 10. In this case the existing neutralizer injection of 0-0.3 gpd (0-1 l/d) proved to be too low to provide protection until 20-30% of the water had condensed and exchanger tubes subject to shock water condensation were seeing pH's of 3 and below.

The corrective action in this case was to adjust the neutralizer injection rate based on routine field Pathfinder® simulations. The first drum temperature and pressure was operated to have small quantities of liquid water. This allows the removal of amine-chloride salts with the drained water and prevents that they are carried into the fractionator tower with the reflux. The corrosion probe readings dropped below 10 mpy and no new exchanger failures have been reported at the time of the writing of this paper.

## Conclusion

With the increase in opportunity crudes processing it is important to rapidly determine all of the impacts on overhead corrosion control that these crudes may have on a CDU OVHD circuit. Computer simulation models can provide the ability to rapidly assess potential impacts on CDU OVHD corrosion.

Computer simulation programs developed to model the ionic nature of CDU overheads and other circuits will provide valuable insights into the process when used with the understanding of the program mechanics and limitations. When properly used, these models can confirm a suspected corrosion mechanism, as well as, identify potential issues before they manifest themselves more dramatically.

There are two different types of models currently available, **DETAILED USE** and **GENERAL USE**, it is important to first define the objectives for using a computer simulation model. Once the objectives are defined then the proper model can be selected with respect to its attributes.

**DETAILED USE** models require an intensive level of data collection and are good for developing and evaluating long-term strategic changes, or post-mortem problem solving, but may not be suitable for frequent monitoring. While some **GENERAL USE** models require less intensive data collection, they provide essentially the same results as the **DETAILED USE** models but are better suited for frequent monitoring and thus can be used in making short-term unit decisions, such as how best to run an opportunity crude.

Equally important, to defining the objectives, is to understand the differences between the various available models and how the model's inherent limitations may affect the output results. No two models in Industry will yield exactly the same answers, however, most will give enough information to make directionally the same conclusions.

## References

1. B. Mason, G. Scattergood, J. Gardenhire, "Handling the Challenge", Hydrocarbon Engineering, March 2008

2. V.K. Braden, P. R. Petersen, L. Bowerbank, J.P. Gorman, M. Malpiedi, "Crude Unit Overhead Corrosion Control", CORROSION/98, Paper 585 (San Diego, CA: NACE 1998)

3. G.G. Dugan and R.G. Richton, "Application of Ionic Equilibrium Process Simulation for Atmospheric Distillation Overhead Systems", CORROSION/98, Paper 586 (San Diego, CA: NACE 1998)

4. P. R. Petersen, "Predicting the Effect of Neutralizing Amines on the pH Profile in Crude unit Overhead Exchangers", CORROSION/88, Paper 166 (St. Louis, MO: NACE 1988)

5. Y. Wu, Oil and Gas Journal, p.38, January 3, 1994

6. T. D. Whitehead, P. A. Ramachandran, "An Analysis of Water Wash Practices Used for Corrosion Control in Condensable Hydrocarbon Systems", CORROSION/91, Paper 318 (Cincinnati, OH: NACE, 1991)

# 227c Case Study: Salt Corrosion Controlled by Discontinuing Water Wash

**George Duggan**, Baker Petrolite Corporation, 12645 West Airport Boulevard, Sugar Land, TX 77478 and Walter Giesbrecht, Flint Hills Resources L.P., Rosemount, MN 55068.

Effective control of overhead system corrosion activity has become more challenging due to the increased prevalence of opportunity crude oils being processed today. Such crude oils can contain significantly higher levels of contaminants (for example chlorides, sulfur, naphthenic acids) than crude oils commonly processed in the past. Combine these more difficult-to-process crude oils with the fact that many refineries are operating well beyond design capacity, and the result is a substantially increased risk of overhead corrosion. Baker Petrolite has developed a comprehensive, engineering-based approach to address these problems using its TOPGUARD Overhead Corrosion Control Program.

For many systems, successful corrosion control can be simply achieved with the use of corrosion inhibitors and neutralizing amines. Alternately, corrosion issues can be addressed with non-chemical methods, such as operational changes, higher-grade metallurgy or water wash. Water washing in distillation column overhead systems is a proven method for controlling under-salt corrosion. However, not every system is a viable candidate for water wash. In fact, in some situations, equipment design can actually render a water wash system ineffective. A TOPGUARD troubleshooting effort on an atmospheric overhead system identified such a case. With a continuous water wash in place, heat exchanger life ranged from six to twelve months due to under-salt corrosion. An engineering evaluation of the system, utilizing the Ionic Model, identified an alternate approach which featured a change in neutralizer strategy coupled with a strict upper limit on overhead chloride levels. With the implementation of these new strategies, water wash injection was discontinued. As a result, heat exchanger life increased dramatically. During the three years of operation with the new strategy, the equipment has experienced no leaks or failures and routine monitoring gives no indication it needs to be replaced. Baker Petrolites TOPGUARD Program ensures that all aspects of equipment mechanical design are considered when selecting a proper corrosion control strategy.

# Caustic Injection Relocation Project

Mohammed S Eid
Saudi Aramco

**ABSTRACT**

Yanbu Refinery Department (YRD) operates a hydroskimming refinery with a capacity of 235,000 Barrels/day. The crude oil passes through three preheat stages (series of heat exchangers) before the entering the crude heaters. After the first pre-heat stage, water is added to the preheated crude oil prior to entering the desalters. A 3°Be caustic is injected upstream of desalters to neutralize acids formed. Also, a 3°Be caustic is injected downstream of the desalters for corrosion control. It was observed since YRD start-up in 1983 that the second preheat stage exchangers have history of active corrosion and fouling problems. It was strongly recommended by Consultant Services Department (CSD) to implement the best practice idea which is relocating the downstream caustic injection from the downstream of the desalters to upstream of the crude heaters to eliminate the corrosion and fouling problem in the second preheat exchangers. On average, this project is saving $80,000/year maintenance cost. The cost avoidance of dumping Diesel to Fuel oil is around $1.5 million. Furthermore, the paper will discuss the challenges faced after the relocating project and the lesson learned.

## PROCESS OVERVIEW

The crude distillation process-flow in plant V04 carries a crude charge through preheating, desalting, charge heating, crude column distillation, and naphtha fractionation. The crude unit consists of two identical pre-heat trains (A & B).

The crude charge passes through the first stage preheat train, which is a series of four heat exchangers:

- ❑ E-1 crude/circulating naphtha exchanger
- ❑ E-2 crude/kerosene product exchanger
- ❑ E-3 crude/LDO product exchanger
- ❑ E-4 crude/no. 1 circulating LDO exchanger

After that, the crude charge flows to the desalters at a temperature of about 120-125°C, and is mixed with wash water before entering the desalters. From the desalters, the crude passes through the second-stage preheat train that heats the crude charge to flash temperature in a series of three heat exchangers:

- ❑ E-5 desalted crude/no. 2 circulating LDO exchanger
- ❑ E-6 desalted crude/HDO product exchanger
- ❑ E-7 desalted crude/cold reduced crude exchanger

Liquid from flash drum, V2 is pumped by flash drum bottoms pump, P2 to the final preheat section:

- ❑ E-8 desalted crude/circulating HDO exchanger
- ❑ E-9 desalted crude/hot reduced crude exchanger

The preheated crude oil finally enters crude charge heaters, H-1A/B where it is heated to crude column flash zone requirements. Heater outlet temperature is controlled at 375°C. **Figure-1**

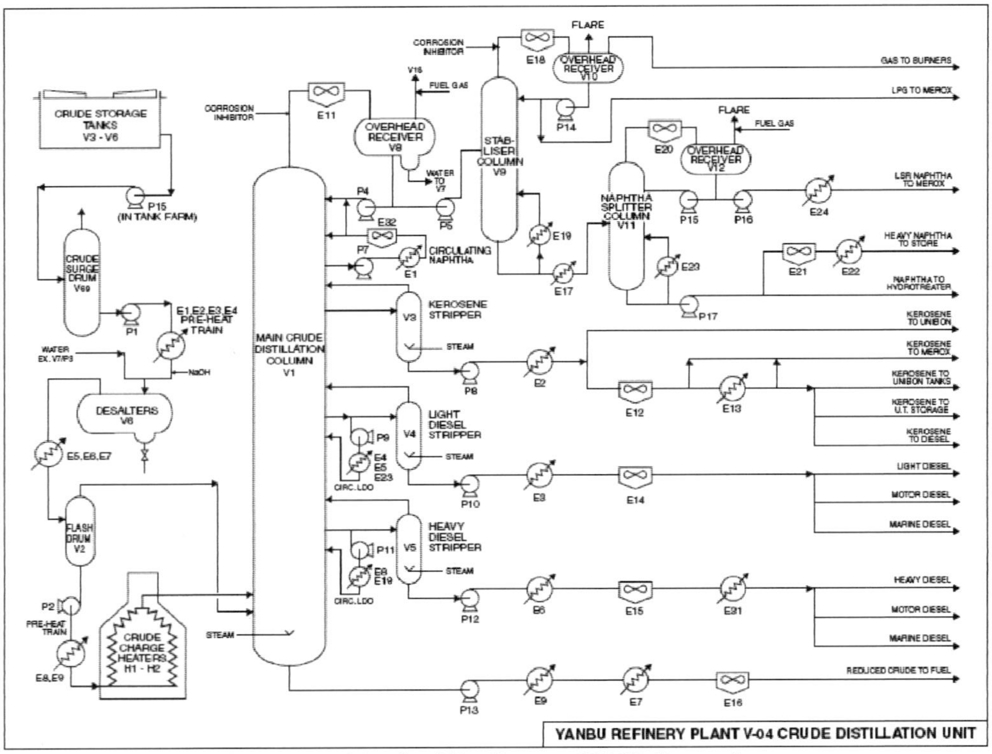


**Figure 1 – Yanbu' Refinery CDU Overview**

## CAUSTIC INJECTION AT CDU

After the first preheat section, water is added to the preheated crude oil by using a pre-set mixing valve and is thoroughly mixed with the crude, prior to entering first stage desalter, V6A. The oil leaves the desalter and is further mixed with freshwater prior to entering the second-stage desalter, V6B. The 3°Be caustic is injected upstream of desalters to neutralize acids formed (during the hydrolysis of salt). The caustic injection is controlled to maintain a pH value of 6-7 of the brine water effluent. The 3°Be caustic is also injected downstream of the desalters for corrosion control, and to control the chloride content in the O/H of the column. **Figure-2**

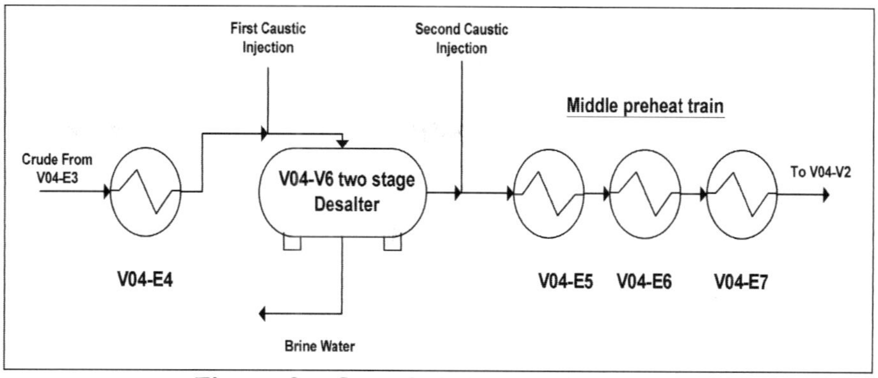


**Figure 2 - Caustic injection locations**

## Fouling & Corrosion Problems

It was observed, since Yanbu' Refinery (YR) start-up in 1983, that middle preheat exchangers V04-E5, E6, and E7 have a history of active corrosion and fouling problems, caused by under-deposit, inorganic deposits, and high temperature sulfide corrosion, which lead to pressure buildup and tube leaks.

The under deposit corrosion products are activated, mainly by the elevated temperatures of the inlet shell sides, leading to under deposit corrosion, due to partial vaporization of the water, and light ends in the crude. The high temperature of the crude passing through the second preheat train is another factor to accelerate such a fouling to occur. With the sulfur content of AL crude at 1.5-2% and with the high operating temperatures, the environment is set for high temperature sulfur corrosion.

In terms of the inorganic deposit inside the tubes, it was noticed, as per the R&D analysis on V04-E7, that around 88 wt% of the plugging material was iron sulfide corrosion product (Pyrite and Pyrrhotiteit), 9 wt% was water soluble salt (Halite-NaCl) and 3 wt% was Calcium Carbonate. These inorganic deposit materials are considered to be the major contributors for fouling inside the tube, which leads to pressure buildup.

The high temperature of the crude passing through the second preheat train is another factor that accelerates fouling. A two-stage desalter after the first preheat train was designed to minimize the salt content in the feed (outlet design for salt is less than 1 PTB). This target could not be achieved due to low desalter efficiency, caused by inadequate water wash rate and a low desalter residence time.

The water wash injection, as per industrial best practice, should be (3-5) volume% to assure that the required salts are dissolved compared to the existing injection rate, which is around 2.9% of volume. The desalter residence time is also important to make sure that no water carry over takes place but the calculated residence time of operating desalter is 11.5 @ 235 MBD, which is under the design (15 min @170 MBD).

It was concluded that downstream caustic injection acts as a catalyst for fouling and under-deposit corrosion, raising the pressure drop and leakage in the middle preheat exchangers

## INSPECTION RECORDS FOR MIDDLE PREHEAT EXCHANGERS

It was found that most of the tube leaks were in the top portion of the bundle, especially for V04-E5 and V04-E6 and this is due to the high heat flux at that location (**Figure-3**). In terms of V04-E7, most of the problems associated with these exchangers are mainly uniform thinning, localized attack and erosion corrosion as per Consulting Services Department (*CSD*).

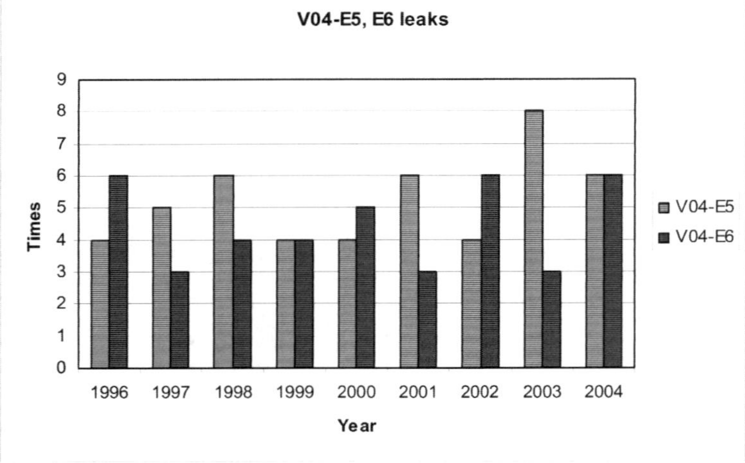


**Figure 3 - Caustic injection locations**

As per inspection records, V04-E5 and V04-E6 tube bundles' metallurgy has been upgraded for both train A and B, from carbon steel and 5 Cr-1/2 Mo, to 12% Cr steel, due to the high frequency of tube failures. Only three exchangers for V04-E7 have been upgraded to 5Cr-1/2Mo while others remain as carbon steel. It was confirmed by CSD that the original carbon had provided satisfactory service life for approximately 20 years and upgrading the subject tube bundles to 5 Cr-1/2 Mo will not fully eliminate the erosion-corrosion problems.

## CAUSTIC INJECTION RELOCATION PROJECT

It was strongly recommended by *CSD* to implement the best practice idea, which was relocating the downstream caustic injection from downstream of the desalters to 66 m upstream of the V04H1A/B heater, to enhance the overhead corrosion control on the chloride content, and to eliminate the corrosion and fouling problem in the second preheat exchangers.

The design includes adding crude slip stream to the caustic stream, before injection to the 14" heater charge line, to eliminate water vaporization and caustic deposition. In addition, interlock system was provided to trip the caustic pumps, once the crude slip stream flow is low. The pre-mixed caustic injection point location is 66 ft upstream of the crude charge heater passes. This location is to allow the premixed caustic to mix well with the bulk crudeIn addition, Monel 400 was used to construct the piping, downstream of the crude slipstream/caustic mixing point, including the quill and the static mixer. The caustic injection relocation was started on May 2006. **Figure-4**

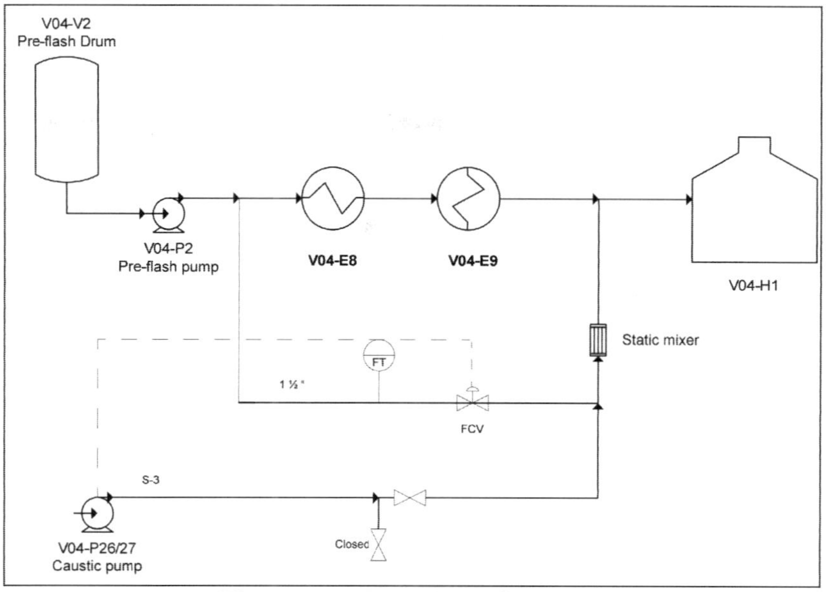


**Figure 4 - Caustic Relocation**

## Post Relocation Project

It was observed that frequent fouling of the second preheat exchangers was eliminated. Testing and Inspection of the desalted crude heat exchangers V04-E5 and E6 were completed on July 2007 (a total of eight exchangers). Initial visual inspection revealed all components were in good condition without any significant mechanical damage. A 100% Magnetic Particle Inspection (MPI), conducted on the internal weld seems of the shells, channel box and covers did not reveal any defects. All exchangers passed the initial and final hydrostatic tests.

YR has major savings after this project. Before, on average, ten exchangers were fouled and leaked every year. Mechanical cleaning, Hydro testing, and maintenance labor cost $8,000. On average, this project annually saves $80,000 in maintenance costs.

Furthermore, this project prevented the frequent isolation of V04-E6 A/B. The isolation of this exchanger resulted in the dumping of Heavy Diesel Oil product to the fuel oil stream, reducing the production of Diesel. **Table-1**

| V04-E6 (Crude and HDO product exchanger) | $m^3/hr$ | bbl/day |
|---|---|---|
| HDO normal flow to both pre-heat trains | 110 | 16,610 |
| HDO flow if one train is isolated (E6A or E6B) | 77 | 11,627 |
| HDO dumping to Fuel Oil | 33 | 4,983 |

**Table1 - V04-E6 flow calculation**

On average, the V04-E6 isolation was experienced 10 times a year, with an average of 3 days of maintenance work. The cost of dumping HDO to F.O. is calculated to be $1.50 MM/year. **Table-2**

| | |
|---|---|
| Average Diesel cost/bbl | $75 |
| Average Fuel oil cost/bbl | $50 |
| Difference | $25 |
| Annual days when isolation | 12 |
| **YR loss per year (million)** | **$1.50** |

**Table 2 - Cost calculation**

On average, this relocation project is annually saving $1.8 million.

## CHALLENGES

After relocation of the caustic injection on May 2006, inspection data at the new caustic injection locations revealed unacceptable corrosion rates in the 14 inch crude lines, upstream of the heaters for both Trains A & B. Estimated localized corrosion rates were 0.6 mm per 5 months for Train A (at the bottom) and 0.4 mm per 5 months for Train B (at the top) - **Table 3**

| Date | Wall thickness (mm) | |
|---|---|---|
| | Location # 1 | Location # 2 |
| | Minimum | Minimum |
| 06/18/2006 | 11.0 | 11.1 |
| 07/03/2006 | 11.0 | 11.1 |
| 09/06/2006 | 11.0 | 11.1 |
| 12/24/2006 | 10.8 | 10.9 |
| 01/22/2007 | 10.6 | 10.8 |
| 01/23/2007 | 10.6 | 10.8 |
| 01/29/2007 | 10.5 | 10.8 |
| 02/11/2007 | 10.4 | 10.8 |
| 02/26/2007 | 10.4 | 10.7 |
| 03/12/2007 | 10.4 | 10.7 |

**Table 3 - New Caustic injection thinning**

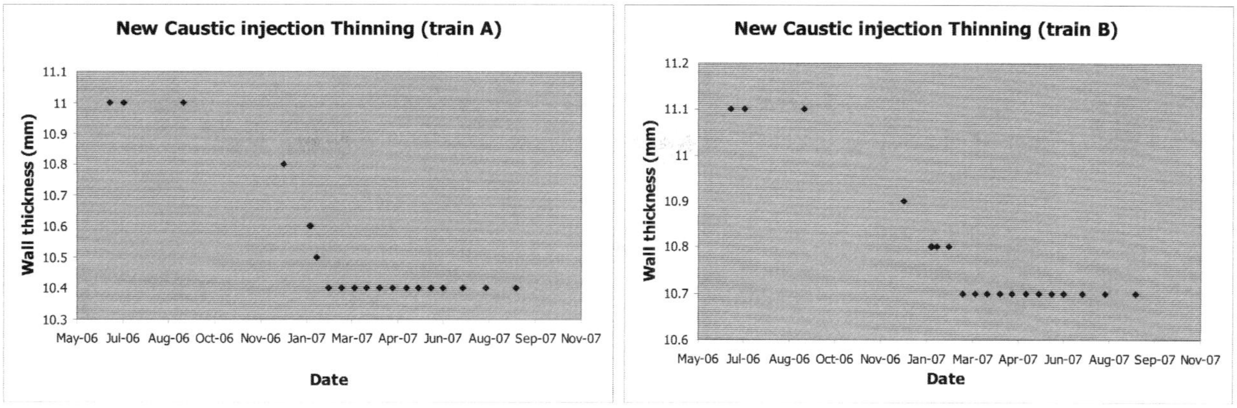


**Figure 5 – New Caustic injection thinning**

A root cause analysis was conducted by *CSD, P&CSD* and the YR plant personnel to focus on potential causes and actions necessary to minimize corrosion. This included looking at the detailed inspection data (UT, Radiographs, Scans, and Infrared) as well as design, fabrication and operational parameters. It was concluded that the damage mechanisms for each train were different and the best scenario for the attack is as follows:

- Metal losses in the 14 inch crude lines are attributed to general corrosion attack by caustic, which attacked the pipe wall prior to the caustic being diluted by the main crude. This would explain the highly localized corrosion seen immediately downstream the injection point in the 6 o'clock position for Train A and in the 12 o'clock position (downstream top-of-the-line weld-o-let) for Train B.

- For **Train A**, it was suspected that there are leaks of partially undiluted caustic on the 14 inch carbon steel line from the root of fillet welds of the quill end plate that is causing corrosion (**Figure 5**). The infrared surveys showed localized areas of much lower temperatures.

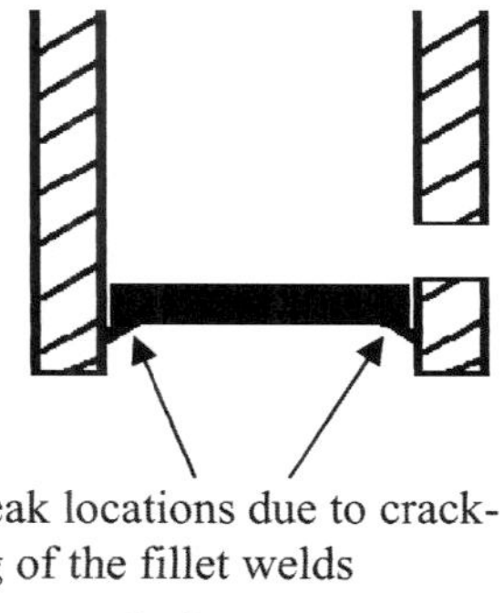


**Figure 5: Train "A" Injection Quill Assembly**

- For **Train B**, it was suspected that improper mixing was allowing some of the caustic to drift upwards to the crude piping, leading to localized attack. The end plate weld in this case appeared to be solid and intact. Also, we cannot be certain if there could be a small leak in the flange weld that could allow top of the line corrosion (**Figure 6**)

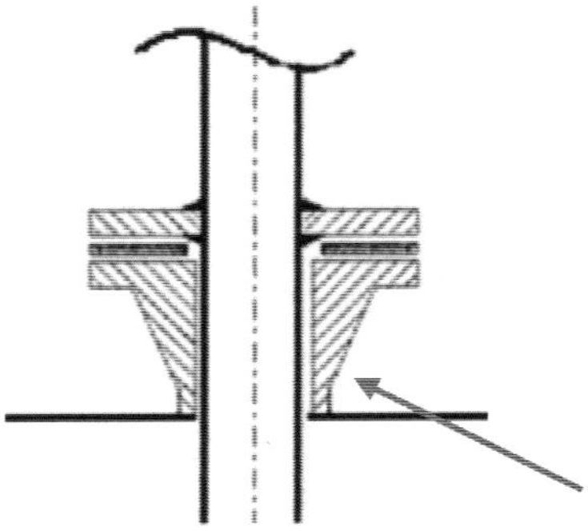


**Figure 6: Train B Injection Quill Assembly**

CONCLUSION

YR had serious fouling and corrosion issues in the second preheat train due to under-deposit, inorganic deposit, and high temperature sulfide corrosion, which lead to pressure buildup and tube leaks. The caustic injection relocation, to upstream of the crude heater project, met the objective of overcoming the heat exchangers fouling and maintenance-related problems. The annual total cost-saving of the relocation was over $1 million vs. Relocation cost less than $100,000. YR faced thinning on the 14" crude line to the heater, where the new injection point was located for several weeks but the thinning stopped and there was no change for the past 7 months. The investigation team believes that the leak on the quill was plugged by itself. In the meantime, YR is conducting UT scanning on the new injection points' bi-weekly, to ensure the integrity of the crude line. Furthermore, YR will take the first coming shutdown opportunity to visually inspect the injection quill for replacement or upgrade.

## References

1- Yanbu' Refinery Instruction Manual.
2- YR Caustic Injection Relocation Project's Package (RFE-2004-291)
3- YR 14" Crude Line Investigation Report

# Corrosion Prevention, Protection & Control In CDU Overhead system

Bhanu Murty. V .Maddala
Mohammed S Eid
Saudi Aramco
P.O. Box 30028 (YA-4367)
Yanbu, Saudi Arabia
Fax: +966-4 397-8118
E-mail: maddala.bhanumurthy@aramco.com

## ABSTRACT

**Saudi Aramco Yanbu Refinery** crude distillation unit (CDU) commissioned in 1983 processes 170 MBD Arabian light crude and expanded over a period to present capacity of 235 MBD. Crude unit overhead line operates under dry condition without any condensation in the line and carries column overhead vapor to air-cooled condensers. It is protected with chemical injections, filming and neutralizing Amine through the reflux line. The Column overhead piping system to the Fin Fan Condensers was in service from the May, 1983 from start-up until January 2, 1996 when the first leak occurred. In 1997 30" vapor line vertical portion has two pinholes with pipe thickness reducing from 9.5 mm to 5.8 mm. The line is converted into two sections as dry and wet sections. The system operated for another four years and inspected in 2001-revealed metal loss in two locations, the first one in the dry section and the second at upstream and downstream of the condenser that was in-services since 1983.

Low thickness in the dry section is due to acid corrosion and under deposit corrosion in addition to the high overhead vapor velocity. Dry section is modified by cladding the part at the upstream of the water injection point to eliminate the under deposit corrosion and moving the water injection further upstream to convert the system to wet system apart from Upgrading the overhead Line to 36". Operating conditions are improved by bettering crude tank supply system with steady crude quality, improving Desalter operation and better monitoring with portable pH analyzers and additional corrosion probes. Wash water quality enhanced by maximizing steam condensate. This has improved the overall system with little corrosion related problems in the overhead line.

## 1. INTRODUCTION

Overhead system can be classified in two sections, dry and wet. The dry section is from the crude column to water injection point where no water is introduced to the system. The wet section is from the water injection point up to overhead receiver including part of overhead line, condenser, and overhead receiver vessel. On 01/02/1996, a pinhole leak was noted at the horizontal pipe near the 3rd elbow from the column top (refer Fig.1).

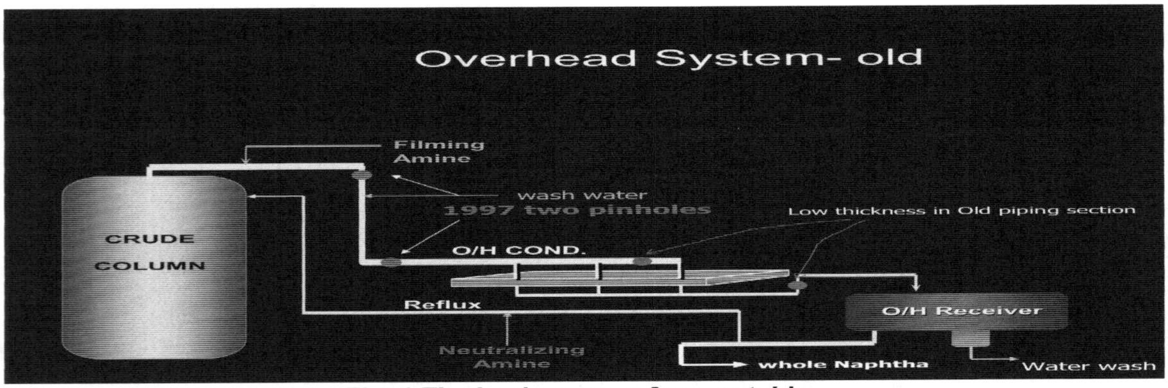


**Fig. 1 The Leak as seen from outside**

The crude unit was shutdown on 01/05/1996 and the 3rd elbow plus about 7.6 meter long pipe section was replaced. Failure is due to acid dew point corrosion during normal operation. On 10/27/1996, another pinhole leak was noted on the vertical pipe just downstream of the 2nd elbow from the column top. The crude unit was shutdown and the 2nd elbow plus about 3.65 meter long vertical pipe section was replaced. The thinned areas on the 1st elbow were externally patched and the unit returned to service on 10/30/1996.Finally in April 1997, the line was completely replaced from the top of the column including the relief header down to fin fan coolers and 24" inlet headers and branches. The most probable cause for this failure is due to acidic dew point corrosion during startup and shutdown of the crude column.
On 11/19/2001, the Refinery Inspection Unit conducted an On Stream Inspection (OSI) program using Radiographic and ultrasonic thickness (UT) surveys on crude overhead line at 13 locations (refer Fig.2). Also high temperature Ultrasonic-scanning used to identify any localized corrosion on the overhead line. Ultrasonic Thickness Surveys are done after removing the external insulation in an approximately 1 meter wide 360$^0$ band around the overhead line at 3 elbows, plus all potential impingement locations, and around the overhead line straight sections near chemical injection points. Overhead line branch connections are also checked so that no dead legs are present. The pipe minimum thickness reduced to 6.8 mm from 8.7 and is the same location where the pinhole leak was detected on 10/27/1996.

| | | | |
|---|---|---|---|
| ON STREAM INSPECTION GAUGE POINT LOCATION | Plant Name | CRUDE DISTILLATION | Line Size.:36" |
| | Operating Temperature: 122 °C | Operating Pressure: | System Name: OVHD VAPOUR FROM COLUMN TO OVERHED CONDENSAERS |

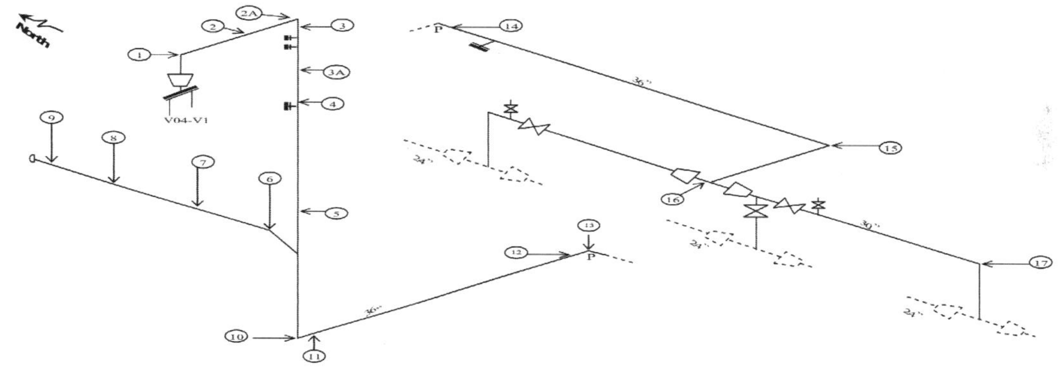


**Fig.2 Overhead Vapor Line Inspected Points**

## 2. PROCESS DESCRIPTION

The Crude Distillation Unit has been revamped for a capacity of 235,000 barrels per stream day (BPSD) of crude oil, equivalent to 31,800 tones per stream day 33.8 API Arabian Light, or 23,857 tones per stream day of 28.6 API Arabian Heavy. To counteract the severe corrosion problems that can occur in the crude overhead system, filming amine, neutralizing amine and water wash injection are provided upstream of the crude column condenser facilities. Processing within the crude unit is divided into four main areas, namely desalting and preheating, crude column distillation, side-stream stripping and naphtha fractionation. The crude oil is distilled in an atmospheric distillation column. The lighter components (gases and naphtha) rise to the top of the column and the heaviest components comprising the reduced crude flow to the bottom of the column. Kerosene and diesel oil are drawn as side products. All of the naphtha and the light hydrocarbon petroleum gases, plus the steam, which is injected, pass overhead from the top of the column and are separated after condensation in Air Cooled Overhead Condensers. The condensed overhead stream flows into the overhead Receiver. Small metering pumps inject neutralizing and filming (Corrosion Inhibitor) into the vapor stream just after it leaves the Column. The Receiver level controller controls the reflux flow rate. The Receiver also has a water boot to collect wash water and stripping steam condensed. The operating and design parameters of the column overhead are:

| | |
|---|---|
| **Service** | Overhead vapor. |
| **Design - Operating Temp** | 149 deg C-122 deg C |
| **Design- Operating Press** | 2.8 kg/cm$^2$g - 0.8 kg/cm$^2$g |

# 3. MATERIALS OF CONSTRUCTION

**Original material:**
Pipe Size: 30"; A155-C55 CLASS 3 (8.7 MM Thick)/Alternate Material: ASTM A-672 Gr.C65
**Upgraded material:**
Dry section Pipe Size: 36"; API-5L Gr.B (9.5 MM Thick)/Hastelloy C-276 (3 MM thick clad)
Total: 9.5 MM Thick.
Wet section Pipe Size: 36"; API-5L Gr.B (9.5 MM Thick) No cladding.

# 4. OBSERVATIONS

## 4.1 The Background

- The 30" overhead line developed a leak at the third elbow in January 1996.
- Another failure occurred at the second elbow from the column in late October 1996.
- Significant losses on the first elbow just off the top of the crude column.

## 4.2 Investigation

An Engineering Services multidisciplinary team was formed to investigate the crude unit overhead corrosion in 1996 following recurring corrosion problems. As part of full inspection survey 13 locations of the overhead system were UT scanned to identify the problem, and locations of low thickness were subjected to radiography shots to confirm the results of UT scanning. Investigations revealed the metal loss of the system is localized in the initial condensation area upstream of the water injection point and no metal loss found in the full condensation area after the wash water injection point. This is due to acid corrosion and/or underdeposit corrosion as revealed by the overhead system operating conditions as some times pH level has dropped below targeted range of 5.5 to 6.5. This is confirmed by severe corrosion in the initial condensation area of the overhead line. Also during the period 1983-1996, neither corrosion nor pitting was noticed in the full condensation area. The system operations were reviewed for the last four years; all corrosion parameters such as Chlorides, Iron and pH of the overhead receiver water were checked and investigated. The possible causes for thinning in crude unit overheads are as follows

1. Acid corrosion
2. Underdeposit corrosion
3. Velocity accelerates corrosion and causes erosion
4. Oxygen accelerates corrosion
5. Total dissolved solids (TDS) accelerate erosion

***4.2.1.Acid corrosion***:

It is due to hydrochloric acid that is formed when HCl comes in contact with condensed water in the overhead line. When HCl dissolves in a small amount of water pH can be very low and the water can be very corrosive.

To control and minimize the HCl content in the overhead system, HCL formation needs to be controlled. HCl is formed when salts in crude such as $MgCl_2$ and $CaCl_2$ hydrolyze under high temperature 160-380 °C according to the following reactions

$CaCl_2 + 2H_2O \rightarrow Ca(OH)_2 + 2HCl$ (1) ( Greater than $400^0$F or $204^0$C)

$MgCl_2 + 2H_2O \rightarrow Mg(OH)_2 + 2HCl$ (2) ( Greater than $250^0$F or $121^0$C)

Therefore, $MgCl_2$ and $CaCl_2$ should be removed from the system before these reactions take place. Caustic injection down stream the desalter is being utilized to convert the unstable salts, $MgCl_2$ & $CaCl_2$, to stable salt, NaCl, according to the following reaction.

$CaCl_2 + 2\ NaOH \rightarrow Ca(OH)_2 + 2\ NaCl$ (4)

$MgCl_2 + 2\ NaOH \rightarrow Mg(OH)_2 + 2\ NaCl$ (5)

$NaCl + H_2O \rightarrow NaOH + HCl$ (6) (Greater than $450^0$F or $232^0$C)

Both products are not hydrolysable and stable. It is worth mentioning that overdosing the caustic will eliminate the chloride issue in the overhead system but could lead to the caustic embitterment of upstream equipment, such as exchangers and heaters.
There was a significant fluctuation of chlorides, sometimes above 25 ppm in overhead receiver water, as illustrated in Fig.3

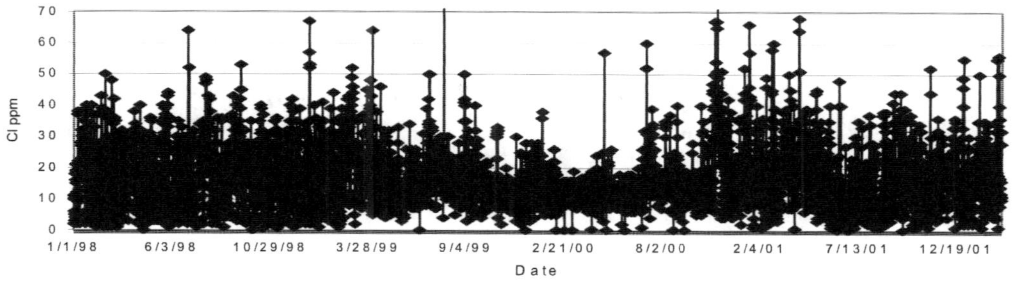


**Figure.3 Chloride levels in water boot of overhead receiver**

This fluctuation is the result of high fluctuation in the salt content of the charge crude and low efficiency of the desalter. The fluctuation in the salt content of charge crude results in unsteady salt content of the desalted crude. This fluctuation requires frequent change in the caustic injection rate, which is not practical and at times cannot be anticipated. Therefore the salt content needs to be as steady as possible through the settling time and mixing in tanks and upstream operation. Meanwhile, it should be noted that low chloride in the overhead system does not mean elimination of acid corrosion, unless the pH is maintained within the target range 5.5-6.5. In addition, introducing water to the system and forcing full condensation upstream of the water dew pint location could dilute the chloride concentration in the initial condensation zone.

***4.2.2* Underdeposit corrosion**:

It is due to any deposit in the overhead system such as an iron sulfide deposit and/or neutralization salt. Neutralization salt acid is formed when naturalizing Amine is introduced to neutralize the hydrogen chloride. Under the condition of an initial condensation area, wet water will be formed and deposit with acid salt, then the solution will hydrolyze to form a low pH (<5) environment. Under this environment underdeposit severe corrosion would take place. As this occurs in the initial condensation area and no wash water is available to wash the system, the dry section of the overhead system is subjected to such corrosion, especially in the horizontal part.

Moreover, overdosing of neutralizing amine and the operating at a pH above 7 will decrease the iron sulfide solubility in the water and ultimately enhance iron sulfide deposit even though wash water is used. Operating conditions of the overhead system showed that overdosing of neutralizing amine resulted in high pH levels, 6.8 to 7.3, and caused higher corrosion rates in the overhead system. The basic solution would result in a loss of the Ferrous Sulfide (FeS) film covering the overhead pipe. When this film is lost, the pipes of the overhead would be susceptible to higher corrosion rates. It was noticed that the pH of the overhead went over its specified maximum range of 6.5 for a considerable period of time. Therefore it is obvious that the overhead system is subjected to under deposit corrosion at the initial condensation area and full condensation areas. YR CDU experienced a significant fluctuation in pH that can lead to acid corrosion and under deposit corrosion. This is due to unsteady salt content in the charge crude and hence the unsteady operation for the overhead system. Therefore, we need to maintain the pH in the overhead system and operate strictly within the pH range of 5.5 to 6.5.

### *4.2.3* Velocity accelerates corrosion and causes erosion:

The overhead line experienced flow-enhanced corrosion (also known as erosion-corrosion) due to HCL condensation. A 30" overhead line was replaced in 1997. High velocity accelerated the corrosion and caused erosion as it removed the protective layer. The fluid velocity across the overhead line was high, around 43 m/sec (141 fps), due to throughput rise.

## 5. DISCUSSION & ANALYSIS

The internal corrosion in a crude overhead line is dependent on process variables, including:

1. Crude quality (salt and water in the crude)
2. Desalter operation
3. Caustic injection type and location
4. Water wash rate, type and injection locations
5. Overhead chemical treatment
6. Crude column operation
7. Program Monitoring and Laboratory Analysis
8. Metallurgy Upgrade
9. Insulation of overhead line
10. Overhead Vapor Velocity

### 5.1) Crude quality (salt and water in the crude):

Usually, Arab Light crude has a BS&W content of 0.1 vol. % or less and a salt content of less than 10 PTB (Pounds per thousand barrels). The crude charge contains natural inorganic salts present within a dispersed water phase. Effective corrosion control begins at the upstream end, paying close attention to crude storage tank operations. Adequate settling time and periodic water drainage are prerequisites to cutting the water (therefore chloride) content in the crude, as well as reducing suspended solids. Hydrolysis of calcium and magnesium chlorides occurs during heating leading to the formation of hydrochloric acid. Significant hydrolysis of magnesium chloride occurs at around 204$^{O}$C (400$^{O}$F), well below the fired heater outlet temperature of 375$^{O}$C (700$^{O}$F). Injection of well-dispersed caustic soda solution (sodium hydroxide) is utilized to neutralize hydrochloric acid (HCl) prior to fractionation in the atmospheric crude column. Any free HCl can enter the overhead as hydrogen chloride vapor, which, upon subsequent condensation at the water dew point, produces a strong acid that is very corrosive.

### 5.2) Desalter operation:

**Configuration**: YR refinery had two-stage desalting equipment, for controlling the amount of hydrolysable chloride appearing as hydrogen chloride in the overhead system. The electrical desalter effectively removes salt-containing water and water-wet solids after thorough mixing with a clean wash water source. For the two-stage configuration, makeup wash water is routed in a counter-current manner to the second stage mix valve and the effluent water, from the second stage, is then pumped to the mix valve at the inlet of the first stage. The three main operating variables affecting water and salt removal efficiencies in the desalter are: wash water %, temperature, and residence time. Wash water quality, mixing valve pressure drop, and pH control affect efficiencies to a lesser extent.

**Wash Water:** To minimize corrosion and scale in the Desalter Vessels, the desalting wash water should meet condensate or stripped sour water qualities. Wash water inlet pH being maintained in the range 5.5 to 7 with oxygen less than 20 ppb and chlorides below15 ppm and ammonia below 20 ppm. At higher pH values and excessive hardness, especially calcium, the potential for scaling increases, and emulsion becomes more stable and hence difficult to break. Oxygen, if present in the system, can cause iron corrosion or/and react with sulfur to form $HSO_3$ and H2$SO_4$, which are very corrosive substances. YR operate with a low wash water rate, <3 vol. % of crude charge, due to good quality wash water availability limitations. Desalter temperatures are maintained between 120-150$^{O}$C. Desalter operations below operating temperatures (<120$^{O}$C) result in less effective desalting while higher than 150$^{O}$C will damage equipment. An optimized desalter operation also reduces salt levels further, as well as bottom sediments. While a residence time of 15 minutes for oil and 60 minutes for water are required, after enhancing throughput, oil residence time dropped to 11.5 minutes. Salt-out of less than 1 PTB and BS&W less than 0.2%w in crude are targeted.

**Desalter Water Effluent pH Control:** YR use injection of caustic upstream of the desalter for pH control of the water effluent and is targeted in the range 5.5-6.5. If this is not done carefully, preheat exchanger fouling and/or emulsion buildup in the desalter will result. If the desalter pH is too high, this might also release ammonia from the water phase into the crude, which would present further concern in the overhead system.
Desalter operations are to be improved further as it is encumbered by crude charge rates in excess of 25-30% above design rates. This represents a considerable reduction in available water settling time. To improve the CDU overhead corrosion and desalting problems, the Desalter water wash rate was increased from about 14 M3/Hr to more than 20 M3/Hr (3% by volume of the crude charge) and as a result, the salt fluctuation in the desalted crude decreased and minimized. The desalting efficiency has increased accordingly.

## 5.3) Caustic Addition Process:

YR using two-stage caustic mixing and injection for overhead chloride control. Fresh Caustic soda is used downstream of the Desalter to neutralize the hydrochloric acid formed from the hydrolysable salts. An effective caustic injection system can reduce neutralizer consumption and help mitigate downstream fouling tendencies. This is accomplished through dispersion and thorough mixing.
Usually, a dilute caustic (3-5 Be") mixture is utilized to avoid stress corrosion cracking at elevated temperatures and to reduce fouling tendencies. The water source for the caustic solution should be of condensate quality with oxygen less than20 ppb. Caustic storage tank is nitrogen blanketed to exclude oxygen. Dilute caustic aids mixing but variations in caustic strength area voided to prevent fouling. To effectively accomplish this mixing, dilute caustic is first mixed with a crude slipstream representing approximately 1%vol. of total charge and the resultant mixture is then injected into the desalted crude charge via a high velocity Monel 400 quill system. The caustic Injection point is relocated from the Desalter down stream to upstream of the heaters to mitigate /eliminate the experienced fouling/corrosion in the second preheat exchangers. The slipstream aids in caustic dispersion and minimizes caustic caused corrosion problems. Caustic injection rates are based on overhead chloride levels. Overdosage of caustic will increase fouling tendencies. It is recommended that caustic injection should not exceed 2 PTB due to the risk of caustic embitterment and fouling concerns.

**5.4)Overhead Wash Water System** :It had a removable atomizing type spray nozzle. The nozzle is located at a suitable distance away from dead legs. The overhead line is insulated upstream of the water spray nozzle location. The receiver boot water is checked for oxygen content. The presence of oxygen under acidic conditions will accelerate corrosion. Target oxygen content is <20 ppb w. Overhead and Desalter makeup wash water source is de-oxygenated and is usually condensate quality (<25 ppm w TDS). Stripped sour water is being used in limited quantities successfully with close monitoring of quality. Under certain circumstances, such as an upset, significant amounts of ammonia and hydrogen sulfide can migrate into the oil phase and cause additional problems in the overhead system as well as disturb the Desalter operation. Overhead wash water rate is at the 3.5 vol. % even though the Ideal rate is 5 vol. % of the total overhead volume. YR using a typical, and probably the most economical approach for recycling overhead receiver as wash water. The water is injected through a removable atomizing style of spray nozzle. The idea is to vaporize the water as quickly as possible by literally producing a fog of water. This also produces a droplet size, which minimizes possible erosion led impingement near the injection point location.

**Corrosion Control**: Overhead water wash stabilizes pH control, helps to solubilize salts, and aids in flushing corrosion products and other solids from exchanger tubes. Solid salt formation is the first step in initiating under-deposit corrosion. If the overhead chloride content is relatively high, then ammonium chloride deposits can form at temperatures above the water dew point. Enough water is added as wash water near the top of the tower to raise the overhead vapor water dew point as quickly as possible to prevent salt deposits. This also results in partial condensation of the hydrocarbon phase, which serves as additional carrier for the inhibitor chemical. No corrosion was observed in the full condensation area after the wash water injection point.

**Tail Water Chlorides**: The combination of good Desalter operation and caustic injection should achieve a chloride level of less than 20-25 ppmw in the tail water. This is somewhat arbitrary, but represents a compromise between practicality and exposure risk to front-end acid corrosion. With effective caustic injection, it is achieved with a desalted crude salt content of 1 PTB.

## 5.5)Overhead Chemical Treatment :

**Neutralizer:** The neutralizer is amine based, completely soluble in water, and forms a non-deposit mobile salt when reacted with hydrogen chloride. This neutralizer is used at top tower conditions. Based on total overhead volume, an addition of approximately 10 ppm of neutralizer would be required to completely neutralize 30 ppm chlorides as hydrochloric acid in the tower generated tail water. Adjustment of neutralizer addition is done based on measured pH in the tail water. This is usually kept in the range of 5.5-6.5 as per our experience. We also relocated the neutralizer injection point from the reflux to the overhead with an engineered quill and it improved the injection efficiency.

**Filming Inhibitor:** Filming inhibitor is aromatic based and is oil soluble, and is effective over a wide range of pH. Also process variables such as pH, temperature, and system velocities may contribute to effectiveness. The filming inhibitor is most effective under steady conditions at a nearly neutral pH and therefore is nearly always used in conjunction with neutralizers. High system velocities may strip the inhibitor barrier especially in areas of high turbulence. This could result in spotty results and higher application rates may be needed. Tail water iron content, corrosion probes, and thickness measurements indicate the effectiveness of the inhibitor program. A target of around 1 ppm w Fe is usually set in overhead receiver boot water. In general, inhibitor consumption should be in the 3-5 ppmV range based on total overhead volumes.

**System Design:** Neutralizer and inhibitor are injected as near as possible to the top of the crude tower, usually in the horizontal run downstream of the first tower elbow. Injection of the amine-based neutralizer into the tower via return reflux or vapor streams should be avoided due to the possibility neither of salt accumulation in stagnant tray areas especially when the top tower section is neither lined nor of Monel trays. YR column top dome is lined with Monel and top trays are Monel too.

To avoid plugging and to aid in vaporizing, the neutralizer is injected with the steam. The inhibitor is usually diluted with naphtha reflux before injection for effective dispersion. The recommended minimum dilution ratio by volume is 50: 1. Removable injection quills were used to affect quick mixing. The neutralizer and inhibitor injection quills are staggered by a meter and at an appropriate distance from elbows and major fittings. Neutralizer should be injected upstream of the inhibitor to prevent the possibility of salt buildup in stagnant areas in the top of the tower. All chemical injection streams are positively metered or gauged and usage is recorded on a daily basis. Diluents slipstream is also quantified.

## 5.6)Crude Column Operation:

The best means of protecting the tower top and overhead system piping is to maintain top temperature at a comfortable margin above water dew point by a minimum of $10^{O}$deg C. The maximum kerosene production (low naphtha end point), combined with a low pressure operation, represents the highest risk for water dew point acid corrosion. This situation produces the lowest dew point margins and the difference between top temperature and water dew point temperature. This might be the case when operating with low crude charge rates. We are presently operating with around $18^{0}$C difference between top temperature and water dew point temperature.

**Process Monitoring:** As one of the CDU Best Practice item YR implemented a tool to monitor the CDU overhead velocity and water dew point. Both parameters were displayed on the operator screen on the DCS for continuous monitoring. This was intended to avoid any acid corrosion and to monitor the velocity in the overhead line during any changes in operating parameters.

### 5.7) Program Monitoring and Laboratory Analysis:

**Program Monitoring:** As part of the corrosion control, the monitoring sheet was developed. The Monitoring cover the following areas:

1. Incoming crude analysis (BS&W, salt content, etc.).
2. Desalter operation (temperature, mixing valve pressure drop, crude and wash water rates, makeup and effluent water rates and qualities (TDS, Cl, pH, etc.), and desalted crude qualities (BS&W, salt content, etc.).
3. Chemical injection (caustic, neutralizer, and inhibitor) daily chemical use, diluents rate, and chemical quality check.
4. Overhead tail water rate, analysis (pH, Cl, Fe).
5. Overhead wash water rate and qualities.
6. Overhead conditions like tower top dew point margin, etc.
7. Corrosion probes data.

The corrosion probes at the upstream and down stream of the vessel can signal the corrosion trend and effectiveness of corrosion inhibitor. There is a plan to upgrade the corrosion monitoring system in this operational area.

**Laboratory testing:** Both regular and desalted crude are tested for salt content, water and sediment. Regular crude is also tested for gravity. Overhead wash water is checked three times a day for pH, chloride and iron.

### 5.8)Metallurgy Upgrade:

The top dry section of the Crude O/H line was exposed to a high acidic corrosive environment, as confirmed by process information review and Inspection records. This is confirmed as the system experienced a corrosion rate of 325 mpy in one month. As a short-term remedy, the severely corroded portions have been sleeved. Later the O/H line dry section was upgraded with 3 mm thick Hastelloy C-276 cladding. The horizontal section and part of vertical section of about 16 meters pipe was clad with Hastelloy.

### 5.9)Insulation of overhead line:

Insulated the horizontal part of the overhead line, upstream of the water injection point, to prevent condensation and hence avoid acid corrosion.

### 5.10) Overhead Vapor Velocity:

High velocity could cause erosion and accelerate corrosion as high velocity fluid can remove the protective layer. The overhead line, replaced in 1997, experienced flow-enhanced corrosion. Therefore the line was resized from 30" to 36" in December 2003. The fluid velocity across the overhead line was brought down from 43 m/sec (141 fps) to 29 m/sec (95 fps) for the gas phase. This helped to overcome the erosion-led problems.

## 6. CORRECTIVE ACTIONS

To preserve the overhead system of crude unit the following actions are implemented as shown in Fig.4

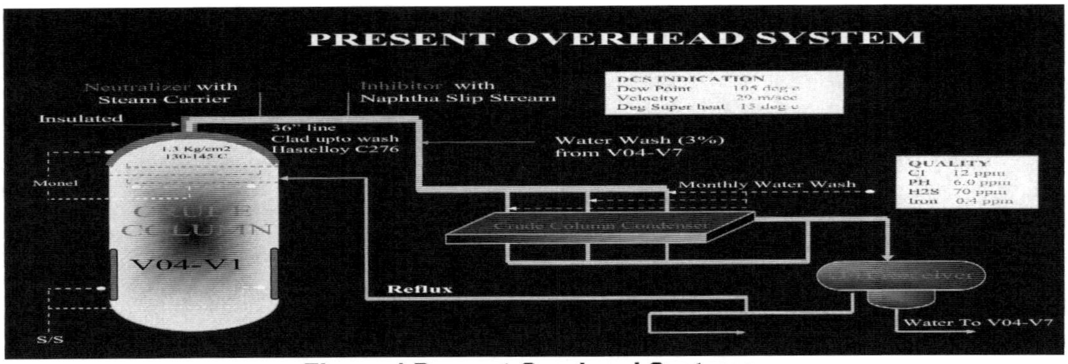


**Figure.4 Present Overhead Systems**

1. Maintained the pH reading of the overhead water in a range 5.5-6.5.
2. Operations provided with field test kit for pH analysis on 4 hour basis and pH meter is calibrated in weekly basis by lab to ensure close monitoring and good control for pH, as the online corrosion monitoring system is not working properly.
3. Fresh caustic in the range of 3 to 4 Be°, based on the operation requirement, is used for injection down stream the desalter to ensure effective neutralization of the salts to maintain the chloride level in the overhead system within the target range 10-30 ppm. The range was fixed based on the salt content of crude to have a good control in the pump injection rate and minimize the fluctuation in the chloride in the overhead receiver boot water. Large or erratic swings in overhead chlorides should be avoided. This in turn causes bounces in pH control, which can strip protective films and expose the system to more corrosion risk.
4. Keep the iron level in the overhead receiver boot water to less than 1 ppm.
5. The major parameters affecting the fluctuation of the salt content in charge crude are not enough settling time in the tank farm and the minimum level in the crude tanks. The minimum level was revised to 3 meters instead of 2 meters.
6. Insulated the horizontal part of the overhead line upstream the water injection point.
7. Relocated the water injection point to be down stream the second elbow of overhead line to force full condensation and eliminate the initial condensation area.
8. O/H line dry section metallurgy was upgraded with 3 mm thick Hastelloy C-276 cladding.

## 7. CONCLUSIONS

The thinning in the overhead system was defined to be in the dry section of the overhead line and was due to acid and under deposit corrosion. Fluctuation in pH and chloride content in the overhead system lead to the environment being conducive to acid and under deposit corrosion. No significant corrosion indication was found in the full condensation section where wash water wash introduced to the system. The thinning locations found in this section were in the old parts of the system that were in-service since the refinery was commissioned in 1983.

An effective corrosion control program for the overhead system always begins at the upstream end, especially with vigilant control of chlorides in the crude feed to the atmospheric crude tower. Close attention to crude storage tank operations, in the form of adequate settling time, and periodic water drainage, and optimization of the Desalter operation, constitute important factors in the minimization of corrosion. Chemical injection requirements (caustic, neutralizer, and filming amine) via engineered injection quills, on the other hand, are dependent upon the effectiveness of upstream operations. To avoid acid corrosion in the crude column overhead line, theoretically, YR operates at an overhead dew point margin of 10°C or more. With diligence, after the metallurgy upgrade, YR enforced tighter operational control on good Desalter operation, caustic injection, neutralization, inhibition and water wash to minimize overhead corrosion and it yielded good results.

On site inspection for last 3 years as shown in Table.1, confirm little thinning in the overhead system.

**Table.1** Inspection results confirming little metal loss

Ultrasonic Examination was carried out on the above mentioned locations on overhead vapor from column top to Air coolers Refer Fig.2.

| **Loc** | **Location Identify** | **WALL THICKNESS (mm.)** | | | | | | | |
|---|---|---|---|---|---|---|---|---|---|
| | | 1996-30" **Dia** | | 8/2004-36" **Dia** | | 2/2006-36" **Dia** | | 3/2007-36" **Dia** | |
| | | Max. | Min. | Max. | Min. | Max. | Min. | Max. | Min. |
| 1 | Elbow | 13 | 14.5 | 10.8 | 10.2 | 10.8 | 10.8 | 10.8 | 10.8 |
| 2 | | | | 10.0 | 9.6 | 10.0 | 10.0 | 10.0 | 10.0 |
| 2A | Elbow | 11 | 13 | 10.8 | 9.8 | 10.8 | 10.8 | 10.8 | 10.8 |
| 3 | | 6.8 | 7.5 | 11.2 | 10.6 | 11.2 | 11.2 | 11.2 | 11.2 |
| 3A | From top to 1st support | 8.0 | 8.5 | 11.2 | 10.6 | 12 | 12 | 12 | 12 |

## 8. ACKNOWLEDGEMENTS

The author wishes to acknowledge the contributions from company's central Consulting Services and Inspection Departments as well as the parent operations engineering unit.

## 9. REFERENCES

1. Saudi Aramco Engineering Report SAER-5941.

# 228a Tutorial: Sulfur Management in Refineries

**Pravin Sheth**, Pravin Consulting, 5012 Devin Green Lane, Fairfax, VA 22030

*Text Not Available.*

## Development of New Generation of Agile Desulfurizers

*Desikan Sundararajan and Abdul-Majeed Azad*
*Department of Chemical and Environmental Engineering*
*The University of Toledo, Toledo, OH 43606.*

## Abstract

The presence of sulfur in logistic fuels (diesel, coal and jet fuels) poses challenge for reforming, as it leads to severe catalyst deactivation and poisoning. Desulfurization is an important and integral component of generating clean power from these abundant fuel sources. Their utilization in high efficiency fuel cells, however, necessitates that the sulfur level in the reformate stream be extremely low, since the SOFC anodes are quickly poisoned by sulfur even in trace levels (~ 1-2 ppm). Thus, the development of effective desulfurizers is important pathway for the efficient utilization of the logistic fuels. Recently, we have developed a series of agile sulfur sorbents with metal oxide(s) dispersed on lightweight, highly ordered nanoporous biomimetic or zeolitic supports. These sorbents have been tested in a wide range of temperatures and dwell times in sulfur-laden gas streams. Their systematic characterization, prior and subsequent to the sulfidation reaction by SEM/EDS and XRD and quantification by gas chromatography and chemical analyses to gauge their efficacy to the desulfurization of fuel streams are presented.

### 1. Introduction

In the quest for cleaner and greener energy, there is great interest in using logistic fuels such as diesel, coal, and jet fuels. One of the challenges that impedes the use of logistic fuels in fuel cells is that they are invariably sulfur-laden to the level of 0.3 to abut 4 wt%. For example, the sulfur content of logistic fuels could be as high 1 wt. %. The presence of sulfur (mainly organosulfur) poisons and deactivates the reforming catalyst. Owing to its high efficacy in utilizing coal, integrated coal gasification combined cycle (IGCC) technology has been widely studied. Unfortunately, sulfur present in the coal poses a major challenge in continuous operation of the IGCC as well [1]. Thus, to obtain continuous uninterrupted clean power, effective sulfur sorbents ought to be developed.

This work reports some preliminary results on the development, characterization and performance of low cost, light weight metal-oxide based desulfurizers. These sorbents were assessed in terms of their sulfur-tolerance under conditions of high sulfur exposure at temperatures typically employed in reforming. To evaluate the performance of these sorbents at various temperatures, desulfurization studies were carried out at 600°C, 700°C and 800°C up to 24 h using gas stream containing 1000 ppm $H_2S$.

### 2. Materials and Methods

Diatomaceous earth (DE) from Eagle-Pitcher (CA) and clinoptilolite (a zeolitic clay isomorph from GSA Resources, AZ) were used as inert support materials for synthesizing the sulfur sorbents. Sorbents were made by dispersing metal oxides onto the support materials using corresponding water soluble precursors: iron (II) chloride, copper (II) nitrate and nickel (II) nitrate hexahydrate (all from Alfa-Aesar) as follows. Slurry of the support material and the desired sorbent metal precursor was homogenized in a rotary evaporator for 4h using ethanol as the dispersion media. The thoroughly mixed and concentrated mass was dried overnight in a vacuum oven at 60°C. The dried cake consisting of soft agglomerates was crushed using mortar and pestle, and pulverized in a ball mill using ethanol and 5-mm diameter high density zirconia spheres. The ball-milled slurry was dried again under ambient conditions and sieved through 325 mesh screen. Each formulation was calcined in air at 700°C for 2h. The clinoptilolite and DE-based sorbents are labeled and will be identified as C1 through C3 and D1 through D3, respectively, the digits 1 to 3 referring to different metal oxide used as sorbent.

The sulfidation experiments were carried out by exposing each sorbent to 1000 ppm H2S (bal. N2) at 600, 700 and 800°C for a period of 12h at each temperature. One gram of the sorbent powder was kept in a stainless steel filter (Swagelok; porosity ~80 mm) and covered with a loosely packed glass wool to allow easy access to the oncoming gas. The $H_2S$ concentration in the exit stream was measure by an online-gas chromatograph (Shimadzu GC 2010), which was programmed to analyze the gas stream and collect data every 3 minutes.

The sorbents were characterized before and after sulfidation, by a host of techniques with respect to their active metal oxide dispersion, structural and morphological features. The electron microscopy was carried out using XL30FEG SEM which is equipped with the capability of performing energy dispersive X-ray spectroscopy (EDS) and Rutherford's back scattering electron (RBSE) mapping as well.

## 3. Results and discussion

Both diatomaceous earth (DE) and zeolites are chemically inert and lightweight materials. Hence they are ideal substrates to perform as the sulfur sorbent supports. Clinoptilolite is the most abundant member of the 48 minerals in the zeolite group. Like DE, it is also naturally-occurring. It is a zeolite comprising a microporous arrangement of silica and alumina tetrahedral, whereas DE is pure silica. Figure 1 shows the morphological features of the two support materials used in the present work.

**Fig. 1.** SEM of clinoptilolite (left) and DE (right) used as sulfur sorbent support matrices.

DE is a lightweight naturally-occurring fossilized remain of the unicellular microorganism called diatom, and possesses highly ordered nano-pores. It is almost an empty structure (> 90% porosity). The diatoms are unicellular, eukryotic (cell or organism with a membrane-bound, structurally discrete nucleus and other well developed subcellular compartments) algae with silica-based cell wall that range in size from less than 1 to greater than 100 μm [2]. The highly periodic nature and uniform size and shape of the pores combined with the thermal and chemical resistance makes them ideal for impregnation with agile sulfur binding species. Moreover, both the support materials are environmentally benign and structurally robust under extreme thermal and chemical conditions.

*Morphological and structural characterization*

The microstructural artifacts and elemental mapping in the as-synthesized clinoptilolite- and diatomite-based sorbent (C1 and D1) are shown in fig. 2.

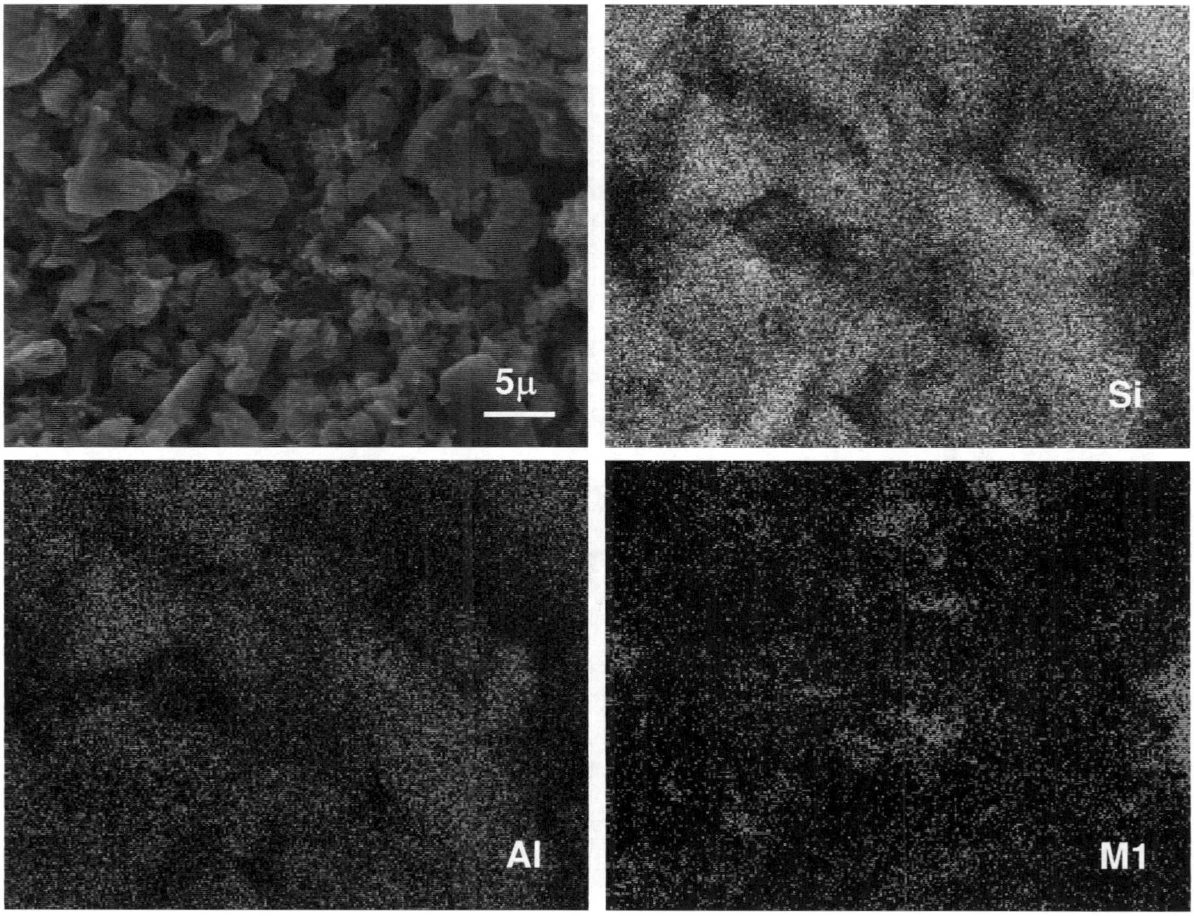


**Fig. 2.** SEM and elemental map of as-prepared sulfur sorbent C1.

As is evident, the sorbent metal (M1) is thoroughly dispersed in the clinoptilolite matrix; uniform distribution of the sulfur scavenging species on the support is a crucial and critical prerequisite for the effective performance of the sorbents.

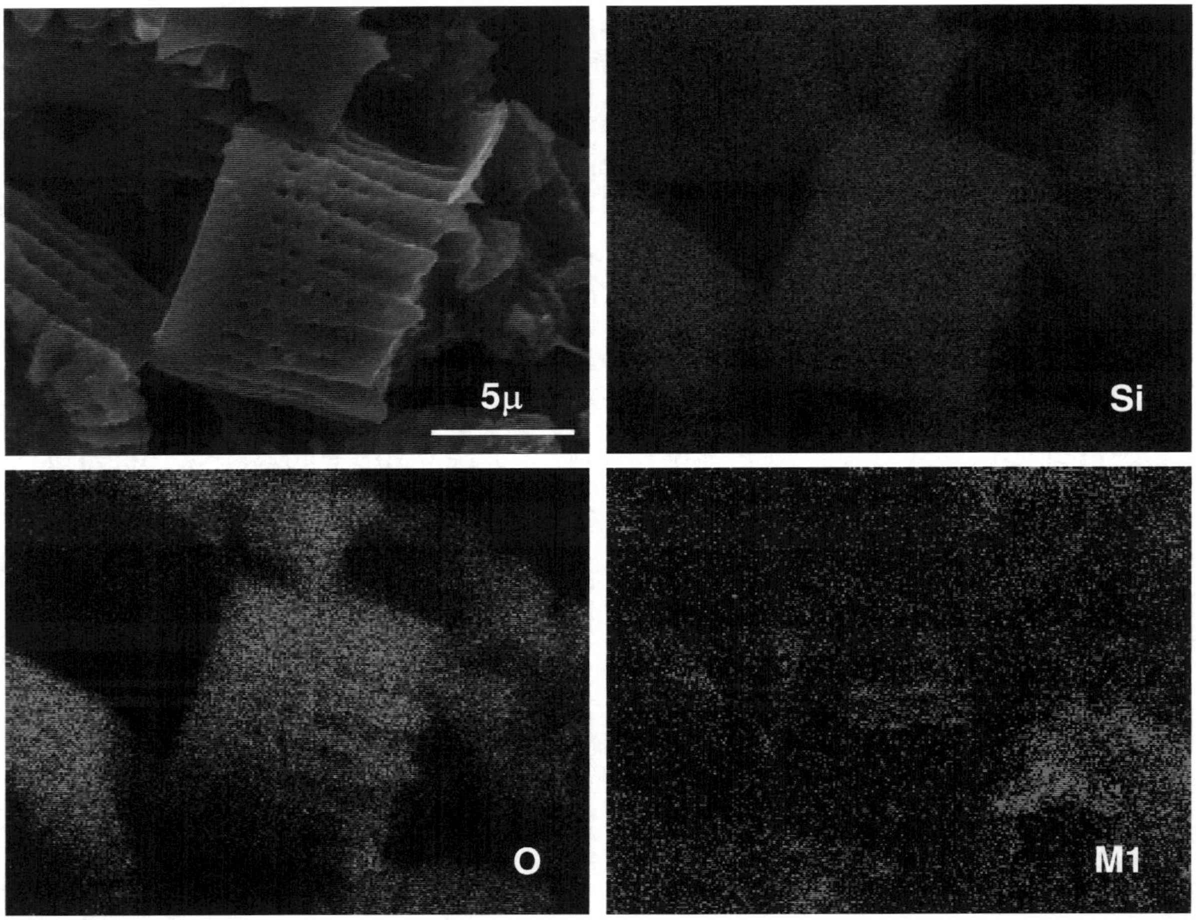


**Fig. 3.** SEM and elemental map of as-prepared sulfur sorbent D1.

Even in the case of diatomite matrix, the dispersion of sulfur-scavenger is highly uniform throughout. Thus, the preparatory technique utilized for making these materials is quite effective for yielding uniform distribution of the sulfur scavengers across the supports.

*Evaluation of sulfur sorption*

As stated above, the sulfidation experiments were carried out by running a gas stream of 1000 ppm $H_2S$/bal. $N_2$ at a flow rate of 100 sccm over each of the formulations for 12h. The $H_2S$ concentration in the exhaust stream was continuously monitored via GC to follow the sorption capacity of each formulation and to discern the breakthrough, which was a measure of the effectiveness of sulfur capture. The time dependence of $H_2S$ concentration in the exit stream for various formulations in the temperature range of 600-800°C is shown in figs. 4 through 6.

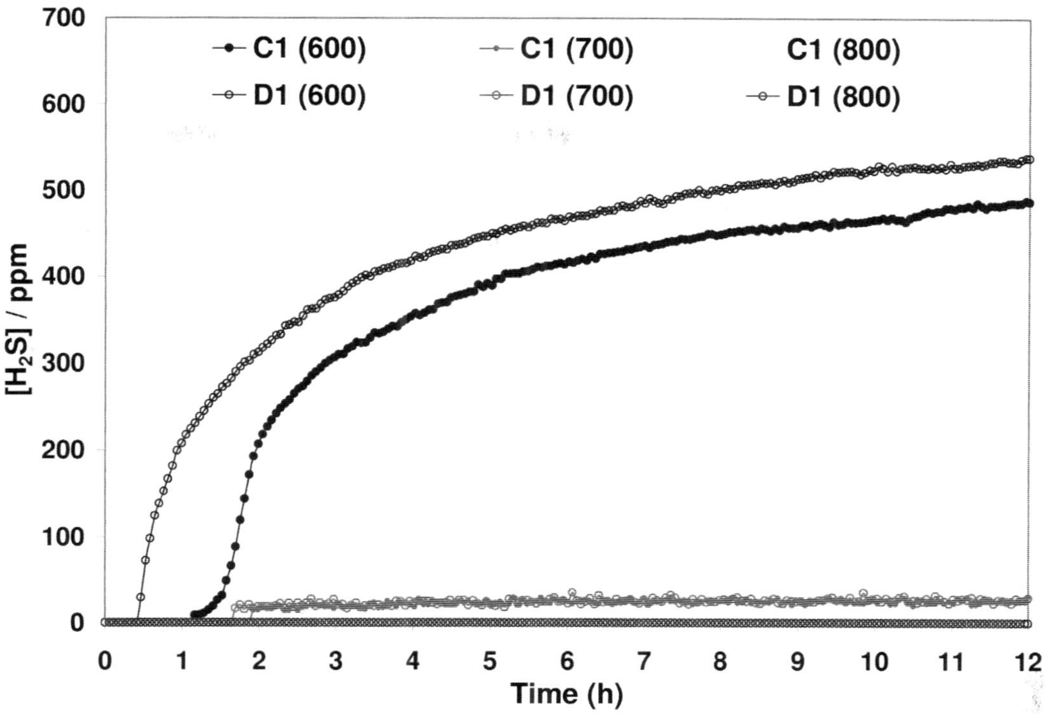


**Fig. 4.** Desulfurization profiles for sorbents C1 and D1.

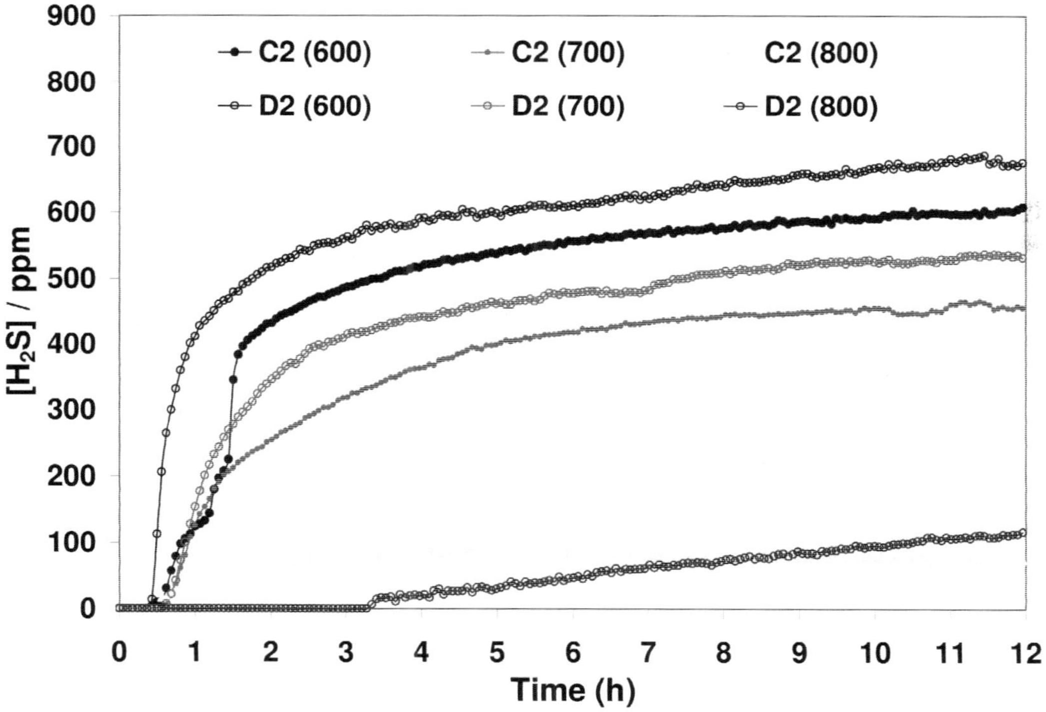


**Fig. 5.** Desulfurization profiles for sorbents C2 and D2.

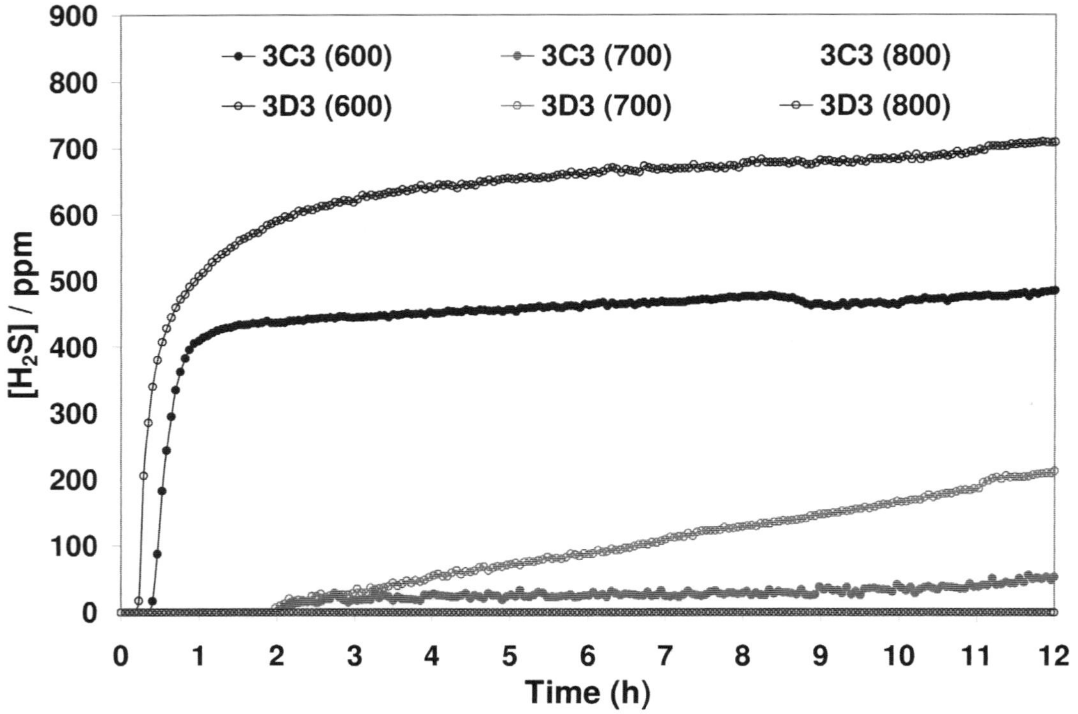


**Fig. 6.** Desulfurization profiles for sorbents C3 and D3.

As can be seen, at a given temperature, the $H_2S$ concentration at the exit stream was lower in the case of C series as compared to the D series. It can also be seen that at 600°C and 700°C the sorbents C1 and D1 exhibited better desulfurization capacity than the other two sorbents. Morphological features and the elemental mapping of the post-sulfided sorbents for sorbent C2 are shown in Fig. 7, which indicate that sulfur species were immobilized and bound to dispersed metal sites on the sorbent surface.

**Fig. 7.** Evidence of sulfur capture on C2 after exposure to 1000 ppm $H_2S$ for 12h at 800°C.

In order to elucidate the likely role of support matrix on the overall performance of the sorbents, a correlation was established between the breakthrough time ($t_b$, defined as the time elapsed before the first indication of sulfur in the exit stream via GC signal) and, the steady-state $H_2S$ concentration in the exit at each temperature, which is shown in Fig. 8. It can be seen that at a given temperature, the breakthrough time is almost independent of temperature in the case of C series sorbents, while it increases almost exponentially in the case of D series.

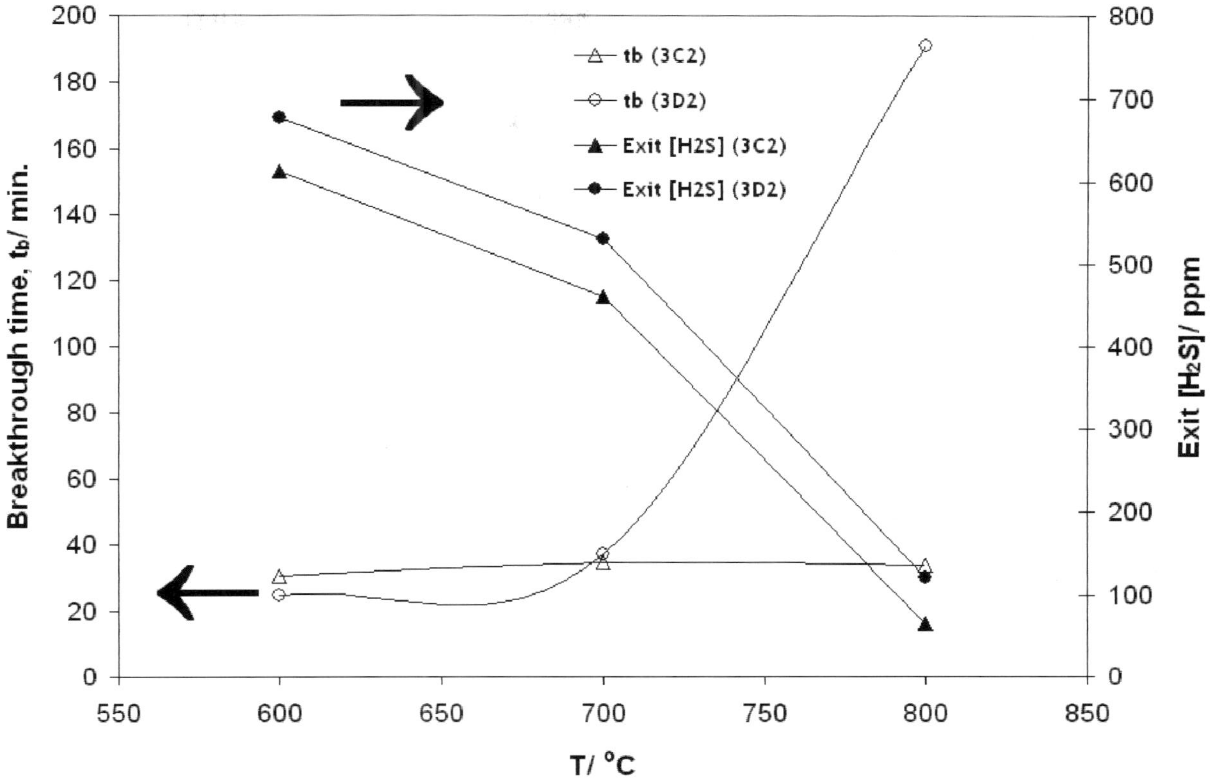


**Fig. 8. Correlation of breakthrough time and temperature**

## 4. Conclusions

Several desulfurizer formulations were synthesized by adding metal oxides to clinoptilolite and diatomaceous earth matrices which showed excellent performance in sulfur stream. Both series showed optimum sorption behavior at 800°C.

## 5. References


1. R.B. Slimane and J. Abbasian, Advances in Environmental Research 4 (2000) 147-162.
2. M.J. Duran, Synthesis and Characterization of Novel Sulfur-Tolerant Catalysts and Desulfurizers for Jet Fuel Reforming, MS Thesis, Department of Chemical and Environmental Engineering, The University of Toledo (2006).

# A Novel FCC Regeneration Process for Reduced $CO_2$ Emission

*Yu Rong, Nan Zhang, Megan Jobson*

**Centre for Process Integration, University of Manchester**

Key word: FCC, Regeneration, $O_2 / CO_2$ feed gas, Boudouard reaction

## Introduction

In the refining industry, Fluid catalytic cracking (FCC) unit is used to crack heavy hydrocarbon molecules. The two basic reaction sections in an FCC unit are the reactor and the regenerator. During operation, cracking reactions take place in reactor to crack feedstock into a range of hydrocarbon products. The reactions also produce carbonaceous deposit on the surface of the catalyst. These materials are collectively referred to as coke. The catalyst is deactivated by the coke deposition. In order to recover its activity, the catalyst is sent to a regenerator.

The major function of the regenerator is to restore the catalyst activity by burning off the coke and supply the heat required for heating and evaporating the FCC feed as well as for the endothermic heat of the cracking reactions. In traditional regeneration process, as illustrated in Figure 1, air is injected into the regenerator reacting with the spent catalyst. Normally, the flue gas from combustion operation is mainly composed of $CO_2$, CO, $SO_x$ and $NO_x$.

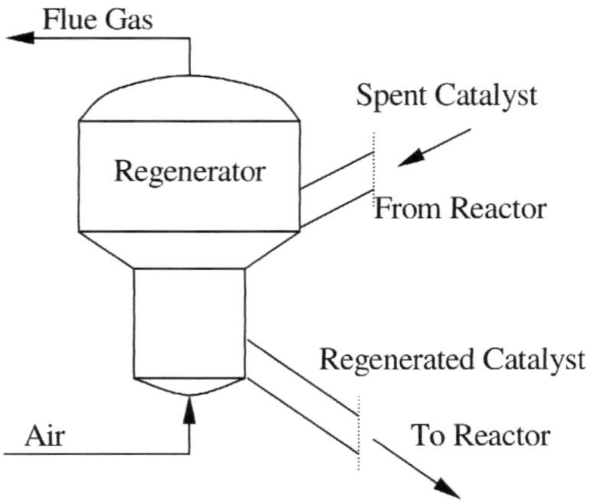


Figure 1 FCC regeneration process

There are two kinds of combustion operations for FCC regeneration – partial combustion and complete combustion. In complete combustion, almost all CO is converted to $CO_2$ in the regenerator with the help of excess oxygen. However, in a partial combustion, there is a significant amount of CO in the flue gas. Before flue gas emission, the entire CO is converted to $CO_2$ in a CO boiler. Therefore, all the coke produced by FCC reaction becomes the source of $CO_2$ emissions, contributing 40 to 45% of total refinery $CO_2$ emissions.

One potential option for minimizing the $CO_2$ emissions is to recover carbon from coke into a usable form. In the chemical industry, CO is a valuable feedstock which could be used as a part of synthesis gas as raw material for producing hydrogen, methanol and other chemicals. It is worth to consider the possibility of CO recovery from FCC regeneration process. However, a large amount of $N_2$ in flue gas makes CO recovery very difficult and expensive. For this reason, various feed gas could be considered in alternative technology.

**Coke gasification with $CO_2$**

Using pure $CO_2$ as reactant, the Boudouard reaction takes place to regenerate the spent catalyst. The reaction is expressed as:

$$C + CO_2 \leftrightarrow 2CO$$

This reaction is an endothermic reaction. In contrast, during the air-firing method the heat released may cause an uncontrolled temperature increase and cause damage to equipment and catalyst. Moreover, the Boudouard reaction converts $CO_2$ into CO which is a valuable feedstock for other chemical industry. Therefore, the use of $CO_2$ for the catalyst regeneration is an attractive option (Furimsky, Houle, et al., 1987).

However, compared with the air-firing method, non-catalytic $CO_2$ -firing method requires longer residence time and higher reaction temperature to achieve high carbon conversion. It has been proved that using catalysts in this process can lower the reaction temperature and increase the reaction rate, by which the energy necessary for the system can be reduced significantly (Nagase and Shimodaira, 1999). Effects of various metals on the Boudouard reaction have been investigated by previous work. (Takenaka and Otsuka, 2001; Carabineiro, McKee, et al., 2001). In the study, they confirmed the rate of reverse Boudouard reaction is enhanced with the presence of certain metal such as Na, Ni and Fe. On the other hand, some of these metals are one major factor causing irreversible deactivation of the catalyst.

To overcome the slow reaction of carbon with $CO_2$ on low temperature, the use of catalysts has been proposed for low temperature by previous work. Carabineiro and McKee (2001) studied the behavior of several metal oxides and binary oxide mixtures as catalysts for reverse Boudouard reaction. In their study, the kinetics of the reaction with metal salt of Mg, Ba, Pb, Cu and Fe has been investigated. Experimental measurements of the catalytic effect of Cr and Ni on the rate of Boudouard reaction have been reported by Karatas (2004). In addition, the carbon gasification with the touch of carbonates and oxide of $Li^+, Na^+, K^+, Cs^+$ and $Ba^{2+}$ has been investigated by Nagase, Shimodaira, Itoh and Zheng (1999).

In the study of FCC catalyst property, the chemical components of the fresh catalyst includes $Al_2O_3, Na_2O$ and $Fe_2O_3$; while the balance catalyst contain some other metal contamination such as Ni, V and Fe. In the study of kinetics and mechanisms of Boudouard reaction over metal catalyst, $Na_2O$ has an advantage on $CO$ formation. Although such metals as Ni, V and Fe are useful catalysts on the Boudouard reaction to increase CO yield, they are contaminants for FCC catalysts.

**Coke gasification with $O_2$/ $CO_2$ mixture**

To avoid long regeneration time and uncontrolled operation temperature, an alternative process, as shown in Figure 2, is proposed to minimizing the $CO_2$ emissions from catalyst regeneration. Pure oxygen mix with a certain amount of recovered $CO_2$ in the feed gas side. The $O_2 / CO_2$ mixture is used as the feed gas for coke combustion in the regenerator instead of air. This technology is named as '$O_2 / CO_2$-firing method'.

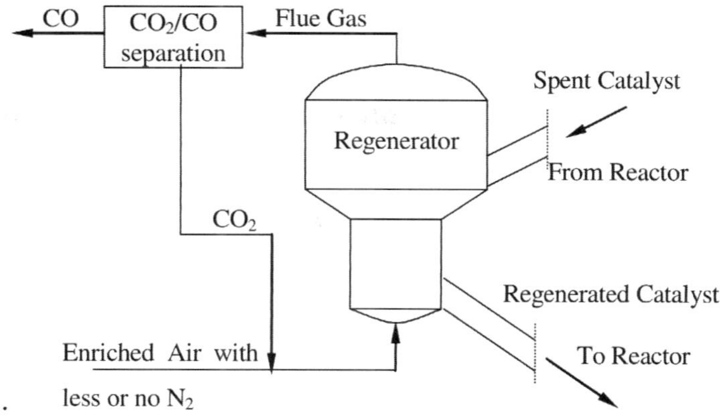


Figure2 Proposed FCC regeneration flowsheet

The carbon dioxide is recycled and reacted with coke to create more carbon monoxide. In this method, $O_2$ is not the only reactant for coke combustion. Part of coke is reacted with carbon dioxide by the reverse Boudouard reaction ($CO_2 + C \rightarrow 2CO$). In addition, the recycled $CO_2$ works as inert gas to control operation temperature of the regenerator.

**Mathematical model for FCC regenerator**

A 'two-region and two-phase' model has been built to simulate the regeneration process. In this existing model, the regenerator has been divided into two regions, a dense bed region, where the bulk of the catalyst is present and a dilute region, where the gas leaving from the dense region along with some catalyst particles moving upwards to the cyclone system is present. The two-phase theory of fluidization (Kunii and Levenspiel, 1991) describes that the dense region could be further divided in two phases, a bubble phase and an emulsion phase. The bubbles move in plug flow and exchange mass and heat with the emulsion phase. The air distributors, the spent catalyst, and the cyclones recycle pipes produce enough turbulence to justify CSTR behavior in the emulsion phase (Ali and Rohani, 1997). Therefore, the dense bed model includes a CSTR reactor model for emulsion phase and a tubular reactor model for the bubble phase, while the dilute region is modelled as a two-phase tubular reactor.

In this 'Two-region and two-phase' model, Boudouard reaction on certain catalyst such as $Na_2O$ and Fe had not been included. In this case, the model fails to simulate the process when the $CO_2$ content in feed gas is high. The assumption of no reaction in bubble phase has not been adopted in this study, since homogeneous $CO$ oxidation reaction has been proved as one significant reaction in the regenerator (Han and Chung, 2001). Therefore, several modifications have been introduced into this existing model to simulate $O_2/CO_2$-firing process.

Overall, the modified model built. The dense bed model includes a CSTR model for energy and coke balances but a tubular reactor model for gas component balance; the dilute region is modeled as a plug flow reactor. The kinetics of coke combustion reaction, $CO$ oxidation reactions and Boudouard reaction are applied into the model.

**Result and Conclusion**

The modified model has been validated by a conventional air feed regeneration process. The molar fractions of various gas components on the exit side of the regenerator is $0.2\%O_2, 1.0\%CO, 16.4\%CO_2$ and $10.12\%H_2O$. The result agrees well the industry data published in the paper of Han and Chung (2001). This modified model and has been applied for a range of operating condition with $CO_2/O_2$ mixture. The CO yield from FCC regeneration can be increased by around 75%, equalized to a 12% reduction of $CO_2$ emission. The model is to be applied for the development of novel process configuration that could further enhance these benefits.

**Reference:**


Adjorlolo, A.A. and Rao, Y.K (1984), Effect of potassium and sodium-carbonate catalysts on the rate of gasification of metallurgical coke, Carbon **22**(2):173-176

Ali, H. and Rohani S. (1997), *Dynamic modeling and simulation of a riser-type fluid catalytic cracking unit.* Chemical Engineering & Technology, **20**(2): p. 118-130.

Carabineiro, S.A., McKee D.W and Silva I.F. (2001), *Uncatalysted and catalysted $CO_2$ reaction using metal catalysts and binary vanadium mixtures supported on activated carbon,* Carbon. **39**(3):p451-463

Han, I.S. and Chung C.B. (2001), *Dynamic modeling and simulation of a fluidized catalytic cracking process. Part II: Property estimation and simulation.* Chemical Engineering Science, **56**(5): p. 1973-1990.

Kunii, D and Levenspiel O. (1991), *Effect of Particles in Bubbles on Fluidized-bed Mass and Heat-transfer Kinetics, Journal of Chemical Engineering of Japan,* **24**(2): p183-188

Kunii, D and Levenspiel O. (1991), *Fluidization Engineering 2nd Ed,* p69, p237-256

Karatas C. (2004), *Catalytic enhancement of graphite–$CO_2$ reaction by in situ reduced chromium and nickel from their oxides $Cr_2O_3$ and NiO over the temperature range 850–1100°C,* Mineral Processing and Extractive Metallurgy, Vol. 113 C19

McKee D.W. (1980), *Catalytic Effects of Alkaline Earth Carbonates in the Carbon-Carbon Dioxide Reaction*. Fuel, **59**(5):308

Nagase K, Shimodaira T, Itoh M, et al. (1999), *Kinetics and mechanisms of the reverse Boudouard reaction over metal carbonates in connection with the reactions of solid carbon with the metal carbonates,* Physical Chemistry Chemical Physics. **1**(24): p5659-5664

## 228e Decarbonisation in Process Sites

**Yuhang Lou**, Robin Smith, and Jhuma Sadhukhan. Centre for Process Integration, Chemical Engineering & Analytical Science, The University of Manchester, Manchester, United Kingdom

This work tackles the carbon dioxide (CO2) emission problem in process sites, particularly in relation to the site utility systems.

There are three basic decarbonisation routes to deal with the CO2 emission problem in energy production. These are pre-combustion, post-combustion and the oxy-combustion routes. For each route, different CO2 separation technologies can be exploited. This work has adopted different decarbonisation routes with both conventional CO2 separation technologies and novel pre-combustion routes.

Unlike the CO2 emission problem of energy generation taken in isolation, the emissions from a utility site can be widely distributed. To challenge this problem, utility sites are integrated with decarbonised combined heat and power generation systems. By doing this, some of the utility products like power, steam and fuel can be substituted by products from the decarbonised power generation system. Consequently, reduced carbon dioxide emissions from the utility system can be achieved.

Thus, a wide range of decarbonisation designs can be applied with power generation. Among them, pre-combustion with novel CO2 separation technology has the best performance. Integrating decarbonised combined heat and power generation systems can give significant potential for CO2 emission reduction to the utility site.

## 228f In Situ Synthesis and Characterization of Bimetallic Mesoporous Silica and Its Applications

**Debasish Kuila**[1], Balaji Tatineni[1], Joslyn Perkins[1], Sergey Yarmolenko[2], Zhigang Xu[2], and Shamsuddin Ilias[3]. (1) Department of Chemistry, North Carolina A&T State University, 1601 East Market Street, Greensboro, NC 71272, (2) Interdisciplinary Research Center, North Carolina A&T State University, (3) Mechanical and Chemical Engineering, North Carolina A&T State University, Greensboro, NC 27411

Bimetallic nanoparticles of Pd-Co / Pd-Ni in MCM type mesoporous silica have been synthesized by hydro-thermal procedure using Pd(NO3)2, CoCl2, Ni(NO3)2, cetyl trimethyl ammonium bromide (CTAB), tetramethoxysilane (TMOS), ammonia and water. These materials have been characterized using XRD, TEM, FT-IR, and BET techniques. The metal oxides in the mesoporous silica were reduced using 4% hydrogen to yield an alloy (Manuscript in preparation). The details of our ongoing characterization and catalytic studies will be presented.

# 228g Ca Based High Temperature CO2 Absorbents for Sorption Enhanced Reforming Process

**Kwang B. Yi**[1], Chang Hyun Ko[1], Jong-Ho Park[1], Su F. Wu[2], and Jong-Nam Kim[1]. (1) Separation Process Research Center, Korea Institute of Energy Research, 71-2 Jang-Dong, Yuseong-gu, Daejeon, 305-343, South Korea, (2) Department of Chemical Engineering, Zhejiang University, Zheda road No. 38, Hangzhou, China

There are several obstacles to overcome to materialize industrial level of sorption enhanced reforming process. One of the major problems of SERP is poor thermal cyclic stability of $CO_2$ absorbent(CaO). To dates, numerous research results showed that the repetitive absorption and desorption of $CO_2$ cause sintering or agglomeration of the absorbent particles resulting in drastic reduction of $CO_2$ absorption capacity along with the increased cycle number. In this study, Ca-based $CO_2$ absorbents were synthesized and modified to achieve the cyclic stability. The cyclic $CO_2$ absorption/desorption tests have been carried out in a thermal gravimetric analyzer using the prepared absorbent. Among the attempts to synthesize the absorbent, liquid phase methods with addition of alumina precursor produced the ultra stable high temperature $CO_2$ absorbents. Absortion/desorption test using the absorbent at 700$^o$C with changing gas atmosphere showed stable $CO_2$ gain of 45 wt% over 100 cycles. $CO_2$ absorption rate was very fast and consistent for each cycle. X-ray diffraction analysis revealed the development of micro porous structure inside of the absorbent during early cycles of absorption/desorption. Focused ion beam (FIB) investigation on cross-section of the absorbent showed major porous structure was developed in the Al rich region. Therefore, it is believed the compound containing Ca and Al formed path ways for fast $CO_2$ diffusion.

Also, various catalyst metal precursors were used to prepare catalyst incorporated absorbents. The prepared catalyst-absorbent materials were tested in a fixed bed reactor to simulated SERP process and its results were compared with that of physical mixture of commercial catalyst and the $CO_2$ absorbent.

# AUTHOR INDEX

# AUTHOR INDEX

**American Institute of Chemical Engineers**
3 Park Avenue
New York, New York 10016-5991

**ISBN: 978-1-60560-213-4**

# Hot Quarks 2010: Workshop for Young Scientists on the Physics of Ultrarelativistic Nucleus-Nucleus Collisions

**Journal of Physics: Conference Series Volume 270**

## La Londe-les-Maures, Cote d'Azur, France
## 21 – 26 June 2010

ISBN: 978-1-61782-372-5
ISSN: 1742-6588

IOP Institute of Physics